Unless Recalled Earlier
Date Due

NOV 11 1987			
NOV 17 1988			
DEC 1 1989			
APR 13 1990			
JUL 21 1993			

BRODART, INC.   Cat. No. 23 233   Printed in U.S.A.

MICROWAVE SEMICONDUCTOR CIRCUIT DESIGN

MICROWAVE SEMICONDUCTOR CIRCUIT DESIGN

W. Alan Davis
The University of Texas at Arlington

 VAN NOSTRAND REINHOLD COMPANY
———————— *New York* ————————

Copyright © 1984 by Van Nostrand Reinhold Company Inc.

Library of Congress Catalog Card Number: 83-6669
ISBN: C-442-27211-1

All rights reserved. No part of this work covered by the copyright hereon may be reproduced or used in any form or by any means—graphic, electronic, or mechanical, including photocopying, recording, taping, or information storage and retrieval systems—without permission of the publisher.

Manufactured in the United States of America

Published by Van Nostrand Reinhold Company Inc.
135 West 50th Street
New York, New York 10020

Van Nostrand Reinhold Company Limited
Molly Millars Lane
Wokingham, Berkshire RG11 2PY, England

Van Nostrand Reinhold Company Limited
480 Latrobe Street
Melbourne, Victoria 3000, Australia

Macmillan of Canada
Division of Gage Publishing Limited
164 Commander Boulevard
Agincourt, Ontario M1S 3C7, Canada

15 14 13 12 11 10 9 8 7 6 5 4 3 2

Library of Congress Cataloging in Publication Data

Davis, W. Alan.
  Microwave semiconductor circuit design.

  Includes index.
  1. Microwave circuits. 2. Microwave devices.
3. Semiconductors. I. Title.
TK7876.D38 1983   621.381′32   83-8669
ISBN 0-442-27211-1

My Father
For the integrity and leadership he gave each task

My Mother
For providing a stable and organized home

My Wife
For giving her family inspiration and encouragement

My Children
For their unbounded optimism

Preface

The microwave frequency range is that portion of the electromagnetic spectrum where the wavelength is of the same order of magnitude as the characteristic size of the circuit carrying the electrical energy. The frequencies most often considered to be in this category lie between approximately 1 and 200 GHz. Microwave circuits usually contain distributed circuit elements. Circuits used at lower frequencies usually have lumped elements, while circuits at higher frequencies use optical techniques.

The microwave frequency range has been applied widely in communication systems and radar systems, and in heating and drying applications. In the last several years, all three of these industries have experienced remarkable growth, and the prospect for dramatic growth in the future is very probable. This growth has fostered the discovery of new improved engineering techniques for the design of microwave circuits and microwave semiconductor devices. Historically, there has been a structural distinction between these two aspects of microwave engineering. This is illustrated by the two IEEE groups, Electron Devices and Microwave Theory and Techniques. The first stresses device research, while the second stresses microwave circuit research. Many university and industrial enterprises have also split these functions into two separate laboratories. However, because of the more stringent performance requirements placed on microwave module designs, this gap between the circuit and device designers has been closing. Today the microwave industry seems to be poised at the beginning of a new development: microwave monolithic circuits. The engineer involved with the design of these circuits will by necessity have to be conversant with both device and circuit design principles. This book provides a basic treatment of the most widely used semiconductor devices and the circuit design techniques required to make useful microwave circuits. These fundamentals apply to both discrete and monolithic circuit designs. This book does not provide a compendium of all the latest microwave devices and semiconductor materials. However, the diodes and transistors that were chosen for discussion do illustrate the fundamental properties inherent in the most widely available semiconductor devices.

This book was written to provide a comprehensive presentation of the fundamentals for the design of active microwave circuits. It has been assumed that the reader is familiar with transmission line theory, electromagnetic theory, the basic ideas of semiconductor materials, and some microwave circuit components. With these assumptions, I have endeavored to provide the needed mathematical derivations within the text. When this did not seem practical, as in Chapter 5, a reference for the source of the derivations is provided. These details have been included because it has been my experience that one cannot fully understand a mathematical expression without going through its derivation. Furthermore, physical insight into the operation of a circuit or device can be greatly enhanced by understanding its mathematical model.

The book has been organized into three parts: (1) passive circuits, (2) active design principles, and (3) semiconductor devices. The first four chapters review the essential microwave circuit concepts and provide detailed design information for some of the most widely used passive microwave components. These designs are based on circuit theory and can be realized in a transverse electromagnetic (TEM) wave transmission line. Up to this point, the specific transmission line medium has been left unspecified. Chapter 5 gives without proofs the transmission line parameters for some of the most widely used transmission line media. In addition, some of the most common discontinuities are discussed. With the aid of this chapter, the circuit components given in Chapters 2, 3, and 4 can be realized. In the second section, Chapters 6 through 10 introduce the concepts needed in active circuit designs. These concepts include the use of the computer in the microwave circuit design process, characteristics of amplifiers and oscillators, noise, statistical thermodynamics, and semiconductor junctions. In the final section, Chapters 11 through 16 each treat a specific semiconductor device. Example circuit designs are provided for each of these devices. This last group of chapters may be read independently of one another.

The wide range of topics presented here would have been almost impossible to assemble together without the aid of many of my past and present colleagues, and the stimulating atmosphere I have found at Raytheon. The research necessary to write this book has been exciting and rewarding. The reader I hope will experience a similar excitement.

I would like to give special thanks to my wife, Margaret, for her many valuable suggestions that have been incorporated in the text. She has not only thoroughly reviewed the grammar and style of the book, but aided in the typing of the final manuscript. I also would like to acknowledge David Freeman for his review of several sections of the book.

ALAN DAVIS

Contents

	Preface	vii
1	Microwave Circuit Analysis	1
	1.1 Introduction	1
	1.2 Fundamental Theorems	2
	1.3 The ABCD Matrix	4
	1.4 The Transmission Line Equation	6
	1.5 Scattering Parameters	9
	1.6 Flow graphs	11
2	Passive Microwave Components	14
	2.1 Introduction	14
	2.2 Directional Couplers	14
	2.3 Attenuators	29
	2.4 Ferrite Devices	31
3	Impedance Transformers and Filters	35
	3.1 Introduction	35
	3.2 Impedance Transformers	36
	3.3 Microwave Filter Design	43
	3.4 Coupled Transmission Lines	62
	3.5 Impedance Inverters	71
	3.6 Coupled Line Filters	78
	3.7 Summary	92
4	Broadband Directional Couplers	94
	4.1 Introduction	94
	4.2 Coupler Synthesis	94
	4.3 Transmission Line Taper	97
	4.4 Asymmetric 180° Coupler Design	100
	4.5 Symmetric 90° Coupler Design	106
	4.6 Conclusions	110

5 Mechanical Realization of Selected Transmission Lines — 112

- 5.1 Transmission Line Synthesis — 112
- 5.2 Rectangular Waveguide — 112
- 5.3 Two Wire Line — 116
- 5.4 Coaxial Line — 117
- 5.5 Stripline — 120
- 5.6 Microstrip — 126
- 5.7 Accommodation of Discontinuities — 132

6 Computer Aided Design, Manufacturing, and Test — 136

- 6.1 Introduction — 136
- 6.2 Computer Aided Design — 137
- 6.3 Computer Aided Manufacturing — 145
- 6.4 Computer Aided Test — 146

7 Characteristics of Amplifiers and Oscillators — 154

- 7.1 Introduction — 154
- 7.2 One-Port and Two-Port Devices — 154
- 7.3 Amplifier Gain — 155
- 7.4 Amplifier Limitations — 156
- 7.5 Multistage Amplifiers — 164
- 7.6 The Energy Source — 167

8 Noise — 170

- 8.1 Introduction — 170
- 8.2 Thermal Noise Theory — 170
- 8.3 Shot Noise Theory — 173
- 8.4 Amplifier Noise Characterization — 175
- 8.5 Oscillator Noise — 179

9 Review of Statistical Thermodynamics — 186

- 9.1 Introduction — 186
- 9.2 Statistical Analysis — 186
- 9.3 Boltzmann Factor — 187
- 9.4 Equipartition Law — 190
- 9.5 Quantum Mechanical Distribution Functions — 192
- 9.6 Some Second Thoughts — 196

10 Review of PN Junction Theory — 200

10.1 Introduction — 200
10.2 PN Junction Theory — 201
10.3 Reverse Biased PN Junction — 205
10.4 Forward Biased PN Junction — 208
10.5 Diffusion Admittance Calculations — 213
10.6 Diffusion and Depletion Admittance — 215

11 Varactor and Step Recovery Diodes — 219

11.1 Introduction — 219
11.2 Voltage Controlled Oscillator — 219
11.3 Parametric Amplifier Design Theory — 222
11.4 Frequency Multipliers — 232

12 Schottky-Barrier Diode Applications — 243

12.1 Schottky-Barrier Device Physics — 243
12.2 Energy-Band Diagrams — 243
12.3 Detector Circuits — 248
12.4 Mixers — 253

13 Circuits Using PIN Diodes — 264

13.1 Introduction — 264
13.2 PIN Diode Operation — 264
13.3 PIN Diode Attenuators — 271
13.4 Switches and Limiters — 272
13.5 Diode Phase Shifters — 277

14 Avalanche Devices and Circuits — 294

14.1 Introduction — 294
14.2 IMPATT Diode Operation — 294
14.3 Read Diode Large Signal Analysis — 297
14.4 Small Signal Analysis of the Read Diode — 306
14.5 Temperature Effects — 311
14.6 Spurious Oscillations — 311
14.7 IMPATT Noise — 312
14.8 Efficiency — 313

14.9	The TRAPATT Mode	315
14.10	Power-Combining Circuits	315

15 Gunn Effect Devices — 325

15.1	Introduction	325
15.2	Properties of GaAs	326
15.3	The Ridley-Watkins-Hilsum Mechanism	329
15.4	Domain Formation	330
15.5	Operating Modes	332
15.6	Large Signal Dipole Domain Analysis	338
15.7	Current (Velocity)-Field Characteristic	341
15.8	Noise in the Gunn Device	345
15.9	Device Equivalent Circuit Model	345
15.10	Waveguide Circuit for Gunn Diodes	350

16 Bipolar and Field Effect Transistors — 355

16.1	Introduction	355
16.2	Bipolar Transistors	356
16.3	The Unipolar MESFET	362
16.4	Circuit Analysis of Two-Port Devices	373
16.5	Amplifier Design	383

17 Present and Future Developments — 396

17.1	Review of Microwave Technology	396
17.2	Future Developments	397

Appendix A Read Diode Avalanche Current Equation — 401

Appendix B Quasistatic Approximation for the Avalanche Current — 405

Index — 411

MICROWAVE SEMICONDUCTOR CIRCUIT DESIGN

Chapter 1
Microwave Circuit Analysis

1.1 INTRODUCTION

The microwave region of the electromagnetic spectrum may be considered to span the frequency range from approximately 1 GHz to 100 GHz. Energy in this frequency range has been used in applications as familiar as the telephone and as remote as bouncing radar signals off our neighboring planets. Microwave energy has been used in drying potatoes in the production of the common potato chip and for producing hot meals quickly in the home kitchen. At the same time, microwave spectroscopy has been used by researchers to study fundamental properties of materials, and physicians have begun using microwave radiation to provide local hyperthermia in the treatment of malignant tumors.

Although microwave energy has been used in the things we use everyday as well as in numerous scientific disciplines, the primary application of microwaves is in the fields of communications and radar. While the devices and techniques that will be described may be used in many applications, it is communications and radar that primarily direct the focus in this book.

Designing a system for either of these two applications requires three broad categories of equipment: antennas, circuits, and active devices. Although antenna theory is an important aspect of design of many microwave systems, circuits and devices will be the primary areas of interest.

Microwave circuit design falls into two distinct divisions. The first of these is obtaining transmission line characteristics such as characteristic impedance, propagation constant, dispersion, loss, and power-handling capability. Finding this information requires a field theory approach. The second division is designing a circuit function embodied in such circuit elements as impedance transformers, filters, multiplexers, power combiners, phase shifters, attenuators, couplers, circulators, isolators, cavity resonators, and connectors. Having obtained the characteristics of the transmission line media, these circuits are usually designed by a circuit theory approach without recourse to field theory.

Active devices are most easily divided into microwave tubes and microwave semiconductors. Several different microwave tubes have been developed for various applications. The earlier devices such as the klystron and magnetron have proven reasonably compact and reliable. The traveling wave tube, despite its expense and high voltage requirement, continues to be unexcelled for broad bandwidth, high power, high gain amplification. Recently, there has been renewed interest in the gyrotron tube because it offers the possibility of high power millimeter wave generation.

However, tubes are considered less reliable than solid state devices, since they do not have to boil electrons off a filament. However, a solid state device won't boil off electrons only as long as it is operated within a safe temperature range. Among the two-terminal solid state devices that have been found useful are the detector, varactor, *PIN,* tunnel, back, IMPATT, TRAPATT, BARITT, and Gunn diodes. Three-terminal devices such as the bipolar transistor and the field effect transistor have made an important contribution to many active circuit designs. One other device, the MASER, although requiring cryogenic temperatures, has been found useful where extremely low noise amplification is required, i.e., radio astronomy and deep space communication. Several solid state devices and some sample circuit designs will be examined and analyzed in this book. The topic of microwave tubes will not be discussed in this text, since it is the author's intention to concentrate on the analysis and design of passive and active microwave circuits that make use of semiconductor devices.

In the present chapter the reader is reminded of some of the fundamental theorems of lumped element circuits that also apply to distributed circuits. Following this discussion, the ABCD matrix description of two-port circuits will be presented. This notation is a particularly convenient way of mathematically cascading lumped or distributed circuits. Since the transmission line equation forms a central role in the analysis of microwave circuits, its origin is discussed in section 1.4. Finally, in section 1.5 a description of the scattering matrix representation is given along with its application to flow graphs describing complicated circuits. These concepts will be used throughout the book and they form a set of basic tools that are very useful in analysis of many microwave circuits.

1.2 FUNDAMENTAL THEOREMS

One of the fundamental theorems of lumped element circuit theory is Foster's reactance theorem. This theorem states that the slope of the reactance with respect to frequency is always positive for a linear, lossless circuit.[1] This theorem is also true for lossless microwave circuits.[2,3] If $\mathscr{E}_m$ represents the magnetic stored energy, $\mathscr{E}_e$ the electrical stored energy, and P_ℓ the dissipated energy, then the input impedance as defined by the power-current definition is

$$Z = \frac{P_\ell + j2\omega(\mathscr{E}_m - \mathscr{E}_e)}{|I|^2/2} \qquad (1.1)$$

where the current in this expression is defined to be proportional to the longitudinal magnetic field intensity. By considering Maxwell's equations, the reactance slope is

$$\frac{\partial X}{\partial \omega} = \frac{4(\mathcal{E}_m + \mathcal{E}_e)}{|I|^2} > 0 \tag{1.2}$$

and similarly the susceptance slope is

$$\frac{\partial B}{\partial \omega} = \frac{4(\mathcal{E}_m + \mathcal{E}_e)}{|V|^2} > 0. \tag{1.3}$$

This implies that like lumped circuits, the poles and zeros of the impedance function of a lossless microwave circuit alternate. For example the input impedance of a shorted transmission line is

$$Z = jZ_0 \tan\left(\frac{\omega \ell}{c}\right) \tag{1.4}$$

which clearly has a positive reactance slope and alternating poles and zeros. Furthermore, the Fourier transforms of the voltage and current in the time domain must be real functions. Consequently their ratio in the frequency domain must be related by the complex conjugate function.[2]

$$Z(-\omega) = Z^*(\omega) \tag{1.5}$$

The real part of the impedance, R, is an even function of frequency

$$R(-\omega) = R(\omega) \tag{1.6}$$

and the imaginary part is an odd function of frequency.

$$X(-\omega) = -X(\omega) \tag{1.7}$$

Reciprocity is another important property of many circuits. If the definition of RF voltage and current is assumed to be proportional to the transverse component of the electric and magnetic fields across the junction, then an N port microwave circuit may be expressed in terms of an equivalent $N \times N$ impedance matrix.

$$[v] = \begin{bmatrix} Z_{11} & Z_{12} & \cdots & Z_{1N} \\ Z_{21} & & & \\ \vdots & & & \\ Z_{N1} & & & Z_{NN} \end{bmatrix} [i] \tag{1.8}$$

Each port represents a single mode of propagation. If more than one mode exists at any one physical port, each mode is represented by an independent voltage and current. If the microwave circuit is passive and contains no non-reciprocal devices (e.g., ferrite devices), then the reciprocity theorem states[2]

$$Z_{ij} = Z_{ji}. \qquad (1.9)$$

It is on the basis of this that the scattering parameter matrix discussed in section 1.5 is also found to be reciprocal.

1.3 THE ABCD MATRIX

When two or more microwave circuits are cascaded together, their overall response may be obtained by simply multiplying the ABCD matrix of each individual circuit together. This matrix relates the input RF voltages and currents to the output RF voltages and currents. It is defined as

$$\begin{bmatrix} v_1 \\ i_1 \end{bmatrix} = \begin{bmatrix} A & B \\ C & D \end{bmatrix} \begin{bmatrix} v_2 \\ i_2 \end{bmatrix}, \qquad (1.10)$$

corresponding to the voltage and current convention shown in Figure 1.1. Sometimes the current i_2 is shown going into the circuit in which case the appropriate change of sign for i_2 would have to be made in (1.10). In addition, for reciprocal circuits the determinate of the ABCD matrix is unity.[4,5]

$$AD - BC = 1 \qquad (1.11)$$

Although the ABCD matrix is a relation between voltages and currents rather than electromagnetic waves, it does express the idea of a voltage propagating down a circuit from left to right. As this voltage wave meets each succeeding circuit element, part of the wave is reflected back toward the source and part continues traveling toward the load. The amount of power that continues to go forward is dependent on both the circuit section itself and the impedances on either side of the circuit section. The image impedances of the two-port circuit shown in Figure 1.2 are those impedances at both the input and output that provide impedance match at both ends. In terms of the ABCD parameters, the image impedances are

$$Z_{I1} = \sqrt{\frac{AB}{CD}} \qquad (1.12)$$

$$Z_{I2} = \sqrt{\frac{BD}{AC}}. \qquad (1.13)$$

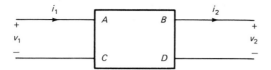

Figure 1.1. Circuit for the definition of the ABCD matrix.

The image impedance is strictly a function of the two-port circuit itself and is independent of the terminating impedances. However, when the terminating impedances are equal to the image impedances, then the circuit is matched. By expressing the ABCD matrix parameters in terms of the z and y parameters, it can be shown that

$$Z_{I1} = \sqrt{z_{11}/y_{11}} \qquad (1.14)$$
$$Z_{I2} = \sqrt{z_{22}/y_{22}} \qquad (1.15)$$

which often provides a convenient method of evaluating the image impedance for a given circuit.

The image impedance is valid for symmetrical as well as unsymmetrical circuits. It is valid for both lumped and distributed circuits. For a transmission line, the image impedance is called the *characteristic impedance*. It is usually designated by the single variable Z_0 since transmission lines are usually symmetrical over the length under consideration. The image propagation constant γ is defined as

$$e^{\gamma} = \sqrt{\frac{v_1 i_1}{v_2 i_2}} = \sqrt{AD} + \sqrt{BC}. \qquad (1.16)$$

The ABCD parameters for specific circuits are found from standard circuit theory of which several examples are shown in Figure 1.3. By cascading these elements together, a wide range of circuits may be generated. The ABCD

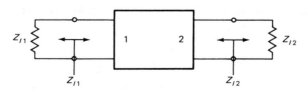

Figure 1.2. Circuit description of the image impedance.

6 MICROWAVE SEMICONDUCTOR CIRCUIT DESIGN

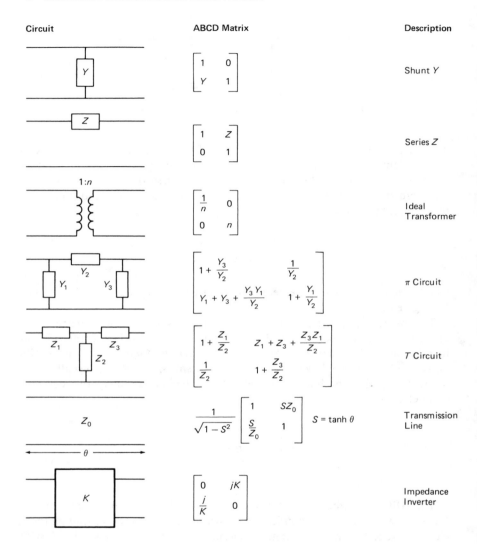

Figure 1.3. ABCD matrices for common 2-port circuits.

matrix of the composite circuit is easily obtained by multiplying their corresponding matrices together.

1.4 THE TRANSMISSION LINE EQUATION

The study of microwave circuits borrows from both electromagnetic field theory and lumped element circuit theroy. In principle, microwave circuits could be analyzed solely in terms of the more exact field theory. However, the sim-

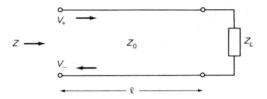

Figure 1.4. Incident and reflected voltages on a transmission line.

plicity and increased intuitive feel associated with the circuit theory approach makes the latter the almost universally accepted method of analysis. Field theory is used for finding the transmission line parameters, such as the characteristic impedance and propagation constant, of specific types of wave-guiding structures. Expressions for these parameters for various types of transmission lines are given in chapter 5. Once these are known, transmission lines may be connected in various ways and the resulting structure analyzed by the circuit theory approach. The key expression for the circuit analysis approach is the transmission line equation.

When a uniform transmission line is excited by a sinusoidal voltage source, the resulting voltages and currents at any position on the transmission line will also vary in a sinusoidal manner. The sum of the incident and reflected voltages and currents at the input of the network in Figure 1.4 are

$$V = V_+ + V_- \qquad (1.17)$$

$$I = \frac{V_+}{Z_0} - \frac{V_-}{Z_0} \qquad (1.18)$$

where Z_0 is the characteristic impedance of the transmission line. By the time the voltage and current waves reach the load, their phases will have been modified by the propagation constant of the line. Consequently at the load,

$$V_L = V_+ e^{-\gamma\ell} + V_- e^{\gamma\ell} \qquad (1.19)$$

$$I_L = \frac{V_+}{Z_0} e^{-\gamma\ell} - \frac{V_-}{Z_0} e^{\gamma\ell}. \qquad (1.20)$$

Solution of these two equations for the incident and reflected voltages gives

$$V_+ = 1/2\, (Z_L + Z_0) I_L\, e^{\gamma\ell} \qquad (1.21)$$
$$V_- = 1/2\, (Z_L - Z_0) I_L\, e^{-\gamma\ell} \qquad (1.22)$$

where $Z_L = V_L/I_L$. Substituting (1.21) and (1.22) into (1.17) and (1.18) gives the input voltage and current in terms of the load impedance and load current.

$$V = \tfrac{1}{2}I_L[(Z_L + Z_0)e^{\gamma\ell} + (Z_L - Z_0)e^{-\gamma\ell}] \qquad (1.23)$$
$$I = \tfrac{1}{2}(I_L/Z_0)[(Z_L + Z_0)e^{\gamma\ell} - (Z_L - Z_0)e^{-\gamma\ell}] \qquad (1.24)$$

The transmission line equation is obtained by taking the ratio of (1.23) and (1.24).

$$Z = Z_0 \frac{Z_L + Z_0 \tanh \gamma\ell}{Z_0 + Z_L \tanh \gamma\ell} \qquad (1.25)$$

For lossless transmission lines, $\gamma\ell = j\beta\ell \triangleq j\theta$*, and (1.25) becomes

$$Z = Z_0 \frac{Z_L + jZ_0 \tan \theta}{Z_0 + jZ_L \tan \theta} \qquad (1.26)$$

There are several important special cases for this equation. First of all it is periodic in θ

$$Z(\theta) = Z(\theta + \pi) \qquad (1.27)$$

so that the input impedance on a transmission line repeats every half wavelength. Secondly, when $\theta = \pi/2$

$$Z = \frac{Z_0^2}{Z_L} \qquad (1.28)$$

which is the basis for a quarter wavelength transformer. If $\theta < \pi/2$ and the transmission line load is a short circuit, then the input impedance is inductive.

$$Z = jZ_0 \tan \theta \qquad (1.29)$$

Similarly, a capacitive reactance is obtained from an open circuited line with $\theta < \pi/2$.

$$Z = -jZ_0 \cot \theta \qquad (1.30)$$

These last three expressions form important building blocks in microwave circuit design and will be discussed more fully in subsequent chapters.

---

*The symbol $\triangleq$ means equal by definition.

1.5 SCATTERING PARAMETERS

Multiport microwave circuits can be described in terms of either y, z, or ABCD parameters. These and other matrix descriptions based on voltages and currents have the disadvantage of relating RF quantities that are not directly measurable. Since reflection and transmission coefficients are directly measurable at microwave frequencies, it has been found convenient to describe a microwave circuit in terms of these coefficients. The scattering matrix does just this. The voltage waves entering and leaving a given port i are

$$a_i = \frac{1}{2\sqrt{Re(Z_i)}} [V + Z_i I] \quad (1.31)$$

$$b_i = \frac{1}{2\sqrt{Re(Z_i)}} [V - Z_i^* I] \quad (1.32)$$

where Z_i is the impedance looking in at the ith port. Furthermore, if Z_{Li} is the source impedance at the ith port the reflection coefficient is

$$\Gamma_i = \frac{b_i}{a_i} = \frac{Z_{Li} - Z_i^*}{Z_{Li} + Z_i} \quad (1.33)$$

and from energy conservation, the magnitude of the transmission coefficient for a two-port circuit is

$$|T| = \sqrt{1 - |\Gamma|^2}. \quad (1.34)$$

The scattering matrix defined for the n-port circuit in Figure 1.5 is

$$\begin{bmatrix} b_1 \\ b_2 \\ \vdots \\ b_n \end{bmatrix} = \begin{bmatrix} S_{11} & S_{12} & \cdots & S_{1n} \\ S_{21} & & & \\ \vdots & & & \\ S_{n1} & \cdots & & S_{nn} \end{bmatrix} \begin{bmatrix} a_1 \\ a_2 \\ \vdots \\ a_n \end{bmatrix}. \quad (1.35)$$

This is a relationship between voltage waves and the vector reflection and transmission coefficients. The scattering matrix may be applied to any n-port circuit. Since the individual S_{ij} components are obtained by setting all except one of the a_i to zero, this implies that physically the circuit must be terminated with a matched load rather than an open circuit for the z parameters or a short circuit for the y parameters.

The scattering matrix may be transformed into one of these other volt-ampere forms. The most useful for microwave analysis is the ABCD matrix which is given in Table 1.1.

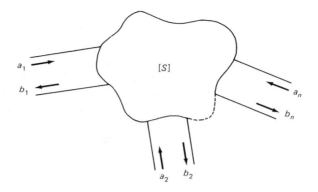

Figure 1.5. Scattering parameter description of an n-port circuit.

Kurokawa[6] has shown that the scattering parameters are related to the actual power delivered to a load from a generator by

$$P_i = \pm \tfrac{1}{2}(|a_i|^2 - |b_i|^2) \tag{1.36}$$

where the $-$ sign is used when $Re(Z_i) < 0$. Just as $Z_{ij} = Z_{ji}$ for reciprocal circuits, $S_{ij} = S_{ji}$. In addition, for a lossless network, the S matrix is unitary, i.e.,

$$[S^*]^T[S] = [I] \tag{1.37}$$

where $[I]$ is the identity matrix, and the superscript T stands for transpose. Consequently, for a lossless, reciprocal two-port circuit

$$|S_{11}| = |S_{22}| \tag{1.38}$$

Table 1.1. Conversion between ABCD and S Parameters.

$$A = \frac{-\Delta + S_{11} - S_{22} + 1}{2S_{21}} \sqrt{\frac{Z_1}{Z_2}}$$

$$B = \frac{\Delta + S_{11} + S_{22} + 1}{2S_{21}} \sqrt{Z_1 Z_2}$$

$$C = \frac{\Delta - S_{11} - S_{22} + 1}{2S_{21}\sqrt{Z_1 Z_2}}$$

$$D = \frac{-\Delta - S_{11} + S_{22} + 1}{2S_{21}} \sqrt{\frac{Z_2}{Z_1}}$$

$$\Delta = S_{11}S_{22} - S_{12}S_{21}$$

$$S_{11} = \frac{AZ_2 + B - CZ_1 Z_2 - DZ_1}{d}$$

$$S_{12} = \frac{2(AD - BC)\sqrt{Z_1 Z_2}}{d}$$

$$S_{21} = \frac{2\sqrt{Z_1 Z_2}}{d}$$

$$S_{22} = \frac{-AZ_2 + B - CZ_1 Z_2 + DZ_1}{d}$$

$$d = AZ_2 + B + CZ_1 Z_2 + DZ_1$$

and

$$|S_{11}|^2 + |S_{21}|^2 = 1. \tag{1.39}$$

The S_{ii} term is therefore the reflection coefficient at the ith port and S_{ji} is the transmission coefficient from the ith to the jth port.

1.6 FLOW GRAPHS

Since the scattering parameters allow one to think more easily in terms of directed waves rather than voltages and currents, a microwave circuit can often be conveniently represented in terms of a flow graph. A flow graph is made up of nodes and directed branches between the nodes, where the label for each branch is a transfer function from the input to the output of that branch. These flow graphs may be reduced to a simple form by means of Mason's nontouching loop rule.[7] An alternative procedure that may be more easily remembered is given by Kuhn.[8] This method consists of four rules that are stated as follows.

Rule 1: The transfer function of two similarly directed branches in series is the product of the two (Figure 1.6a).

Rule 2: The transfer function of two similarly directed branches in parallel is the sum of the two (Figure 1.6b).

Rule 3: The transfer function with a self loop S_2 is found by dividing all the input branches by $(1 - S_2)$ (Figure 1.6c).

Rule 4: The transfer function remains unchanged if a node with an input branch and N output branches is split into two nodes. The input branch goes to each of the new nodes. There will be $k < N$ output branches leaving one node and the remaining $N - k$ branches leaving the second node. The dual graph with one output branch and several input branches also holds (Figure 1.6d).

Rules 1 and 2 are self evident, and rule 3 is easily proved by letting the voltage wave V_b at node b be given in terms of the input voltage wave V_a at node a by

$$V_b = V_a S_1 + V_b S_2. \tag{1.40}$$

Solving this for the ratio V_b/V_a provides the overall transfer function.

$$\frac{V_b}{V_a} = \frac{S_1}{1 - S_2} \tag{1.41}$$

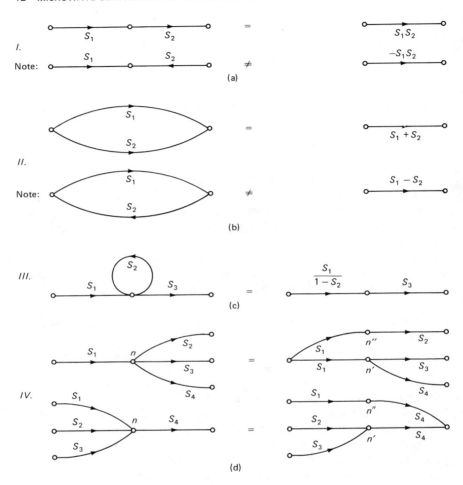

Figure 1.6. Four rules for flow graph reduction.

For rule 4, it is obvious that the outputs will be unchanged after splitting the node. Simplifying flow graphs in this manner is especially helpful in finding the transfer function of a circuit. It is used, for example, in error correction for an automatic network analyzer or analyzing a two-port device like the field effect transistor.

REFERENCES

1. R. M. Foster, "A Reactance Theorem," *Bell System Technical Journal,* Vol 3, pp. 259–267, 1924.

2. R. E. Collin, *Foundations for Microwave Engineering.* New York: McGraw-Hill, 1966.
3. J. L. Altman, *Microwave Circuits.* New York: D. Van Nostrand, 1964.
4. E. A. Guillemin, *Synthesis of Passive Networks.* New York: John Wiley & Sons, 1957.
5. G. L. Matthaei, L. Young, and E. M. T. Jones, *Microwave Filters, Impedance-Matching Networks, and Coupling Structures.* New York: John Wiley & Sons, 1962.
6. K. Kurokawa, "Power Waves and the Scattering Matrix," *IEEE Trans. on Microwave Theory and Techniques,* Vol. MTT-13, pp. 194–202, March 1965.
7. S. J. Mason, "Feedback Theory—Further Properties of Signal Flow Graphs," *Proceedings of the IRE,* Vol. 44, pp. 920–926, July 1956.
8. N. Kuhn, "Simplified Signal Flow Graph Analysis," *The Microwave Journal,* Vol. 6, pp. 59–66, November 1963.

Chapter 2
Passive Microwave Components

2.1 INTRODUCTION

Both active and passive microwave modules used in communication and radar systems make use of various passive components. Some of the more important of these are directional couplers, matched attenuators, ferrite circulators and isolators, impedance transformers, filters, broadband directional couplers, all-pass circuits, and power combiners. In this chapter the first three of these components will be discussed while the subsequent chapter will be devoted to transformers and filters. The subject of broadband directional couplers is based on impedance transformer design and is discussed in chapter 4. The topic of power combiners is deferred to the discussion of IMPATT diodes where these power combiners find their major application.

The subject of directional couplers contained in section 2.2 is restricted to the branch line and rat race couplers. The analysis of these couplers makes use of even and odd mode excitation of a symmetrical circuit. This illustrates an important concept that may be applied to a wide variety of circuits. The fixed attenuator described in section 2.3 is a widely used laboratory component and may, when needed, be incorporated in the design of miniature microwave circuits. Since attenuators are usually designed with quasi-lumped element resistors, the concept of image impedance is used in understanding their design (chapter 1). Finally a short discussion of ferrite circulators and isolators is found in section 2.4.

2.2 DIRECTIONAL COUPLERS

A directional coupler is a lossless, reciprocal, four-port circuit consisting of two pairs of ports in which (a) the ports of each pair are mutually isolated from one another, and (b) one pair of ports are matched. Later it will be shown that these conditions imply that the other pair of ports are also matched. The ideal directional coupler illustrated in Figure 2.1 shows that power incident at port 1 is transmitted to ports 2 and 3 with no power transmitted to port 4. The quality of a coupler is measured by the insertion loss from port 1 to port 2

$$L = 10 \log \frac{P_2}{P_1} \tag{2.1}$$

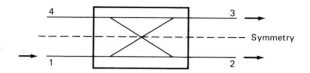

Figure 2.1. Directional coupler symbol.

the coupled power from port 1 to port 3

$$C = 10 \log \frac{P_3}{P_1} \qquad (2.2)$$

the isolation between ports 1 and 4

$$I = 10 \log \frac{P_4}{P_1} \qquad (2.3)$$

and the directivity between ports 3 and 4

$$D = 10 \log \frac{P_4}{P_3}. \qquad (2.4)$$

This latter quantity is useful when dealing with couplers where the coupled output at port 3 is small. It makes little sense to have a coupler with a coupling $C = -20$ dB and isolation $I = -20$ dB. For this reason couplers with low coupled output power are often specified in terms of their directivity. In this example a coupler with $D = -20$ dB would be an excellent coupler since its isolation would be

$$I = C + D = -40 \text{ dB}. \qquad (2.5)$$

Directional couplers operate on the principle of constructive and destructive interference of two waves. A signal at the input splits into two waves that arrive at the isolated port 180° out of phase with one another and therefore cancel one another. At the direct and coupled output ports, these waves arrive in phase with one another and interfere constructively.

A directional coupler may be used as a power level monitor, a local oscillator injection device, an attenuator, a power combiner/divider, or a device to provide a fixed relative phase angle between two signals. A hybrid coupler is one whose two outputs have a relative phase angle of 90°. This coupler is often used in combining two and sometimes four amplifiers or locked oscillators by means of either the reflection or transmission type circuits shown in Figure 2.2.

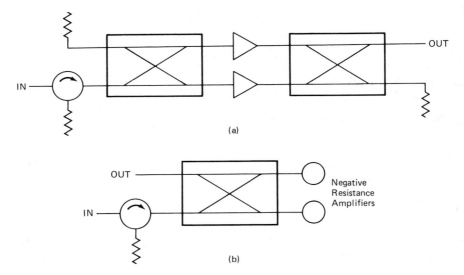

Figure 2.2. Transmission (a) and reflection (b) combining circuits using 90° directional couplers.

The reflection circuit is useful for combining negative resistance amplifiers and requires only a single coupler and one isolator at the input. The transmission circuit requires two couplers, an isolator, two loads, and when reflection amplifiers are used, two circulators. On the basis of the parts list, the reflection circuit is clearly the preferred method. However, the transmission circuit provides superior isolation between amplifiers and better return loss at the input port. This may be seen intuitively in Figure 2.3 where the underlined terms in the expression for power leakage from amplifier 2 to 1 are small. The only thing separating the two amplifiers in the reflection circuit is the isolation of the coupler, while in the transmission circuit the amplifiers are isolated by both the isolation of the second coupler and the return loss of the first coupler. This result has been verified by measured data taken on two circuits using short slot waveguide 3 dB hybrid couplers. The choice between the two circuits in Figure 2.2 reduces to a compromise between the isolation needed between the amplifiers and the number of parts required for the combining circuit.

The definition of the directional coupler provides sufficient information to obtain the scattering matrix for the coupler. The isolation of the fourth port from the input implies $S_{14} = S_{32} = 0$ and the matched requirement implies $S_{11} = S_{22} = 0$. For a lossless circuit the S matrix is unitary

$$[S][S^*]^T = [I] \qquad (2.6)$$

and reciprocity implies

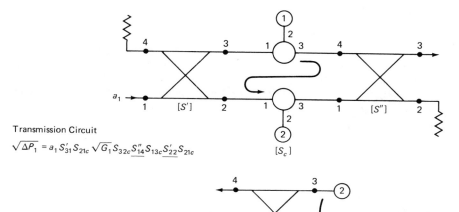

Figure 2.3. Transmission and reflection combining circuits. Underlined S parameters are small quantities.

$$[S] = [S]^T. \tag{2.7}$$

Doing the indicated multiplication in (2.6) shows the row 1, column 3 element is $S_{13}S_{33}^* = 0$, and the row 2, column 4 element is $S_{24}S_{44}^* = 0$. Since S_{13} and S_{24} are assumed to be nonzero, then $S_{33} = S_{44} = 0$ and the remaining ports 3 and 4 must also be matched. The equation (2.6) can then be written as

$$\begin{bmatrix} 0 & S_{12} & S_{13} & 0 \\ S_{12} & 0 & 0 & S_{24} \\ S_{13} & 0 & 0 & S_{34} \\ 0 & S_{24} & S_{34} & 0 \end{bmatrix} \begin{bmatrix} 0 & S_{12}^* & S_{13}^* & 0 \\ S_{12}^* & 0 & 0 & S_{24}^* \\ S_{13}^* & 0 & 0 & S_{34}^* \\ 0 & S_{24}^* & S_{34}^* & 0 \end{bmatrix} = \begin{bmatrix} 1 & 0 & 0 & 0 \\ 0 & 1 & 0 & 0 \\ 0 & 0 & 1 & 0 \\ 0 & 0 & 0 & 1 \end{bmatrix}$$

(2.8)

The row 1, column 4 and row 2, column 3 multiplications provide the following relations.

$$S_{12}S_{24}^* + S_{13}S_{34}^* = 0 \tag{2.9}$$
$$S_{12}S_{13}^* + S_{24}S_{34}^* = 0 \tag{2.10}$$

Dividing (2.9) by (2.10) and taking the absolute value of the result shows that

$$|S_{24}| = |S_{34}| \tag{2.11}$$

and in a similar fashion it can be shown that

$$|S_{12}| = |S_{34}|. \tag{2.12}$$

This states that the magnitudes of the coupling terms are the same and the magnitudes of the insertion loss terms are the same. If the reference planes are chosen so that $S_{12} = S_{34} = C_1$ where C_1 is real, and $S_{13} = C_2 e^{j\theta}$ where C_2 is real, then (2.9) can be rewritten as

$$C_1 S_{24}^* + C_1 C_2 e^{j\theta} = 0 \tag{2.13}$$

and solving for S_{24} yields

$$S_{24} = C_2 e^{j(\pi - \theta)} \tag{2.14}$$

The scattering matrix for the coupler then is

$$[S] = \begin{bmatrix} 0 & C_1 & C_2 e^{j\theta} & 0 \\ C_1 & 0 & 0 & C_2 e^{j(\pi-\theta)} \\ C_2 e^{j\theta} & 0 & 0 & C_1 \\ 0 & C_2 e^{j(\pi-\theta)} & C_1 & 0 \end{bmatrix} \tag{2.15}$$

This expression says that in-phase power entering at ports 1 and 4 will have a coupled output to port 3 that is 180° out of phase with the coupled power leaving port 2. The 90° coupler is one in which $\theta = \pi/2$ and the 180° coupler is one in which $\theta = 0$ or π. The relationship between C_1 and C_2 is found by multiplying the row 1, column 1 element of (2.8).

$$|C_1|^2 + |C_2|^2 = 1 \tag{2.16}$$

The design of 90° 3 dB directional couplers using coupled lines often requires unrealizable line widths and spacings. The realizability difficulty can be relieved substantially by cascading two identical 8.34 dB couplers to achieve the desired 3 dB result. This coupling factor is obtained by cascading the two identical couplers shown in Figure 2.4. The output of the first coupler, b_i, is obtained by multiplying (2.15) by the input vector $[a_1, 0, 0, a_4]^T$. The inputs to the second coupler are

$$a_1' = b_2 = C_1 a_1 + jC_2 a_4 \tag{2.17}$$
$$a_4' = b_3 = jC_2 a_1 + C_1 a_4 \tag{2.18}$$

The outputs for the second coupler are obtained by again multiplying the S matrix in (2.15) by the vector $[a_1', 0, 0, a_4']^T$.

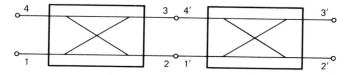

Figure 2.4. Cascade of two couplers.

$$b'_2 = (C_1^2 - C_2^2)a_1 + j2C_1C_2a_4 \qquad (2.19)$$
$$b'_3 = j2C_1C_2a_1 + a_4(C_1^2 - C_2^2) \qquad (2.20)$$

If the inputs to the first coupler are $a_1 = 1$ and $a_4 = 0$, the magnitudes of the outputs are $[b'_2] = [b'_3] = 1/\sqrt{2}$.

$$C_1^2 - C_2^2 = 1/\sqrt{2} \qquad (2.21)$$
$$2C_1C_2 = 1/\sqrt{2} \qquad (2.22)$$

Simultaneous solution of (2.21) and (2.22) yields

$$C_1^2 = \tfrac{1}{2}(1 + 1/\sqrt{2}) \qquad (2.23)$$
$$\text{or } -0.68 \text{ dB} \qquad (2.24)$$
$$C_2^2 = \tfrac{1}{2}(1 - 1/\sqrt{2})$$
$$\text{or } -8.34 \text{ dB}$$

Thus 3 dB power split is obtained by cascading two identical 8.34 dB hybrid couplers.

2.2.1 Analysis of Symmetrical 4-Port Circuits

The analysis of branch line as well as many other symmetrical networks has been greatly facilitated by the method of superposition of even and odd mode excitations. This method was applied by Reed and Wheeler[1] to obtain the fre-

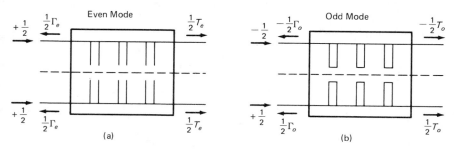

Figure 2.5. Even and odd mode excitation of a symmetrical circuit.

quency response of several symmetrical networks. Two of the most commonly used of these are the branch line coupler and the rat race coupler.

The procedure begins by exciting ports 1 and 4 with the even and odd mode voltage amplitudes $+\frac{1}{2}$ and $\pm\frac{1}{2}$ respectively (Figure 2.5). For the even mode circuit, currents at the line of symmetry are zero. These lines may be open circuited without affecting the circuit response. For the odd mode circuit, voltages at the line of symmetry are zero. These lines may be short circuited to ground without affecting the circuit response. In this way the 4-port circuit is reduced to a 2-port circuit. By superposition the outputs at the four ports are

$$b_1 = \tfrac{1}{2}(\Gamma_e + \Gamma_o) \tag{2.25}$$
$$b_2 = \tfrac{1}{2}(T_e + T_o) \tag{2.26}$$
$$b_3 = \tfrac{1}{2}(T_e - T_o) \tag{2.27}$$
$$b_4 = \tfrac{1}{2}(\Gamma_e - \Gamma_o). \tag{2.28}$$

The reflection and transmission coefficients in (2.25) to (2.28) may be expressed in terms of the circuit ABCD parameters. This relationship is obtained from Table 1.1 where the input and output impedances are matched ($Z_0 = Z_1 = Z_2$).

$$\Gamma = \frac{A + B/Z_0 - CZ_0 - D}{A + B/Z_0 + CZ_0 + D} \tag{2.29}$$

$$T = \frac{2}{A + B/Z_0 + CZ_0 + D} \tag{2.30}$$

In turn the ABCD matrix elements are calculated by cascading individual elements in the circuit under consideration. From Figure 1.3, the ABCD matrix for a shunt admittance of $1/Z$ is

$$\begin{bmatrix} 1 & 0 \\ 1/Z & 1 \end{bmatrix}. \tag{2.31}$$

The ABCD matrix for a transmission line with characteristic impedance Z_0 and electrical length θ, where $\Omega \stackrel{\Delta}{=} \tan\theta$, is

$$\frac{1}{\sqrt{1+\Omega^2}} \begin{bmatrix} 1 & j\Omega Z_0 \\ \dfrac{j\Omega}{Z_0} & 1 \end{bmatrix}. \tag{2.32}$$

The matrix parameters for the individual sections in the following examples will consist of shunt admittances and transmission lines. These are cascaded together to obtain ABCD matrices for the even and odd mode circuits. Next the even and odd mode reflection and transmission coefficients are found from (2.29) and (2.30). Finally the outputs from each port are obtained from (2.25)–(2.28). The design parameters are constrained to insure that port 1 is matched, i.e., $b_1 = 0$.

2.2.2 The Branch Line Coupler

For the even and odd mode excitations the branch line coupler may be reduced to the two 2-port circuits shown in Figure 2.6. Since it is convenient to express all line lengths in terms of multiples of the shortest line, let $\theta' = \dfrac{\omega \pi}{\omega_0 4}$ and let

$$\Omega' = \tan \theta'. \qquad (2.33)$$

Then for the quarter wavelength transmission line of length $2\theta'$

$$\tan 2\theta' = \frac{2 \tan \theta'}{1 - \tan^2 \theta'} \qquad (2.34)$$
$$= \frac{2\Omega'}{1 - \Omega'^2}$$

and the ABCD matrix for the transmission line is

$$\begin{bmatrix} 1 - \Omega'^2 & j2\Omega' Z_0 \\ \dfrac{j2\Omega'}{Z_0} & 1 - \Omega'^2 \end{bmatrix}. \qquad (2.35)$$

For the even mode circuit in Figure 2.6a, the ABCD matrix is a cascade of a shunt open circuit stub with $Z_0 = a$, a transmission line with $Z_0 = b$, and another open circuit stub with $Z_0 = a$.

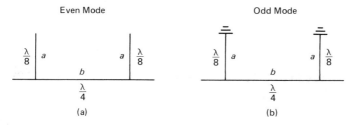

Figure 2.6. Even and odd mode equivalent circuits for the branch line coupler.

$$M_e(\theta) = \begin{bmatrix} 1 & 0 \\ \dfrac{j\Omega'}{a} & 1 \end{bmatrix} \dfrac{1}{1+\Omega'^2} \begin{bmatrix} 1-\Omega'^2 & j2\Omega'b \\ \dfrac{j2\Omega'}{b} & 1-\Omega'^2 \end{bmatrix} \begin{bmatrix} 1 & 0 \\ \dfrac{j\Omega'}{a} & 1 \end{bmatrix} \quad (2.36)$$

$$= \dfrac{1}{1+\Omega'^2} \begin{bmatrix} 1-\Omega'^2 - \dfrac{2\Omega'^2 b}{a} & j2\Omega'b \\ j2\Omega'\left[\dfrac{1-\Omega'^2}{a} + \dfrac{1}{b} - \dfrac{\Omega'^2 b}{a^2}\right] & \dfrac{2b}{a} + 1 - \Omega'^2 \end{bmatrix}$$

For synthesis purposes, this should be evaluated at the center frequency where $\Omega' = 1$.

$$M_e(1) = \begin{bmatrix} -\dfrac{b}{a} & jb \\ j\left(\dfrac{1}{b} - \dfrac{b}{a^2}\right) & -\dfrac{b}{a} \end{bmatrix} \quad (2.37)$$

In similar fashion the odd mode composite *ABCD* matrix is found.

$$M_o(\theta) = \begin{bmatrix} 1 & jb \\ \dfrac{-j}{a\Omega'} & -\dfrac{b}{a} \end{bmatrix} \dfrac{1}{1+\Omega'^2} \begin{bmatrix} 1-\Omega'^2 & j2\Omega'b \\ \dfrac{2\Omega'}{b} & 1-\Omega'^2 \end{bmatrix} \begin{bmatrix} 1 & 0 \\ \dfrac{-j}{a\Omega'} & 1 \end{bmatrix}$$

$$= \dfrac{\begin{bmatrix} 1-\Omega'^2 + \dfrac{2b}{a} & j2\Omega'b \\ j2\left[\dfrac{\Omega'}{b} - \dfrac{1-\Omega'^2}{a\Omega'} - \dfrac{b}{a^2 \Omega'}\right] & \dfrac{2b}{a} + 1 - \Omega'^2 \end{bmatrix}}{1+\Omega'^2} \quad (2.38)$$

At center frequency where $\Omega' = 1$

$$M_o(1) = \begin{bmatrix} \dfrac{b}{a} & jb \\ j\left(\dfrac{1}{b} - \dfrac{b}{a^2}\right) & \dfrac{b}{a} \end{bmatrix}. \quad (2.39)$$

The even and odd mode reflection coefficients are obtained from (2.29). Since in this case $A = D$, then $\Gamma_e = \Gamma_o = 0$ implies

$$B/Z_0 = CZ_0. \quad (2.40)$$

The constraints on the characteristic impedances a and b are obtained from (2.40) and either (2.37) or (2.39).

$$\frac{1}{a^2} = \frac{1}{b^2} - \frac{1}{Z_0^2} \tag{2.41}$$

The even and odd mode transmission coefficients obtained from (2.30) are simplified with the aid of (2.41).

$$T_e = -\frac{b}{a} - j\frac{b}{Z_0} \tag{2.42}$$

$$T_o = \frac{b}{a} - \frac{jb}{Z_0} \tag{2.43}$$

Substituting these into (2.26) and (2.27), the output amplitudes at ports 2 and 3 are found to be

$$b_2 = \frac{-jb}{Z_0} \tag{2.44}$$

$$b_3 = \frac{-b}{a} \tag{2.45}$$

which have a phase difference of 90°. For a 3 dB coupler, $b_2 = b_3 = 1/\sqrt{2}$, so that $b = Z_0/\sqrt{2}$ and $a = Z_0$. Using these values, the ideal frequency dependence of the branch line coupler is plotted in Figures 2.7 and 2.8.

2.2.3 The Rat Race Coupler

Just as was done for the branch line coupler, the rat race coupler shown in Figure 2.9 is divided into the even and odd mode equivalent circuits (Figure 2.10). The last stub is three times longer than the fundamental line length $\theta' = \frac{\omega\pi}{\omega_0 4}$, so that

$$\tan 3\theta' = \frac{\Omega'(3 - \Omega'^2)}{1 - 3\Omega'^2} . \tag{2.46}$$

If the characteristic impedance of the ring is a, the even mode ABCD matrix is

24 MICROWAVE SEMICONDUCTOR CIRCUIT DESIGN

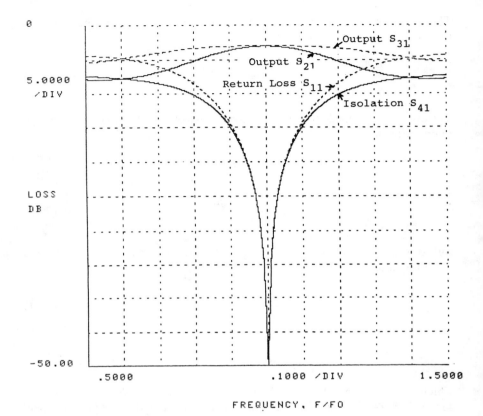

Figure 2.7. Frequency response of branch line coupler.

$$M_e(\theta) = \begin{bmatrix} 1 & 0 \\ \dfrac{-j\Omega'}{a} & 1 \end{bmatrix} \dfrac{1}{1+\Omega'^2} \begin{bmatrix} 1-\Omega'^2 & j2\Omega'a \\ \dfrac{j2\Omega'}{a} & 1-\Omega'^2 \end{bmatrix} \begin{bmatrix} 1 & 0 \\ \dfrac{j\Omega'(3-\Omega')}{a(1-3\Omega')} & 1 \end{bmatrix}$$

or

$$M_e(\theta) = \dfrac{\begin{bmatrix} \dfrac{1-10\Omega'^2+5\Omega'^4}{1-3\Omega'^2} & j2\Omega'a \\ \dfrac{2\Omega'}{a}(3-\Omega'^2) & 1-3\Omega'^2 \end{bmatrix}}{1+\Omega'^2} \quad (2.47)$$

PASSIVE MICROWAVE COMPONENTS 25

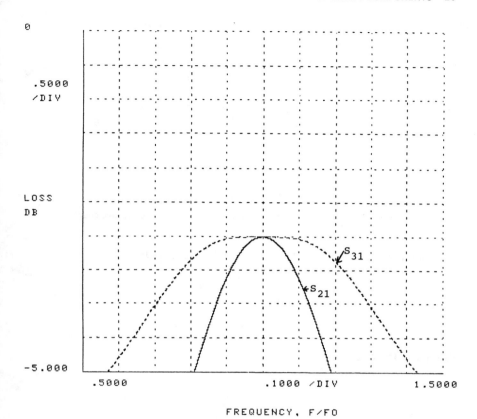

Figure 2.8. Outputs of the branch line coupler with expanded loss scale.

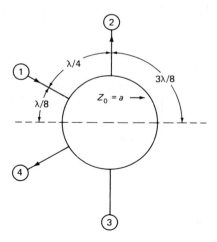

Figure 2.9. The rat race coupler.

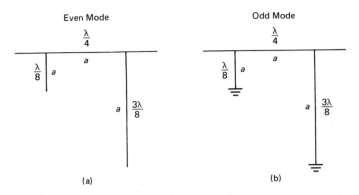

Figure 2.10. Even and odd mode equivalent circuits for the rat race coupler.

At center frequency where $\Omega' = 1$ this matrix reduces to

$$M_e(1) = \begin{bmatrix} 1 & ja \\ \dfrac{j2}{a} & -1 \end{bmatrix} \qquad (2.48)$$

Similarly the odd mode matrix is

$$M_o(\theta) = \begin{bmatrix} 1 & 0 \\ \dfrac{-j}{\Omega'a} & 1 \end{bmatrix} \dfrac{1}{1+\Omega'^2} \begin{bmatrix} 1-\Omega'^2 & j2\Omega'a \\ \dfrac{j2\Omega'}{a} & 1-\Omega'^2 \end{bmatrix} \begin{bmatrix} 1 & j2\Omega'a \\ \dfrac{-j(1-3\Omega'^2)}{a\Omega'(3-\Omega'^2)} & 1 \end{bmatrix},$$

$$M_o(\theta) = \dfrac{\begin{bmatrix} \dfrac{5-10\Omega'^2+\Omega'^4}{3-\Omega'^2} & j2\Omega'a \\ \dfrac{j2(3\Omega'^2-1)}{a\Omega'} & 3-\Omega'^2 \end{bmatrix}}{1+\Omega'^2} \qquad (2.49)$$

and at center frequency

$$M_o(1) = \begin{bmatrix} -1 & ja \\ \dfrac{j2}{a} & 1 \end{bmatrix}. \tag{2.50}$$

In this case $A \ne D$, in contrast to the branch line coupler, so conditions have to be found that ensure

$$b_1 = \tfrac{1}{2}(\Gamma_e + \Gamma_o) = 0. \tag{2.51}$$

From (2.48) and (2.50) the reflection coefficients at center frequency are

$$\Gamma_e = \frac{2 + j(a/Z_0 - 2Z_0/a)}{j(a/Z_0 + 2Z_0/a)} \tag{2.52}$$

$$\Gamma_o = \frac{-2 + j(a/Z_0 - 2Z_0/a)}{j(a/Z_0 + 2Z_0/a)}. \tag{2.53}$$

Substituting these into (2.51) gives

$$a = \sqrt{2}Z_0. \tag{2.54}$$

For this value of a the outputs at the four ports at the center frequency are

$$b_1 = 0 \tag{2.55}$$
$$b_2 = -j/\sqrt{2} \tag{2.56}$$
$$b_3 = 0 \tag{2.57}$$
$$b_4 = -j/\sqrt{2}. \tag{2.58}$$

In this case, the two output signals are in phase with one another. If the input were chosen to be at port 2, the outputs at ports 1 and 3 would be 180° out of phase with one another. Figures 2.7, 2.8, 2.11, and 2.12 show that the rat race coupler has somewhat broader bandwidth than the branch line coupler. The coupling and insertion loss remain within the range of 3 ± 0.05 dB for a 17% fractional bandwidth for the branch line coupler and for a 28% bandwidth for the rat race coupler. Similarly, the input VSWR < 1.1 over a 5.5% bandwidth for the branch line coupler and over a 16.33% bandwidth for the rat race coupler.

Reed and Wheeler[1] pointed out that broader bandwidth designs can be obtained by adding sections to the branch line coupler. However, the characteristic impedances of the lines become increasingly difficult to realize. Hughes and Wilson[2] described a variation of the rat race coupler (Figure 2.13). Their

28 MICROWAVE SEMICONDUCTOR CIRCUIT DESIGN

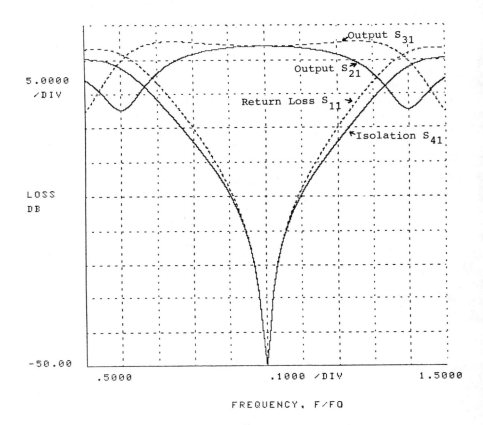

Figure 2.11. Frequency response of the rat race coupler.

circuit provided more bandwidth because of its inherent symmetry. In this case, the transmission coefficients to each of the two output ports is the same at off-resonant frequencies, in contrast to the rat race coupler shown in Figure 2.9. Although Hughes and Wilson use 100 ohm loads for the isolation resistors, the magnitude of the outputs are affected very little by the choice of the resistor for bandwidths up to 50%. Analysis of this circuit based on the above described bisection method of Reed and Wheeler shows that for bandwidths greater than 50%, load resistors of 35 ohms yield flatter transmission coefficients. Called the symmetric hybrid, this circuit is really the same as a 2-way Gysel power divider.[3] The Gysel power divider will be discussed in chapter 16.

The broadband balun described by Bex[4] is another modification of the rat race coupler (Figure 2.14). It has more bandwidth than the rat race coupler and more circuit parameters to vary in order to achieve the desired performance. The original paper showed no isolation resistors, but under balanced

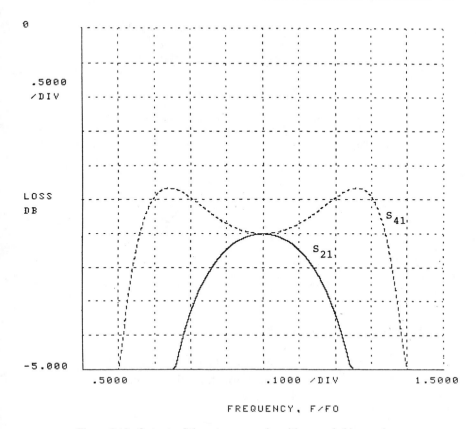

Figure 2.12. Outputs of the rat race coupler with expanded loss scale.

conditions, they could be added to the circuit at the end of the dashed lines with no affect on the circuit operation (Figure 2.14). The line lengths ℓ_1 and ℓ_2 are design variables that might be chosen as quarter wavelength and half wavelength lines respectively for an initial trial solution.

2.3 ATTENUATORS

Fixed microwave attenuators are used to provide a desired insertion loss while at the same time maintain a minimum reflection coefficient. The equivalent circuit for an attenuator is the T circuit given in Figure 2.15. The image impedance defined in chapter 1 for this circuit is

$$Z_I = \sqrt{R_1 R_2 + R_1^2/4} \qquad (2.59)$$

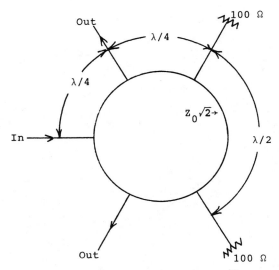

Figure 2.13. Symmetric hybrid circuit.[2,3]

and the transducer power loss is

$$\alpha^2 = \frac{\text{Power delivered to the load}}{\text{Available power from the source}} \qquad (2.60)$$
$$= \frac{R_0|i|^2/2}{v^2/(8R_0)}.$$

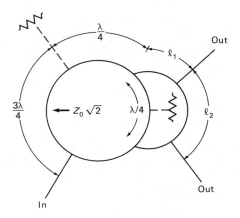

Figure 2.14. Broadband balun circuit.[4]

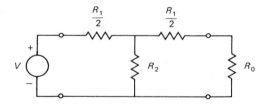

Figure 2.15. Equivalent circuit for the fixed attenuator.

Application of Kirchoff's current law to the circuit in Figure 2.15 provides the relationship between v and i, so that

$$\alpha^2 = \frac{4R_0^2 R_2^2}{[R_0^2 + R_0(R_1 + 2R_2) + R_1^2/4 + R_1 R_2]^2}. \quad (2.61)$$

For the attenuator to be reflectionless, the image impedance must be equal to the source resistance, i.e., $Z_I = R_0$. Substituting (2.59) into (2.61) yields

$$\alpha^2 = \frac{4R_2^2}{(2R_0 + R_1 + 2R_2)^2}. \quad (2.62)$$

The design objective is to find the values for R_1 and R_2 when the attenuation α and the image impedance R_0 are known. The value for R_2 in terms of R_1 is obtained from (2.59).

$$R_2 = \frac{R_0^2 - R_1^2/4}{R_1} \quad (2.63)$$

Substituting this into (2.62), the resulting expression may be solved for R_1.

$$R_1 = \frac{2R_0(1 - \alpha)}{1 + \alpha} \quad (2.64)$$

Table 2.1, which summarizes the resistance values needed for a single section attenuator, was generated based on this equation.

2.4 FERRITE DEVICES

Certain ferrite materials when biased by a DC magnetic field will exhibit nonreciprocal properties. A wave traveling in one direction will have a different propagation constant than a wave propagating in the opposite direction. Con-

Table 2.1. Attenuator Resistance Values for $R_o = 50$ ohms.

ATTENUATION, dB	SERIES RESISTANCE $R_1/2$, ohms	SHUNT RESISTANCE R_2, ohms
3	8.579	141.421
6	16.667	66.667
10	25.975	35.136
20	40.909	10.101
30	46.935	3.165
40	49.010	1.000

sequently the relationship between the RF magnetic flux density and the magnetic field intensity is the tensor permeability.[5]

$$\overset{\leftrightarrow}{\mu} = \mu_o \begin{bmatrix} 1 + \chi_{xx} & \chi_{xy} & 0 \\ \chi_{yx} & 1 + \chi_{yy} & 0 \\ 0 & 0 & 1 \end{bmatrix} \quad (2.65)$$

$$\omega_o = \gamma B_o \quad (2.66)$$

$$\chi_{xx} = \chi_{yy} = \gamma \mu_o M_s \frac{\omega_o}{\omega_o^2 - \omega^2} \quad (2.67)$$

$$\chi_{xy} = -\chi_{yx} = \gamma \mu_o M_s \frac{j\omega}{\omega_o^2 - \omega^2} \quad (2.68)$$

The quantity γ is the gyromagnetic constant, M_s is the DC saturation magnetization of the material, ω is the radian RF frequency, and B_o is the DC magnetic flux density going in the $\hat{z}$ direction. The asymmetry of the permeability makes it possible to construct useful nonreciprocal microwave elements such as circulators and isolators.

An ideal circulator is a three-port device in which all power entering port 1 is transmitted to port 2 and none to port 3; all power entering port 2 is transmitted to port 3, and none to port 1, etc. as indicated in Figure 2.16. An isolator is often made by simply terminating one of the ports with a matched load. The minimum requirements for an ideal circulator, found by Carlin[6], are that the three-port device is nonreciprocal, matched, and lossless. The second condition implies $S_{ii} = 0$ so the S matrix is reduced to

$$[S] = \begin{bmatrix} 0 & S_{12} & S_{13} \\ S_{21} & 0 & S_{23} \\ S_{31} & S_{32} & 0 \end{bmatrix} \quad (2.69)$$

The lossless feature implies the S matrix is unitary, i.e.,

$$[S][S^*]^T = [I] \quad (2.70)$$

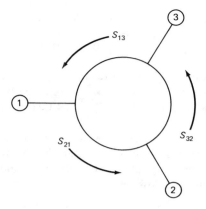

Figure 2.16. Three-port circulator.

where $[I]$ is the identity matrix. Performing the indicated multiplication yields the following results.

row	column		
2	3	$S_{21}S_{31}^* = 0$	(2.71)
3	3	$|S_{31}|^2 + |S_{32}|^2 = 1$	(2.72)
1	3	$S_{12}S_{32}^* = 0$	(2.73)
1	1	$|S_{12}|^2 + |S_{13}|^2 = 1$	(2.74)
1	2	$S_{13}S_{23}^* = 0$	(2.75)
2	2	$|S_{21}|^2 + |S_{23}|^2 = 1$	(2.76)

If it is assumed the transmission coefficient S_{21} from port 1 to port 2 is nonzero then

$$(2.71) \rightarrow S_{31} = 0$$
$$(2.72) \rightarrow |S_{32}| = 1$$
$$(2.73) \rightarrow S_{12} = 0$$
$$(2.74) \rightarrow |S_{13}| = 1$$
$$(2.75) \rightarrow S_{23} = 0$$
$$(2.76) \rightarrow |S_{21}| = 1$$

Since the phases of the nonzero elements are dependent on the choice of the reference plane, they may be chosen real. The scattering matrix for the lossless, matched, three-port, circuit has the properties of a circulator.

$$[S] = \begin{bmatrix} 0 & 0 & 1 \\ 1 & 0 & 0 \\ 0 & 1 & 0 \end{bmatrix} \qquad (2.77)$$

Achieving this result with a ferrite disk and a magnet was discussed by Bosma and later by Fay and Comstock.[8] More recently Wu and Rosenbaum[9] designed a wideband circulator from 7 to 12.4 GHz with some degradation in maximum isolation.

Design of a circulator requires choosing several design parameters. Among these is the ferrite disk radius, the disk height, the DC magnetic field intensity, and the impedance matching into the circulator. In addition, the desired frequency range and insertion loss tolerance govern the type of ferrite material.

REFERENCES

1. J. Reed and G. J. Wheeler, "A Method of Analysis of Symmetrical Four-Port Networks," *IRE Transactions on Microwave Theory and Techniques*, Vol. MTT-4, pp. 246–252, October 1956.
2. J. Hughes and K. Wilson, "High Power Multiple IMPATT Amplifiers," *Proc. European Microwave Conference*, pp. 118–122, 1974.
3. V. H. Gysel, "A New N-Way Power Divider/Combiner Suitable for High Power Applications," *1977 IEEE MTT-S—International Microwave Symposium Digest*, p. 116, May 1975.
4. H. Bex, "New Broadband Balun," *Electronics Letters*, Vol 11, pp. 47–48, January 23, 1975.
5. R. E. Collin, *Foundations for Microwave Engineering*. New York: McGraw-Hill, 1966.
6. H. J. Carlin, "Principles of Gyrator Networks," *Proc. of the Symposium on Modern Advances in Microwave Techniques*, Polytechnic Institute of Brooklyn, Vol. 4, pp. 175–204, November 8–10, 1954.
7. H. Bosma, "On Stripline Y-Circulation at UHF," *IEEE Transactions on Microwave Theory and Techniques*, Vol. MTT-12, pp. 61–72, January 1964.
8. C. E. Fay and R. L. Comstock, "Operation of the Ferrite Junction Circulator," *IEEE Transactions on Microwave Theory and Techniques*, Vol. MTT-13, pp. 15–27, January 1965.
9. Y. S. Wu and F. J. Rosenbaum, "Wideband Operation of Microstrip Circulators," *IEEE Transactions on Microwave Theory and Techniques*, Vol. MTT-22, pp. 849–856, October 1974.

Chapter 3
Impedance Transformers and Filters

3.1 INTRODUCTION

A microwave module may contain amplifiers, oscillators, phase shifters, switches, etc. It will most likely require a semiconductor device to be matched to the system impedance level, usually 50 ohms. The module is required to operate over a specified bandwidth. Moreover, certain frequency bands must often be kept out of the module. These requirements lead to the necessity of building impedance transformers and filters.

The purpose of the present chapter is to review some of the design techniques that are used in synthesizing impedance transformers and filters. Specific design procedures, along with the pertinent derivations, are given for a few of the most common techniques. A few examples are given in terms of ideal, lossless, dispersionless transmission lines. The problems of physical realization in various transmission line media along with accommodating discontinuity reactances are deferred until chapter 5.

The next section in this chapter contains a discussion of some simple narrowband impedance transformers. Following this, section 3.2.2 presents a complete design procedure for broadband quarter wavelength impedance transformers. The next major section is concerned with microwave filter theory. Lumped element filter synthesis is provided initially because microwave filter design borrows heavily from the well-developed lumped element theory. Subsequently, two microwave filter design techniques are discussed. The first in section 3.3.2 is termed *redundant filter synthesis*. It is based on Richard's transformation. In this transformation a lumped element capacitance is substituted by an open circuit stub, and a lumped element inductance is substituted by a short circuit stub. The second method, termed *non-redundant filter synthesis*, is discussed in section 3.3.3. In this method, unit elements (quarter wavelength transmission lines) play an essential role in the filtering action of the circuit. Section 3.4 presents the theory of coupled transmission lines and provides equivalent circuits for a coupled line whose four ports may be terminated in many different ways. Section 3.5 considers a third microwave filter technique based on the impedance inverter concept. Usually, this technique is limited to narrower filter bandwidths than the two previously mentioned methods. Two examples are given for this approximate technique: one based on coaxial line disks and the other on the widely used stripline coupled line.

3.2 IMPEDANCE TRANSFORMERS

The usual impedance transformation problem is to match a complex load or a real load to a real source impedance. For the complex load, maximum bandwidth is obtained if the impedance transformation is done with a small lumped element circuit positioned close to the load. The distributed element design is divided into two techniques: (1) narrowband matching of a complex load, and (2) broadband matching of a real load. These two basic approaches by no means exhaust the available design techniques for matching structures.

3.2.1 Narrowband Impedance Transformers

From the many techniques that have been developed to transform the impedance level of a microwave circuit, two will be described here. The classical laboratory approach is the double stub tuner. Two open circuit or short circuit stubs can of course be designed into a microstrip or stripline circuit. The desired impedance matching of a complex load can be achieved over a narrow range of frequencies.[1] An alternative approach is to choose a transmission line length and characteristic impedance that will provide the desired real resistance. In the circuit in Figure 3.1, the load impedance $R_L + jX_L$ is to be transformed to some desired real resistance R. The required values Z_0 and θ may be obtained by solving the transmission line equation.

$$R = Z_0 \frac{R_L + j(X_L + Z_0 \tan \theta)}{Z_0 - X_L \tan \theta + jR_L \tan \theta} \tag{3.1}$$

This can be arranged as two equations with two unknowns by separating the real and imaginary parts.

$$Z_0(1 - R_L/R) = X_L \tan \theta \tag{3.2}$$

$$\tan \theta = \frac{Z_0 X_L}{RR_L - Z_0^2} \tag{3.3}$$

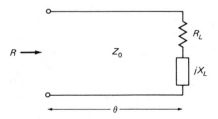

Figure 3.1. Terminated transmission line.

IMPEDANCE TRANSFORMERS AND FILTERS

Solving these expressions for θ and Z_0 gives the required transmission line parameters for matching a complex load to the desired real resistance R.

$$Z_0 = \frac{\sqrt{RR_L - R_L^2 - X_L^2}}{\sqrt{1 - R_L/R}} \tag{3.4}$$

$$\tan \theta = \frac{\sqrt{(1 - R_L/R)(RR_L - R_L^2 - X_L^2)}}{X_L} \tag{3.5}$$

By this approach, only those impedances can be matched that give a real value for Z_0. Furthermore, since there is no control over the bandwidth, this approach is not useful for broadband designs.

3.2.2 Quarter Wavelength Transformers

The most widely used method of broadband impedance matching between two real impedances is the quarter wavelength multisection impedance transformer. The design of these circuits is based on the assumption that multiple reflections can be neglected. From the diagram of the two discontinuities in Figure 3.2, the voltage reflection and transmission coefficients are given by the following relations.*

$$\Gamma_1 = \frac{Z_2 - Z_1}{Z_2 + Z_1} \tag{3.6}$$

$$\Gamma_2 = \frac{Z_1 - Z_2}{Z_1 + Z_2} = -\Gamma_1 \tag{3.7}$$

$$T_{21} = 1 + \Gamma_1 = \frac{2Z_2}{Z_1 + Z_2} \tag{3.8}$$

$$T_{12} = 1 + \Gamma_2 = \frac{2Z_1}{Z_1 + Z_2} \tag{3.9}$$

The reflection coefficient at the input to the circuit of Figure 3.3 will be the sum of a series of multiple reflections arising at each of the impedance discontinuities. If this circuit is excited by a reflectionless source with a wave of unit amplitude and zero phase, then the first discontinuity will reflect a wave Γ_1 and transmit the rest of the wave T_{21} to node 2. At node 2, part of this wave is reflected back toward node 1 with a reflection coefficient of Γ_2. Upon reaching node 1 part of this wave is transmitted to the source, thereby joining the orig-

---

*The voltage transmission coefficient is obtained by noting that the sum of the incident and reflected voltages (V^+ and V^-) is equal to the voltage across the load V_L. Hence, $V^+ + \Gamma V^+ = TV^+$ or $1 + \Gamma = T$.

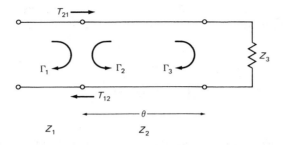

Figure 3.2. Multiple reflections in a single-section transformer.

inal reflected wave. The remainder is re-reflected back toward node 2 again. After accounting for the phase shift of the traveling wave, the output to the source at the left side of the circuit is given below. After each pass of the wave,

1) Γ_1
2) $T_{21}\Gamma_2 T_{12} e^{-j2\theta}$
3) $T_{21}\Gamma_2 \Gamma_a \Gamma_2 T_{12} e^{-j4\theta}$
4) $T_{21}\Gamma_2 \Gamma_a \Gamma_2 \Gamma_a \Gamma_2 T_{12} e^{-j6\theta}$
⋮

The reflection coefficient is simply the sum of all these output waves.

$$\Gamma = \Gamma_1 + T_{12}T_{21}\Gamma_2 e^{-j2\theta} \sum_{k=0}^{\infty} \Gamma_2^k \Gamma_a^k e^{-j2k\theta} \quad (3.10)$$

$$= \Gamma_1 + \frac{T_{12}T_{21}\Gamma_2 e^{-j2\theta}}{1 + \Gamma_a \Gamma_2 e^{-j2\theta}} \quad (3.11)$$

This last result is obtained by making use of the formula for the sum of an infinite geometric series. Replacing the transmission coefficient with the corresponding reflection coefficients in (3.7) to (3.9),

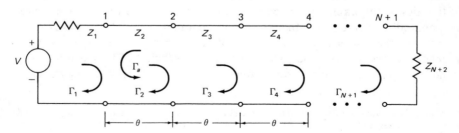

Figure 3.3. Multiple section impedance transformer.

IMPEDANCE TRANSFORMERS AND FILTERS 39

$$\Gamma = \frac{\Gamma_1 + \Gamma_2 e^{-j2\theta}}{1 + \Gamma_a \Gamma_2 e^{-j2\theta}}. \qquad (3.12)$$

For small reflections, $\Gamma_a\Gamma_2 \ll 1$ so that (3.12) is approximated by

$$\Gamma \cong \Gamma_1 + \Gamma_2 e^{-j2\theta}. \qquad (3.13)$$

Consequently, for small reflections the total reflection coefficient is simply that obtained by taking first order reflection coefficients into account. When this is extended to the multiple section impedance transformer shown in Figure 3.3, the input reflection coefficient is

$$\Gamma = \rho_1 + \rho_2 e^{-j2\theta} + \rho_3 e^{-j4\theta} + \cdots + \rho_{N+1} e^{-j2N\theta} \qquad (3.14)$$

where an N step transformer will have $N + 1$ reflection steps and $\rho_i = |\Gamma_i|$. Transformers are usually designed with symmetrical steps in the sense that $\rho_i = \rho_{N+2-i}$. This leaves an extra ρ_i when N is even. Consequently, the last term, A, will be different depending on whether N is even or odd.

$$\Gamma = \rho_1(e^{jN\theta} + e^{-jN\theta})e^{-jN\theta} + \rho_2(e^{j(N-2)\theta} + e^{-j(N-2)\theta})e^{-jN\theta} + \cdots + A$$

$$(3.15)$$

where

$$A = \begin{cases} \rho_{(N+1)/2}(e^{j\theta} + e^{-j\theta})e^{-jN\theta} & N \text{ odd} \\ \rho_{(N/2+1)}e^{-jN\theta} & N \text{ even} \end{cases} \qquad (3.16)$$

This is transformed into an expression in terms of cosines.

$$\Gamma = 2e^{-jN\theta}[\rho_1 \cos(N\theta) + \rho_2 \cos(N-2)\theta + \cdots + B] \qquad (3.17)$$

where

$$B = \begin{cases} \rho_{(N+1)/2} \cos\theta & N \text{ odd} \\ \tfrac{1}{2}\rho_{(N/2+1)} & N \text{ even} \end{cases} \qquad (3.18)$$

When these reflection coefficients ρ_i are chosen to have ratios corresponding to the like terms of the Nth order Chebyshev polynomial, the transformer will provide a minimum passband VSWR for a given output/input impedance ratio and bandwidth. This design technique was utilized by Cohn[2] to give a design for quarter wavelength transformers with an arbitrary number of sections. The reflection coefficients are chosen to be in the ratio of a set of yet to be determined numbers, a_i, that are related to the Chebyshev polynomial.

40 MICROWAVE SEMICONDUCTOR CIRCUIT DESIGN

$$\rho_1:\rho_2:\rho_3:\ldots:\rho_{N+1} = a_1:a_2:a_3:\ldots:a_{N+1} \tag{3.19}$$

To find a relationship between the ρ_i and the a_i, consider the mth reflection coefficient.

$$\rho_m = \frac{Z_{m+1} - Z_m}{Z_{m+1} + Z_m} \cong \frac{1}{2} \ln \frac{Z_{m+1}}{Z_m} \tag{3.20}$$

This approximation is achieved by retaining only the first term in a Taylor series expansion of a logarithm. The ratio (3.19) can be written in terms of an unknown proportionality constant K.

$$a_1 = K \ln \frac{Z_2}{Z_1}$$

$$a_2 = K \ln \frac{Z_3}{Z_2}$$

$$\vdots$$

$$a_{N+1} = K \ln \frac{Z_{N+2}}{Z_{N+1}}$$

The proportionality constant, K, may be eliminated by forming the ratio

$$\frac{a_m}{\sum_{i=1}^{N+1} a_i} = \frac{K \ln \dfrac{Z_{m+1}}{Z_m}}{K\left[\ln \dfrac{Z_2}{Z_1} + \ln \dfrac{Z_3}{Z_2} + \cdots + \ln \dfrac{Z_{N+2}}{Z_{N+1}}\right]}. \tag{3.21}$$

$$m = 1, 2, \ldots, N+1$$

(3.21)

By combining the logarithms in the right hand side of (3.21), the value for the $(m+1)$th impedance in the transformer may be found when the mth impedance is known.

$$\ln \frac{Z_{m+1}}{Z_m} = \frac{a_m \ln \dfrac{Z_{N+2}}{Z_1}}{\sum_{i=1}^{N+1} a_i} \tag{3.22}$$

The Chebyshev polynomial is chosen to have a magnitude ≤ 1 in the passband defined from θ_1 to $\theta_2 = \pi - \theta_1$ as shown in Figure 3.4. This is related to the fractional bandwidth by

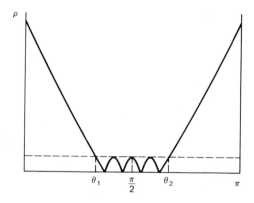

Figure 3.4. Chebyshev transformer reflection coefficient.

$$w = \frac{\Delta f}{f_0} = \frac{\theta_2 - \theta_1}{\pi/2} \tag{3.23}$$

or solving for θ_1

$$\theta_1 = \pi/2 - w\pi/4. \tag{3.24}$$

The Chebyshev polynomial defined as

$$T_n(x) = \cos[n(\arccos x)] \quad x \leq 1 \tag{3.25}$$
$$T_n(x) = \cosh[n(\operatorname{arccosh} x)] \quad x > 1 \tag{3.26}$$

will have the desired property for the transformer if the argument is changed to $x = \cos\theta/\cos\theta_1$. By equating the Chebyshev function to Γ in (3.17),

$$\Gamma = K'T_N(\cos\theta/\cos\theta_1)e^{-jN\theta} \tag{3.27}$$

the proportionality constants, a_i, may be found. A simple recursion formula for these a_i values is found in Cohn's paper.[2] These have been rearranged in tabular form in Table 3.1. Even though it is rare that a quarter wavelength transformer would need more than $N = 4$ sections, other applications might require more impedance steps. The table may be extended by using the following recursion formulas where A_{jk} is the jth row and kth column of the table. When j is even the last column is

$$A_{j,j/2+1} = 2x_0 A_{j-1,j/2} - A_{j-2,j/2} \quad N = j/2 + 1 \tag{3.28}$$

Table 3.1. Proportionality Constants a_i for the Chebyshev Transformer.

N	a_1	a_2	a_3	a_4
1	x_0			
2	x_0^2	$2x_0^2 - 2$		
3	x_0^3	$3x_0^3 - 3x_0$		
4	x_0^4	$4x_0^4 - 4x_0^2$	$6x_0^4 - 8x_0^2 + 2$	
5	x_0^5	$5x_0^5 - 5x_0^3$	$10x_0^5 - 15x_0^3 + 5x_0$	
6	x_0^6	$6x_0^6 - 6x_0^4$	$15x_0^6 - 24x_0^4 + 9x_0^2$	$20x_0^6 - 36x_0^4 + 18x_0^2 - 2$

$x_0 = \sec \theta_1$
$\theta_1 = \pi/2 - w\pi/4$
$w = $ fractional bandwidth

where $x_0 = \sec \theta_1$ and $A_{01} \triangleq 2$. For the rest of the table where j is odd or j is even with $k < j/2$

$$A_{jk} = (A_{j-1,k} + A_{j-1,k-1})x_0 - A_{j-2,k-1} \tag{3.29}$$

The entries in Table 3.2 illustrate the correspondence between these a_i values and the reflection coefficients of the symmetrical impedance transformer.

Consider as an example a two-section impedance transformer that matches 10 ohms to 100 ohms over a bandwidth of 8 to 12 GHz. The fractional bandwidth $w = 0.4$, so the angle θ_1 obtained from (3.24) is 1.257 and $x_0 = \sec \theta_1 = 3.236$. For $N = 2$, Table 3.1 gives $a_1 = 10.472$ and $a_2 = 18.944$. Since $Z_1 = 10$, (3.22) gives $Z_2 = 18.304$, then $Z_3 = 54.634$, and finally $Z_4 = 100$. A plot of the reflection coefficient for this design is shown in Figure 3.5. A practical realization of the transformer will require accounting for discontinuities associated with the abrupt changes in the impedances. The question of discontinuities will be considered later in chapter 4.

Table 3.2. Reflection Coefficients for Symmetrical Transformer.

N	ρ_1	ρ_2	ρ_3	ρ_4	ρ_5	ρ_6	ρ_7
1	a_1	a_1					
2	a_1	a_2	a_1				
3	a_1	a_2	a_2	a_1			
4	a_1	a_2	a_3	a_2	a_1		
5	a_1	a_2	a_3	a_3	a_2	a_1	
6	a_1	a_2	a_3	a_4	a_3	a_2	a_1

$a_i = a_{N+2-i}$

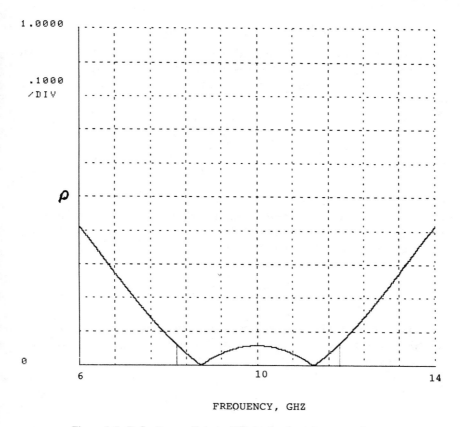

Figure 3.5. Reflection coefficient of Chebyshev impedance transformer.

3.3 MICROWAVE FILTER DESIGN

An electrical filter is a two-port circuit that has a desired specified response to a given input signal. Most filters are used to allow certain frequencies to be transmitted to the output load while rejecting the remaining frequencies. Many designs demand not only a specified amplitude response, but a specified phase response as well. Filters can be designed in either the frequency domain or the time domain, although the former approach is the more fully developed. The design of microwave filters owes much to the progress made in the design of lumped parameter filters. Consequently, this section provides a brief review of lumped element filter theory. Then two techniques that depend on lumped element filters will be described for the design of microwave filters.

3.3.1 Filter Approximation Problem

Modern filter synthesis is concerned with synthesizing a polynomial that approximates a desired transducer power gain function. This is the ratio of the power delivered to the load and the available power from the source as shown in Figure 3.6.

$$G = \frac{|V_L|^2/(2R_L)}{|V_g|^2/(8R_g)} \tag{3.30}$$

$$= \frac{P_{inc}(1 - |\Gamma|^2)}{P_{inc}} \tag{3.31}$$

The power P_{inc} is the incident power and Γ is the voltage reflection coefficient. The transducer gain for a lowpass filter that has a maximally flat passband is known as the Butterworth filter and is given by

$$G = \frac{1}{1 + (\omega/\omega_c)^{2n}}. \tag{3.32}$$

If steeper passband skirts are required for a given order n, then the Chebyshev response is often used. The penalty is small ripples in the passband. This gain function is given by

$$G = \frac{1}{1 + \epsilon^2 T_n^2(\omega/\omega_c)}. \tag{3.33}$$

Since the Chebyshev polynomial oscillates between ± 1 for $\omega \leq \omega_c$, the gain has a ripple in the passband with a maximum amplitude of $1/(1 + \epsilon^2)$. When $\omega > \omega_c$, the Chebyshev function becomes large and the gain quickly approaches zero as the frequency increases.

There are several choices for the approximation function. Figures 3.7 and 3.8 show examples of the insertion loss and transmission phase for the maximally flat, 0.1 dB ripple Chebyshev, linear phase, and Gaussian filters. The Chebyshev filter seems to provide the best response for most applications.

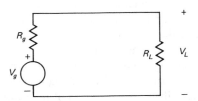

Figure 3.6. Circuit for transducer power gain calculation.

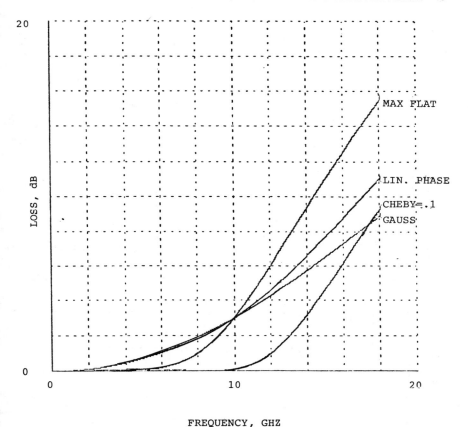

Figure 3.7. Insertion loss for different filter approximations.

Microwave filters are often based on lumped lowpass filters of the form shown in Figure 3.9. The relationship between these circuits in Figure 3.9 and the approximation gain function is obtained by equating the response to the filter with that of the gain function (3.31). For the Chebyshev function

$$1 + \epsilon^2 T_n^2(\omega/\omega_c) = \frac{1}{1 - |\Gamma|^2} \tag{3.34}$$

$$= \frac{Z_g + Z_L}{4R_g R_L}. \tag{3.35}$$

Equating coefficients of like powers of ω/ω_c, the g_i values for the lowpass prototype circuit may be found. Fortunately, these g_i values are extensively tabulated[3] for a large number of filter responses. In addition a simple recursion

46 MICROWAVE SEMICONDUCTOR CIRCUIT DESIGN

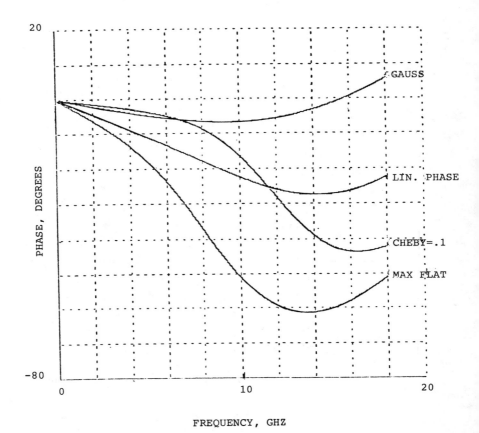

Figure 3.8. Insertion phase for different filter approximations.

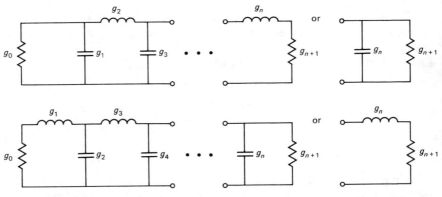

Figure 3.9. The lumped low-pass prototype filter starting with (a) shunt capacitor or (b) series inductor.

formula given below may be used to find the g_i values. For the Butterworth filter these values are

$$g_0 = g_{n+1} = 1$$
$$g_k = 2 \sin [(2k - 1)\pi/2n], \quad k = 1,2,3,\ldots,n$$

and for the Chebyshev filter

$$g_0 = 1$$
$$g_{n+1} = \begin{cases} 1 & n \text{ odd} \\ \tanh^2 (\beta/4) & n \text{ even} \end{cases}$$
$$g_1 = 2a_1/\gamma$$
$$g_k = \frac{4a_{k-1}a_k}{b_{k-1}g_{k-1}} \quad k = 2,3,\ldots,n$$
$$a_k = \sin [(2k - 1)\pi/2n] \quad k = 1,2,\ldots,n$$
$$b_k = \gamma^2 + \sin^2 (k\pi/n) \quad k = 1,2,\ldots,n$$
$$\beta = \ln [\coth (A_m/17.37)]$$
$$A_m = 10 \log (\epsilon^2 + 1)$$
$$\gamma = \sinh (\beta/2n).$$

Because the input and output impedances for the even order Chebyshev filter are different, most designs use the odd order polynomial. However, if an even order polynomial is required, Lind[4] has developed a Chebyshev-like approximation function that gives equal terminations.

Transformation of this lowpass prototype to the desired impedance level is accomplished by multiplying all resistances R and inductances L_k by the desired impedance level R_L, and dividing all capacitances C_k by this factor.

$$L_k' = R_L L_k \quad (3.36)$$
$$C_k' = C_k/R_L \quad (3.37)$$
$$R' = R_L R \quad (3.38)$$

The cutoff frequency is adjusted from 1 to ω_c by dividing the frequency sensitive reactances by ω_c.

$$\omega = \omega'/\omega_c \quad (3.39)$$
$$L_k' = L_k/\omega_c \quad (3.40)$$
$$C_k' = C_k/\omega_c \quad (3.41)$$
$$R' = R \quad (3.42)$$

Thus the lumped, lowpass circuit, designed to have a cutoff frequency of 1 radian/sec and a terminating resistance of 1 ohm, can be readily transformed to a circuit with any terminating impedance and cutoff frequency. If s_n represents the complex normalized radian frequency for the lowpass prototype circuit, then the highpass circuit is obtained by replacing s_n in the lowpass circuit with

$$s \leftarrow s_c/s_n \tag{3.43}$$

The obvious consequence of this is that reactances that varied directly with frequency in the lowpass circuit now vary inversely with frequency in the highpass circuit and vice versa. The lowpass to bandpass transformation is obtained by replacing s_n with

$$s \leftarrow \frac{1}{w}\left(\frac{s}{\omega_o} + \frac{\omega_o}{s}\right) \tag{3.44}$$

where the fractional bandwidth is

$$w = \frac{\omega_2 - \omega_1}{\omega_o} \tag{3.45}$$

and ω_o is the geometric mean of the passband edges.

$$\omega_o = \sqrt{\omega_1 \omega_2} \tag{3.46}$$

Finally the transformation for the bandstop filter is the reciprocal of the bandpass transformation.

$$s \leftarrow w\left(\frac{s}{\omega_o} + \frac{\omega_o}{s}\right)^{-1} \tag{3.47}$$

The results of these transformations are illustrated in Figure 3.10 where the circuit elements for the highpass bandpass, and bandstop filters are all derivable from the lowpass L and C.

Many microwave filters are designed with commensurate line lengths. Owing to the periodicity of the tangent function in the transmission line equation, the simple lowpass or highpass filter provides a bandpass characteristic. Also, many microwave filter designs make direct use of the lumped bandpass filter prototype as a stepping stone for the microwave realization. The first technique will be discussed in the following pages while the latter will be discussed in section 3.5.

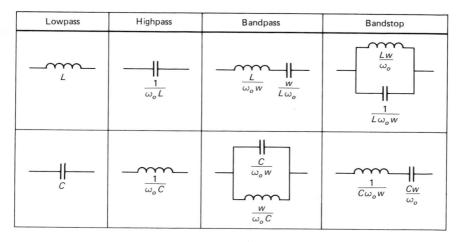

Figure 3.10. Element transformation from the lowpass prototype circuit.

3.3.2 Redundant Element Synthesis

Filters could be realized almost directly from lumped element prototype circuits except that the distance between circuit elements is not negligible at microwave frequencies. In this section a technique for transforming a lumped element circuit to a microwave circuit with the aid of Richard's transform is discussed. To circumvent the problem of nonzero distances between circuit elements, Kuroda's identities are employed to physically separate circuit elements without compromising the function of the filter. The term *redundant* arises from the necessity of introducing into the design extra quarter wavelength lines known as unit elements. These redundant unit elements do not contribute to the filtering action.

3.3.2.1 Richard's Transformation.
Richard[5,6] showed in 1948 that distributed networks composed of commensurate open and short circuit transmission lines could be synthesized like a lumped RLC network. Richard's transformation uses the complex frequency variable

$$S = \Sigma + j\Omega = \tanh(\ell s/v). \tag{3.48}$$

The length ℓ is the shortest commensurate line length, v is the wave velocity in the transmission media, and $s = \sigma + j\omega$ is the complex frequency. When $\sigma = 0$, then

$$\Omega = \tan\left(\frac{\omega \pi}{\omega_0 2}\right) \tag{3.49}$$

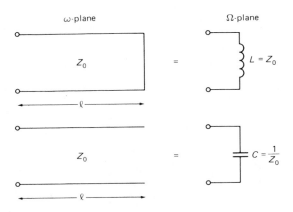

Figure 3.11. Ω-plane inductance and capacitance.

The transmission lines are a quarter wavelength long ($\ell = \lambda_o/4$) at the frequency ω_o. Consequently, a short circuit stub may be represented as an S-plane inductance L and an open circuit stub as an S-plane capacitance C (Figure 3.11).

$$j\Omega L = jZ_0 \tan\left(\frac{\omega\pi}{\omega_o 2}\right) = jZ_0\Omega \qquad (3.50)$$

$$j\Omega C = j \tan\left(\frac{\omega\pi}{\omega_o 2}\right)/Z_o = j\Omega/Z_0 \qquad (3.51)$$

With this transformation the analytical techniques found in lumped element synthesis techniques such as described by Guillemin[7] may be used in the S-plane to synthesize distributed networks. For example, a simple lowpass Butterworth filter obtained by conventional lumped element synthesis can be used as the basis for a microwave filter (Figure 3.12a). The lumped lowpass circuit frequency passband is $0 < \omega < 1$, and the corresponding passband for the microwave filter in Figure 3.12b and 3.12c is $0 < \Omega < 1$. The microwave filter will have a passband $0 < \omega < \omega_o/2$, so the stub length will be $\ell = \lambda/8$ at the cutoff frequency. The problem of the proximity of the various elements with one another may be alleviated by introducing redundant unit elements and making use of the Kuroda identities.

3.3.2.2 Kuroda Identities. The unit element has no equivalent in lumped element circuit theory and therefore introduces a new design variable in microwave filter design. A unit element is a quarter wavelength transmission line with a specified characteristic impedance. The S-plane inductors and capaci-

IMPEDANCE TRANSFORMERS AND FILTERS   51

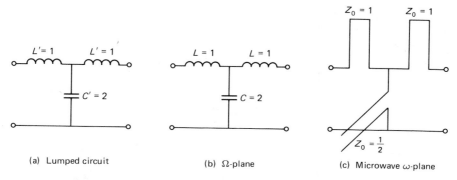

Figure 3.12. Microwave Butterworth filter synthesis from a lumped element design.

tors can be represented by a unit element in cascade with a short or open circuit respectively, as indicated in Figure 3.13. The four Kuroda identities are listed in Figure 3.14. These identities may be verified by showing that the matrices of each pair of circuits are equal. For the first pair the left-hand circuit is described by the ABCD matrix that follows.

$$\begin{bmatrix} 1 & 0 \\ S/Z_2 & 1 \end{bmatrix} \frac{1}{\sqrt{1-S^2}} \begin{bmatrix} 1 & SZ_1 \\ S/Z_1 & 1 \end{bmatrix}$$

$$= \frac{1}{\sqrt{1-S^2}} \begin{bmatrix} 1 & SZ_1 \\ S\left(\dfrac{1}{Z_1} + \dfrac{1}{Z_2}\right) & S^2 \dfrac{Z_1}{Z_2} + 1 \end{bmatrix} \quad (3.52)$$

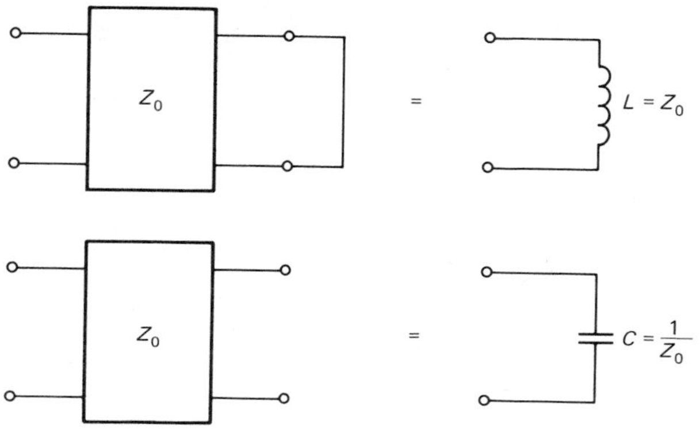

Figure 3.13. Equivalence of circuit with unit element Z_0 and S-plane L or C.

52 MICROWAVE SEMICONDUCTOR CIRCUIT DESIGN

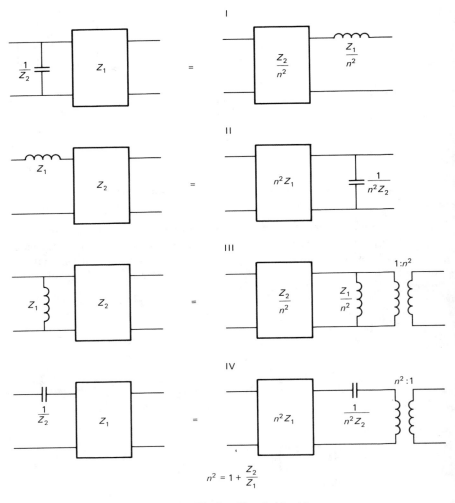

Figure 3.14. The four Kuroda identities.

The corresponding matrix for the right-hand circuit is

$$\frac{1}{\sqrt{1-S^2}} \begin{bmatrix} 1 & \dfrac{SZ_2}{n^2} \\ \dfrac{Sn^2}{Z_2} & 1 \end{bmatrix} \begin{bmatrix} 1 & \dfrac{SZ_1}{n^2} \\ 0 & 1 \end{bmatrix}$$

$$= \frac{1}{\sqrt{1-S^2}} \begin{bmatrix} 1 & S\left(\dfrac{Z_1}{n^2} + \dfrac{Z_2}{n^2}\right) \\ \dfrac{Sn^2}{Z_2} & S^2\dfrac{Z_1}{Z_2} + 1 \end{bmatrix}. \quad (3.53)$$

IMPEDANCE TRANSFORMERS AND FILTERS   53

The corresponding matrix elements of (3.52) and (3.53) are equal if $n^2 = 1 + Z_2/Z_1$. The second Kuroda identity can be verified in the same way or can be seen directly by inspection of the first identity. The left-hand circuit of the third identity is described by the following ABCD matrix.

$$\begin{bmatrix} 1 & 0 \\ \frac{1}{SZ_1} & 1 \end{bmatrix} \frac{1}{\sqrt{1-S^2}} \begin{bmatrix} 1 & SZ_2 \\ \frac{S}{Z_2} & 1 \end{bmatrix}$$

$$= \frac{1}{\sqrt{1-S^2}} \begin{bmatrix} 1 & SZ_2 \\ \frac{1}{SZ_1} + \frac{S}{Z_2}\frac{Z_2}{Z_1} + 1 \end{bmatrix} \quad (3.54)$$

The corresponding ABCD matrix for the right-hand circuit is

$$\frac{1}{\sqrt{1-S^2}} \begin{bmatrix} 1 & \frac{SZ_2}{n^2} \\ \frac{Sn^2}{Z_2} & 1 \end{bmatrix} \begin{bmatrix} 1 & 0 \\ \frac{n^2}{SZ_1} & 1 \end{bmatrix} \begin{bmatrix} \frac{1}{n^2} & 0 \\ 0 & n^2 \end{bmatrix}$$

$$= \frac{1}{\sqrt{1-S^2}} \begin{bmatrix} \frac{1}{n^2} + \frac{Z_2}{Z_1 n^2} & SZ_2 \\ \frac{S}{Z_2} + \frac{1}{SZ_1} & n^2 \end{bmatrix} \quad (3.55)$$

Equations (3.54) and (3.55) are identical if $n^2 = 1 + Z_2/Z_1$. Finally the fourth Kuroda identity is derived in the same fashion as the third.

Since the inductors and capacitors in Figure 3.14 are S-plane circuit elements, they represent either short or open circuit stubs. The first identity then states that a shunt open circuit stub in cascade with a unit element has the same electrical characteristics as a unit element cascaded with a short circuit series stub. The Kuroda identities are useful in transforming a circuit design that may be unrealizable in the given circuit media into one that is realizable. They are also used to insert unit elements between S-plane reactances and to separate them from one another.

3.3.2.3 Filter Design Using Redundant Unit Elements.
The realization of the ladder network in Figure 3.12 is complicated by the requirement that all the microwave stubs are located at one point on the transmission line. This realization problem may be alleviated for an n element ladder network by introducing $(n - 1)$ unit elements at the terminals of the circuit. The characteristic impedance of the unit elements must all be equal to the termination resistance.[8] These added unit elements do not affect the amplitude response of

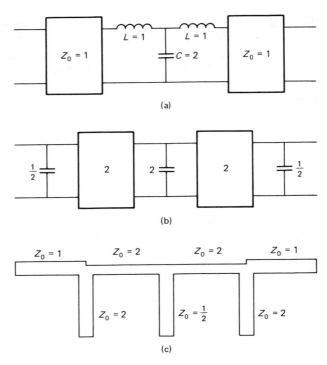

Figure 3.15. Microwave filter realization by (a) adding a redundant unit element, (b) applying Kuroda identity, and (c) stripline fabrication.

the circuit. Using the first two Kuroda identities in Figure 3.14, the unit elements are transformed inside the ladder network until all the S-plane L's and C's are separated by a unit element. In this way the circuit becomes physically realizable.

The simple normalized filter in Figure 3.12 provides a good example of the use of unit elements. A unit element of value 1 is introduced at each end of the circuit (Figure 3.15a). Using the first Kuroda identity, the unit elements are transferred inside the ladders, and the S-plane inductors are transformed into S-plane capacitors. This circuit is therefore realized as three open circuit stubs separated by quarter wavelength transmission lines as shown in Figure 3.15c.

3.3.3 Non-Redundant Circuit Synthesis

Microwave circuit synthesis may also be carried out by making use of unit elements as an integral part of the filter design. This non-redundant synthesis method relies on extracting unit elements from the impedance function using Richard's theorem[6] rather than L's and C's. There is no analogous procedure in the theory of lumped circuit synthesis.

Richard's theorem states that if the driving point impedance $Z(S)$ is positive real and rational in S, then a unit element of value $Z(1)$ may be extracted from the impedance function. This leaves a factor of $(S - 1)$ in the numerator and denominator that cancels out, but does not reduce, the order of the impedance function. If however $Z(1) = -Z(-1)$, then an added factor $(S + 1)$ also cancels from the numerator and denominator. This does reduce the order of the impedance function. The non-redundant approximation problem is therefore to provide an impedance function $Z(S)$ with $Z(1) = -Z(-1)$.

Consider the circuit in Figure 3.16 which consists of a unit element with characteristic impedance Z_0 terminated with a load Z_L. The input impedance is given by the transmission line equation.

$$Z(S) = Z_0 \frac{Z_L(S) + Z_0 S}{Z_0 + Z_L(S)S} \qquad (3.56)$$

At $S = 1$, the input impedance is equal to the characteristic impedance of the unit element.

$$Z(1) = Z_0 \frac{Z_L(1) + Z_0}{Z_0 + Z_L(1)} = Z_0 \qquad (3.57)$$

Solving (3.56) for the load impedance $Z_L(S)$ and substituting $Z(1) = Z_0$,

$$Z_L(s) = Z(1) \frac{Z(S) - SZ(1)}{Z(1) - SZ(S)}. \qquad (3.58)$$

Consequently when a unit element of value $Z(1)$ is extracted from the impedance function, the remainder is that given by (3.58). If $Z(1) = -Z(-1)$ then the order of the remaining impedance function is reduced. Reduction of the impedance function continues by reapplication of Richard's theorem or by extraction of S-plane L's and C's—whichever is appropriate.

A unit element can be extracted from an impedance function based on the Butterworth or Chebyshev approximations given by (3.32) or (3.33), respectively. However, the degree of the resulting impedance function is not reduced and the extracted unit element is a redundant element. Horton and Wenzel[9]

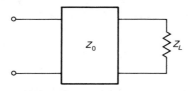

Figure 3.16. Unit element terminated with a load, Z_L.

derived generalized Butterworth and Chebyshev functions that give a maximally flat or equal ripple response.

The choice of these functions are obtained in such a way that both LC circuit elements and unit elements are incorporated into the approximation function. The S matrix discussed in chapter 2 can be rearranged in a form that allows representation of a cascade of circuit elements by the matrix of each individual element.

$$\begin{bmatrix} b_1 \\ a_1 \end{bmatrix} = \frac{1}{S_{21}} \begin{bmatrix} -\Delta & S_{11} \\ -S_{22} & 1 \end{bmatrix} \begin{bmatrix} a_2 \\ b_2 \end{bmatrix} \qquad (3.59)$$

The determinate of the S matrix used in (3.59) is Δ. A highpass filter is designed to have m S-plane series C's and shunt L's, and to have n unit elements. Horton and Wenzel[9] showed that if care is taken to keep the filter from being redundant, such as having two adjacent shunt L's, the square of the row 1 column 2 element of (3.59) is of the form

$$\left| \frac{S_{11}}{S_{21}} \right|^2 = \left[\frac{S_c}{S} \right]^{2m} \left[\frac{1 - S_c^2}{1 - S^2} \right]^n P_{m+n} \left(\frac{S^2}{S_c^2} \right) \qquad (3.60)$$

where P_{m+n} is an $m + n$ degree polynomial. Now consider the frequency transformations*

$$Z^2 = 1 + (S/\Omega_c)^2 \qquad (3.61)$$
$$Z^2 = 1 + (\Omega_c/S)^2 \qquad (3.62)$$

for the highpass and lowpass circuits respectively. These functions map the passband in the S plane into the imaginary axis in the Z plane and the stopbands in the S plane into the real axis in the Z plane. For the highpass circuit when $S = j\Omega > j\Omega_c$, Z is imaginary while for $\Omega < \Omega_c$, Z is real.

The function defined by

$$e^{j2\phi(Z)} = \frac{\Pi(Z_i + Z)}{\Pi(Z_i - Z)} \qquad (3.63)$$

is an allpass transfer function that is defined to have real roots or conjugate pair roots on the imaginary axis of Z. Also, the function

$$\tfrac{1}{2}(1 + e^{j2\phi(Z)}) \qquad (3.64)$$

---

*In this discussion Z should not be confused with impedance.

IMPEDANCE TRANSFORMERS AND FILTERS   57

is guaranteed to be positive real. Moreover since the roots Z_i are either real or occur in complex conjugate pairs, the exponent in (3.63) is an odd function.

$$e^{j2\phi(Z)} = e^{-j2\phi(-Z)} \tag{3.65}$$

The filter function formed by the product of (3.64) with its complex conjugate can be written in three equivalent forms.

$$|F_H(Z)|^2 = \tfrac{1}{4}(1 + e^{j2\phi(Z)})(1 + e^{j\phi(-Z)}) \tag{3.66}$$
$$= \cos^2 \phi(Z) \tag{3.67}$$
$$= \frac{[\Pi(Z_i + Z) + \Pi(Z_i - Z)]^2}{4\Pi(Z_i^2 - Z^2)} \tag{3.68}$$

The denominator of (3.68) when transformed back into the S plane is

$$Z_i^2 - Z^2 = \frac{S_i^2 - S^2}{\Omega_c^2} \tag{3.69}$$

so that by choosing $S_i = 0$, the poles of (3.68) are the same as those originating from the highpass elements of $|S_{11}|^2/|S_{21}|^2$ in (3.60). By choosing $S_i = 1$, the poles in (3.68) are the same as those in (3.60) originating from the unit elements.

The ith component of the phase angle in (3.63) is given by

$$\exp(j2\phi_i) = \frac{Z_i + Z}{Z_i - Z} \tag{3.70}$$

so that

$$F_H(Z) = \cos(\Sigma \phi_i). \tag{3.71}$$

The individual contributions to the phase angle in (3.70) can be expressed in the S plane as follows.

$$\cos \phi_i = \frac{Z_i}{\sqrt{Z_i^2 - Z^2}} \tag{3.72}$$
$$= \left[\frac{S_i^2 - S_c^2}{S_i^2 - S^2}\right]^{1/2} \tag{3.73}$$

For each of the m highpass elements, $S_i = 0$

$$\cos \phi_{hp} = \frac{jS_c}{jS} \tag{3.74}$$

and for each of the n unit elements, $S_i = 1$

$$\cos \phi_{ue} = \left[\frac{1 - S_c^2}{1 - S^2}\right]^{1/2} \tag{3.75}$$

From (3.71) the highpass approximation function is

$$F_H = \cos(m\phi_{hp} + n\phi_{ue}) \tag{3.76}$$

and after substitution of (3.74) and (3.75)

$$F_H = \cos\left[m \arccos\left(\frac{S_c}{S}\right)\right] \cos\left\{n \arccos\left[\frac{1 - S_c^2}{1 - S^2}\right]\right\}$$
$$- \sin\left[m \arccos\left(\frac{S_c}{S}\right)\right] \sin\left\{n \arccos\left[\frac{1 - S_c^2}{1 - S^2}\right]\right\} \tag{3.77}$$

The gain G or transmission coefficient T is

$$|T|^2 = \frac{1}{1 + |\Gamma|^2/|T|^2} \tag{3.78}$$

where the S parameter notation in (3.60) has been replaced with Γ and T. The filter function for the highpass filter in (3.77) has a maximum amplitude of 1 and can be rewritten as

$$\frac{|\Gamma|^2}{|T|^2} = \epsilon^2 \left[T_m\left(\frac{S_c}{S}\right) T_n\left(\sqrt{\frac{1 - S_c^2}{1 - S^2}}\right) \right.$$
$$\left. - U_m\left(\frac{S_c}{S}\right) U_n\left(\sqrt{\frac{1 - S_c^2}{1 - S^2}}\right) \right]^2 \tag{3.79}$$

where

$$T_n(x) = \cos[n \arccos x] \qquad x \leq 1 \tag{3.80}$$
$$U_n(x) = \sin[n \arccos x] \qquad x \leq 1. \tag{3.81}$$

The small number ϵ defines the passband ripple. The generalized Butterworth function was derived in similar fashion by Horton and Wenzel.[9]

IMPEDANCE TRANSFORMERS AND FILTERS   59

$$\frac{|\Gamma|^2}{|T|^2} = \left(\frac{S_c}{S}\right)^{2m} \left(\sqrt{\frac{1 - S_c^2}{1 - S^2}}\right)^{2n} \quad (3.82)$$

The lowpass S-plane functions are obtained by replacing S with $1/S$ and S_c with $1/S_c$ in (3.79) and (3.82).

The function $U_n(x)$ is not the Chebyshev polynomial of the second kind as normally defined and should not be confused with it. The $T_n(x)$ and $U_n(x)$ polynomials can be easily obtained from the following recursion formulas. Beginning with

$$T_0(x) = 1$$
$$U_0(x) = 0$$
$$T_1(x) = x$$

these recursion formulas are

$$T_n(x) = 2xT_{n-1}(x) - T_{n-2}(x) \qquad n = 1,2,\ldots \quad (3.83)$$
$$U_n(x) = xU_{n-1}(x) + \sqrt{1 - x^2}\,T_{n-1}(x) \qquad n = 1,2,\ldots \quad (3.84)$$

These polynomials for $n = 0$ to 5 are shown in Table 3.3.

Since these filters are constructed from commensurate transmission lines, the passbands repeat every 180 electrical degrees. The highpass functions in (3.79) and (3.82) are thus used to build bandpass filters from θ_c to $180° - \theta_c$. The lowpass forms of (3.79) and (3.82) are used to build bandstop filters from θ_c to $180° - \theta_c$ where $S = j \tan \theta$.

The synthesis procedure begins by choosing the filter fractional bandwidth w and from this θ_c.

$$\theta_c = \frac{\pi}{2}(1 - w/2) \quad (3.85)$$

Table 3.3. The Polynomials $T_n(x)$ and $U_n(x)$

n	$T_n(x)$	$U_n(x)$
0	1	0
1	x	$\sqrt{1 - x^2}$
2	$2x^2 - 1$	$\sqrt{1 - x^2}\,2x$
3	$4x^3 - 3x$	$\sqrt{1 - x^2}\,(4x^2 - 1)$
4	$8x^4 - 8x^2 + 1$	$\sqrt{1 - x^2}\,(8x^3 - 4x)$
5	$16x^5 - 20x^3 + 5x$	$\sqrt{1 - x^2}\,(16x^4 - 12x^2 + 1)$

The total number of poles is $m + n$, where m is the number of highpass elements and n is the number of nonredundant unit elements. The choice of m and n is based on realizability constraints. Horton and Wenzel[9] point out that for narrowband designs, physical layout dictates the choice of m and n, while for broadband designs, physical size for a given selectivity may be the limiting factor. Since stub networks are difficult to build for narrow bandwidth filters, coupled line structures are often used. In this case m is small and n is large. The familiar coupled line filter uses $m = 1$. For wideband designs, the filter response differs with different choices of m and n. Usually stub filters are superior, the best being obtained with $n = 0$. However, such a filter is difficult to fabricate because of the proximity of all the elements to a single junction. A large n/m ratio leads to large circuits and poor response because the unit elements are not as good contributors to filter response as the L and C elements. Broadband designs may still need the addition of some nonredundant unit elements to achieve a practical design.

The next step in the synthesis procedure is to obtain the magnitude squared of the reflection coefficient from (3.78).

$$\Gamma(S)\Gamma(-S) = |\Gamma|^2 = 1 - |T|^2 \tag{3.86}$$

The poles of $|\Gamma|^2$ are found by obtaining the roots of the denominator polynomial, usually requiring a computer program. The left-hand plane poles are chosen to make $\Gamma(S)$ realizable. The only restriction on the numerator is that the zeros be in complex conjugate pairs such that $\Gamma(S)\Gamma(-S) = |\Gamma(S)|^2$. Usually the left-hand plane zeros are chosen so that they do not lie on the imaginary axis. The normalized impedance function is then

$$Z(S) = \frac{1 + \Gamma(S)}{1 - \Gamma(S)} \tag{3.87}$$

from which the L's, C's, and unit elements may be extracted. Use of the function (3.79) or (3.82) with $n \neq 0$ will guarantee that the application of Richard's theorem n times to (3.87) will reduce the order of the polynomials in the numerator and denominator.

As an example consider the 3-pole Chebyshev filter. In previous examples in this chapter, no mention was made of the desired bandwidth during the synthesis process because bandwidth scaling was easily accomplished after the normalized circuit had been synthesized. Now, however, bandwidth scaling is not simple, so a choice of the desired bandwidth must be made at the outset of the design procedure. For this example, the fractional bandwidth is chosen to be 50%, the passband ripple is 0.1 dB, the number of unit elements is 2 and the number of highpass elements is 1. It is convenient to choose $n = 2$ rather than 1 so as to not end up with different impedance levels at the input and output of the circuit. For the choices of m, n, ϵ, and w given above,

IMPEDANCE TRANSFORMERS AND FILTERS   61

$$S_c = j \tan(3\pi/8) \triangleq jx. \qquad (3.88)$$

Substituting this into (3.79)

$$\frac{|\Gamma|^2}{|T|^2} = -\frac{\epsilon^2[S^2(x + 2\sqrt{1+x^2}) + (2x^2+1)x + 2x^2\sqrt{1+x^2}]^2}{S^2(1-S^2)^2 - \epsilon^2[S^2(x+2\sqrt{1+x^2}) + x(2x^2+1) + 2x^2\sqrt{1+x^2}]^2}$$

(3.89)

From (3.78) $|\Gamma(S)|^2$ can be obtained. Putting in the actual numbers gives

$$|\Gamma|^2 = \frac{[S^2 1.166 + 9.312]^2}{-S^6 + S^4 3.360 + S^2 20.718 + 86.722}. \qquad (3.90)$$

This and the following numerical work was done using 7 significant figures, but the numbers shown have been truncated to three decimal places. In general a root-solver program is needed to factor the denominator, but in this case one root in S^2 can be found from Newton's method and the two remaining roots by synthetic division. Choosing the left-hand plane roots,

$$\Gamma(S) = \pm \frac{S^2 1.166 + 9.312}{(S + 2.755)(S + 0.795 + j1.658)(S + 0.795 - j1.658)}$$
$$= \pm \frac{S^2 1.166 + 9.312}{S^3 + S^2 4.345 + S 6.218 + 5.066} \qquad (3.91)$$

By choosing the negative sign, the circuit topology can be of the form of a unit element, a shunt short circuit stub, and a unit element all in cascade. From (3.87) and (3.91)

$$Y_1(S) = \frac{S^3 + S^2 5.511 + S 7.759 + 18.625}{S^3 + S^2 3.179 + S 7.759}. \qquad (3.92)$$

Since $Z_1(1) = -Z_1(-1) = 0.363 = 1/Y_1(1)$, the application of Richard's theorem reduces the numerator and denominator polynomial of (3.92) to give

$$Y_2(S) = Y_1(1) \frac{S^2 Y_1(1) + S 7.759 + 18.625}{S[S + Y_1(1)]}. \qquad (3.93)$$

The next step is to remove a shunt S-plane inductance. The remaining admittance is

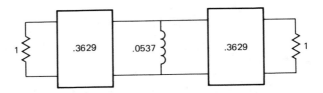

Figure 3.17. The 50% bandwidth Chebyshev filter example.

$$Y_3(S) = Y_2(S) - 1/LS. \qquad (3.94)$$

By choosing $L = 0.0537$, the constant term in the numerator of (3.93) is removed leaving

$$Y_3(S) = Y_1(1) \frac{SY_1(1) + 1}{Y_1(1) + S}. \qquad (3.95)$$

Consequently the final circuit element is a unit element with a value equal to the first. The final circuit is shown in Figure 3.17.

3.4 COUPLED TRANSMISSION LINES

Structures considered thus far have been open circuit stubs, short circuit stubs, and unit elements. When two transmission lines are physically close together, the electromagnetic field on one will affect the other so that energy can be coupled from one line to the other. The coupled lines are assumed to support only transverse electromagnetic (TEM) waves so that a DC capacitance equivalent circuit can model the coupled transmission line circuit. A pair of coupled lines with a common ground are shown in Figure 3.18a and an equivalent circuit of the static lumped capacitances between the two transmission lines and each line to ground is shown in Figure 3.18b. From the short circuit admittance parameter matrix formulation,

$$[C] = \begin{bmatrix} C_{11} + C_{12} & -C_{12} \\ -C_{12} & C_{22} + C_{12} \end{bmatrix}. \qquad (3.96)$$

This model can be transformed into a circuit consisting of unit elements and ideal transformers by the circuit equivalence shown in Figure 3.19. The relationship between the elements of these two circuits is provided by the equivalence of their y matrices.

$$\begin{bmatrix} y_a + y_b & -y_a \\ -y_a & y_a + y_c \end{bmatrix} = \begin{bmatrix} y_p + y_s & -ny_s \\ -y_s n & y_s n^2 \end{bmatrix} \qquad (3.97)$$

IMPEDANCE TRANSFORMERS AND FILTERS   63

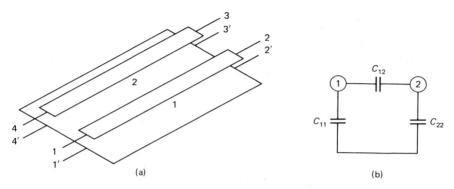

Figure 3.18. (a) Coupled lines above a common ground plane, and (b) equivalent circuit at a given position along the TEM coupled line. (*from J. A. G. Malherbe*, Microwave Transmission Line Filters, *by permission of Artech House, Inc., Dedham, Mass., 1979.*)[10]

By simultaneous solution of the three independent equations represented by (3.97) the left-hand circuit elements are found to be

$$y_a = ny_s \tag{3.98}$$
$$y_b = y_p + y_s(1 - n) \tag{3.99}$$
$$y_c = n(n - 1)y_s \tag{3.100}$$

and from (3.98) and (3.100)

$$n = 1 + y_c/y_a. \tag{3.101}$$

Thus the capacitance matrix in (3.96) can be rewritten in terms of a new C' matrix represented in Figure 3.20, whose element values are

$$C_{11} = C'_{11} + C'_{12}(1 - n) \tag{3.102}$$
$$C_{12} = nC'_{12} \tag{3.103}$$
$$C_{22} = n(n - 1)C'_{12} \tag{3.104}$$

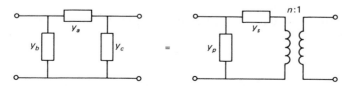

Figure 3.19. Equivalence of pi circuit with L circuit and ideal transformer.

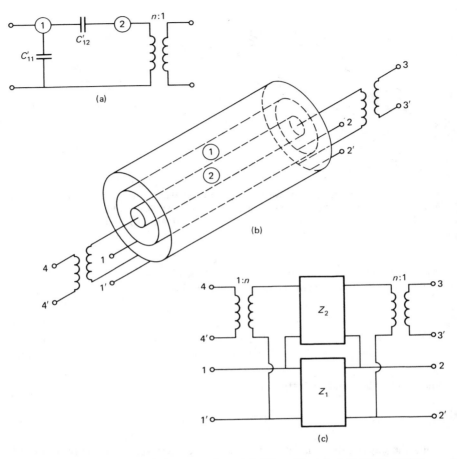

Figure 3.20. Derivation of equivalent circuit for coupled lines. (*from J. A. G. Malherbe, Microwave Transmission Line Filters, by permission of Artech House, Inc., Dedham, Mass., 1979.*)[10]

which imply

$$n = \frac{C_{22} + C_{12}}{C_{12}}. \tag{3.105}$$

These equations may be inverted to give the primed element values.

$$C'_{12} = \frac{C_{12}^2}{C_{12} + C_{22}} \tag{3.106}$$

$$C'_{11} = C_{11} + \frac{C_{12}C_{22}}{C_{12} + C_{22}} \tag{3.107}$$

IMPEDANCE TRANSFORMERS AND FILTERS   65

This new equivalent C' matrix represents two transmission lines, the second of which is completely uncoupled from ground. This could be realized as a coaxial line in which the center conductor of line 2 lies within the center conductor of line 1[10] (Figure 3.20b). Between ports 1-1' and 2-2' is a unit element with characteristic impedance

$$Z_1 = \frac{1}{vC'_{11}}, \qquad (3.108)$$

and between lines 4-4' and 3-3' is a unit element with characteristic impedance

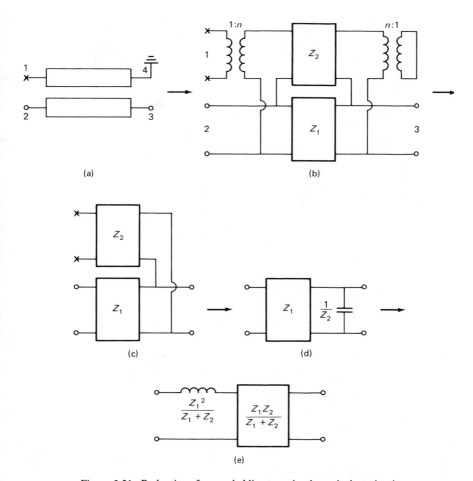

Figure 3.21. Reduction of a coupled line to a simple equivalent circuit.

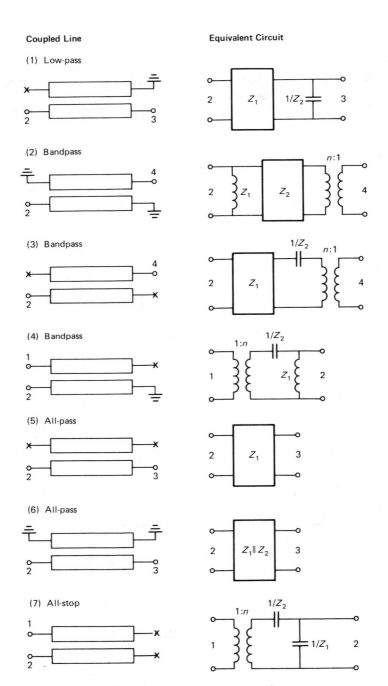

Figure 3.22. Equivalent circuits for coupled line circuits where $n^2 = 1 + Z_2/Z_1$ for symmetrical coupled lines. (*from J. A. G. Malherbe*, Microwave Transmission Line Filters, *by permission of Artech House, Inc., Dedham, Mass., 1979.*)[10]

IMPEDANCE TRANSFORMERS AND FILTERS 67

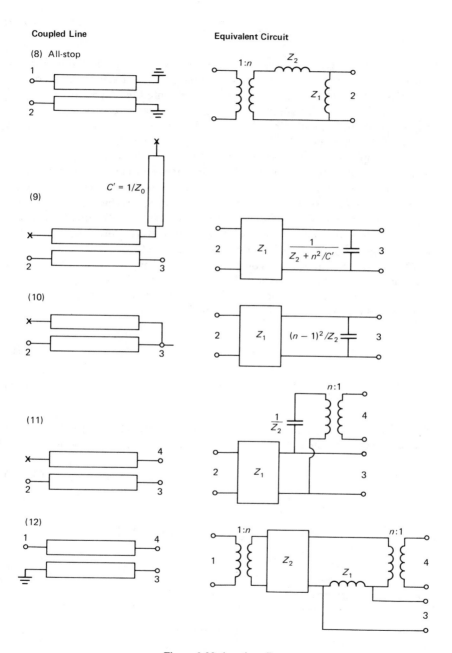

Figure 3.22 (continued).

$$Z_2 = \frac{1}{vC'_{12}}, \tag{3.109}$$

where v is the propagation velocity of the TEM wave in the circuit media. The ideal transformer changes the voltages on line 2 to $1/n$ above the ground level. The coupled line in Figure 3.18a may be represented by the equivalent circuit in Figure 3.20c. By designating two of the four available ports as the input/output ports and terminating the remaining ports with a given impedance, many circuit functions may be obtained. Some of these, however, may have no practical application. A useful example of a circuit function is the coupled line with port 1 open circuited, port 4 short circuited, and ports 2 and 3 chosen as the input/output ports. The graphical reduction of this circuit to one that is easily analyzable is shown in Figure 3.21. Other equivalent circuits shown in Figure 3.22[10] can be obtained using the same technique. The equivalent circuit in Figure 3.21 may be used to realize the Butterworth filter with redundant unit elements (Figure 3.15b) as a coupled line structure (Figure 3.23).

The line widths and spacings of the coupled line may be found once the even and odd mode characteristic impedances are known. These may be found in terms of the Z_1 and Z_2 of the equivalent circuit. By substituting (3.106) and (3.107) into (3.108) and (3.109), the impedances are

$$Z_1 = \frac{1}{v}\left[C_{11} + \frac{C_{12}C_{22}}{C_{12} + C_{22}}\right]^{-1} \tag{3.110}$$

$$Z_2 = \frac{C_{12} + C_{22}}{vC_{12}^2}. \tag{3.111}$$

If equal voltages are applied to each of the two conductors in Figure 3.24a, the lines are excited in their even mode. The total even mode capacitance is

$$C_e = C_{11} = C_{22} = 2C_f + 2C_p + 2C_{fe} \tag{3.112}$$

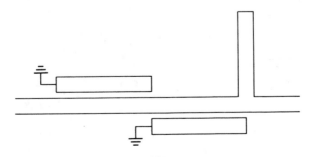

Figure 3.23. Simple Butterworth filter realization with coupled lines.

IMPEDANCE TRANSFORMERS AND FILTERS   69

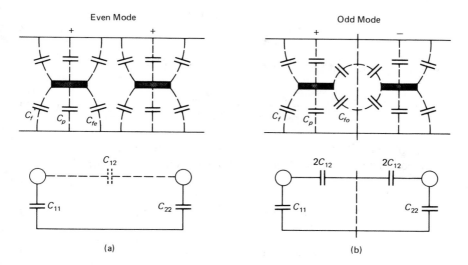

Figure 3.24. Static capacitances for even and odd mode excitation.

where equal line widths are assumed. When voltages of equal magnitude but opposite sign are applied to the center conductors as shown in Figure 3.24b, the lines are excited in their odd mode. The total odd mode capacitance is

$$C_o = 2C_{12} + C_{11} = 2C_f + 2C_p + 2C_{fo}. \quad (3.113)$$

It may be noted that $C_{12} = \tfrac{1}{2}(C_o - C_e) = C_{fo} - C_{fe}$. The even and odd mode characteristic impedances are

$$Z_{0e} = \frac{1}{vC_e} \quad (3.114)$$

$$Z_{0o} = \frac{1}{vC_o} \quad (3.115)$$

In terms of the even and odd mode capacitances, the unit element impedances are

$$Z_1 = \frac{1}{v}\left[C_e + \frac{\tfrac{1}{2}(C_o - C_e)C_e}{\tfrac{1}{2}(C_o - C_e) + C_e}\right]^{-1}$$
$$Z_1 = \tfrac{1}{2}(Z_{0e} + Z_{0o}) \quad (3.116)$$

and

$$Z_2 = \frac{\frac{1}{2}(C_o - C_e) + C_e}{\frac{1}{4}v(C_o - C_e)^2} \tag{3.117}$$

To invert (3.116) and (3.117) it may be shown that

$$\frac{Z_1 Z_2}{Z_1 + Z_2} = \frac{2 Z_{0e} Z_{0o}}{Z_{0e} + Z_{0o}}. \tag{3.118}$$

By combining this expression with (3.116), explicit formulas for the even and odd mode impedances are found in terms of the unit element impedances.

$$Z_{0e} = Z_1 \left[1 + \sqrt{\frac{Z_1}{Z_1 + Z_2}} \right] \tag{3.119}$$

$$Z_{0o} = Z_1 \left[1 - \sqrt{\frac{Z_1}{Z_1 + Z_2}} \right] \tag{3.120}$$

The value of the turns ratio n in the equivalent circuit may be obtained explicitly in terms of Z_1 and Z_2 when equal width coupled lines are used. In this case $C_{11} = C_{22}$ so that from (3.102) and (3.104)

$$C'_{11} + C'_{12}(1 - n) = n(n - 1) C'_{12}. \tag{3.121}$$

Substituting the impedances in (3.108) and (3.109) into (3.121), the value for n is found.

$$n^2 = 1 + \frac{Z_2}{Z_1} \tag{3.122}$$

$$= \left[\frac{Z_{0e} + Z_{0o}}{Z_{0e} - Z_{0o}} \right]^2 \tag{3.123}$$

The even and odd mode characteristic impedances may now be written in terms of the turns ratio n.

$$Z_{0e} = Z_1 \left[1 + \frac{1}{n} \right] \tag{3.124}$$

$$Z_{0o} = Z_1 \left[1 - \frac{1}{n} \right] \tag{3.125}$$

A filter design procedure may be summarized as follows. A circuit structure is derived by a synthesis technique resulting in a circuit consisting of unit elements, S-plane inductors, S-plane capacitors, and ideal transformers. When a coupled line realization is needed for one or more sections of the circuit, the equivalent circuit models in Figure 3.22 are consulted for the coupled line realization. It should be noted that the list of circuits in this figure is not exhaustive. The value for Z_{0e} and Z_{0o} are then given by (3.119) and (3.120) from which the line widths and spacings for the coupled lines may be found. These mechanical parameters of course depend on the chosen transmission line media. Explicit formulas for the commonly used microstrip and stripline will be given in chapter 5.

3.5 IMPEDANCE INVERTERS

Impedance (or K) inverters and admittance (of J) inverters were first used by Cohn[11,12] in the design of waveguide and coupled line filters. The impedance and admittance inverters are defined in Figure 3.25 and are useful in filter designs requiring bandwidths no wider than approximately 20%. This bandwidth limitation may be increased to 40% if K and J inverters are used alternately. A simple realization of an immittance inverter is a quarter wavelength transmission line. If a lowpass filter requires shunt capacitances and series inductances, either one of the reactance types may be replaced by an immittance inverter. For example, the lowpass filter may be realized by series inductances separated by impedance inverters or shunt capacitances separated by admittance inverters. Similarly, a bandpass filter may be realized by series LC series resonant circuits separated by impedance inverters or parallel LC circuits in shunt separated by admittance inverters. The impedance inverter realization is illustrated in Figure 3.26. The values for $K_{i,\,i+1}$, L'_i, and C'_i for the K inverter circuit in Figure 3.26b are obtained by equating the response of this filter with that of the lumped prototype circuit in Figure 3.26a. A section of each of these circuits is shown in Figures 3.27 and 3.28. To make the input series LC part equivalent in the two circuits, the impedance level of the lumped prototype is adjusted by the factor L'_{i-1}/L_{i-1} (Figure 3.29). For this circuit to be equivalent to that in Figure 3.28, the indicated input admittances Y_i in both circuits must be equal.

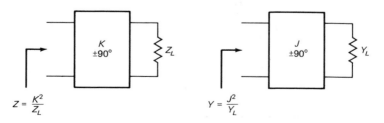

Figure 3.25. Definition of impedance and admittance inverters.

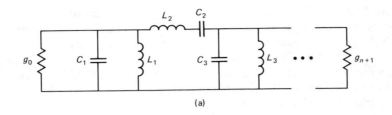

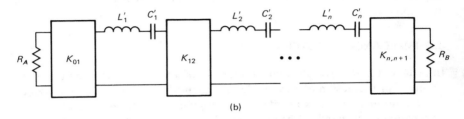

Figure 3.26. (a) Lumped bandpass prototype filter and (b) realization with K-inverters.

$$\frac{K_{i-1,i}^2}{j[\omega L'_i - 1/(\omega C'_i)] + Z'_{i+1}}$$
$$= \left[\frac{j\omega C_i L_{i-1}}{L'_{i-1}} - \frac{jL_{i-1}}{\omega L_i L'_{i-1}} + \frac{Y_{i+1}L_{i-1}}{L'_{i-1}}\right]^{-1} \quad (3.126)$$

Equating like coefficients of ω and using $L_i C_i = L'_i C'_i = 1/\omega_o^2$,

$$K_{i-1,i} = \sqrt{\frac{L'_i L'_{i-1}}{L_{i-1} C_i}} \quad (3.127)$$

Incrementing the subscripts i by 1, and expressing reactive elements in terms of the lowpass elements obtained by the transformation given in Figure 3.10

$$K_{i,i+1} = \omega_o w \sqrt{\frac{L'_i L'_{i+1}}{g_i g_{i+1}}}. \quad (3.128)$$

Figure 3.27. Lumped prototype section.

IMPEDANCE TRANSFORMERS AND FILTERS   73

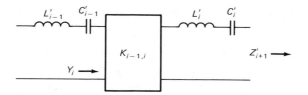

Figure 3.28. Impedance inverter section.

In similar fashion the first and last K inverters are obtained.

$$K_{0,1} = \sqrt{\frac{\omega_o w L_1' R_g}{g_0 g_1}} \qquad (3.129)$$

$$K_{n,n+1} = \sqrt{\frac{\omega_o w L_n' R_L}{g_n g_{n+1}}} \qquad (3.130)$$

The values L_i' have as yet not been specified. The resonant frequency ω_o is used to eliminate the C_i', and the reactance slope at ω_o is used to find L_i'. The reactance slope parameter χ is given by

$$\chi = \frac{\omega_o}{2} \frac{dX(\omega)}{d\omega}\bigg|_{\omega_o} \qquad (3.131)$$

which is related to the circuit Q by

$$Q = \frac{\chi}{R}. \qquad (3.132)$$

Similarly, a shunt resonator will have a susceptance slope parameter

$$b = \frac{\omega_o}{2} \frac{dB(\omega)}{d\omega}\bigg|_{\omega_o}. \qquad (3.133)$$

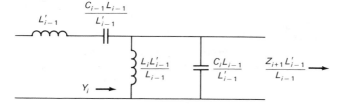

Figure 3.29. Prototype with adjusted impedance level.

For a series lumped LC resonant circuit,

$$\chi = \omega_o L'. \tag{3.134}$$

A series resonant LC circuit can be approximated by a half wavelength transmission line. If the transmission line is terminated by some resistance R, then the input reactance is

$$X = Z_c \frac{(Z_c^2 - R^2)\tan\phi}{Z_c^2 + R^2 \tan^2\phi} \tag{3.135}$$

where Z_c is the characteristic impedance of the transmission line between the K inverters. By differentiating the reactance and substituting $\phi = \pi$ for the chosen half wavelength transmission line, the slope is

$$\chi = \frac{\pi Z_c (Z_c^2 - R^2)}{2 Z_c^2}. \tag{3.136}$$

Assuming that $Z_c \gg R$, the reactance slope is

$$\chi \cong \frac{\pi Z_c}{2}. \tag{3.137}$$

Thus the series resonant circuits between the K inverters may be realized with a transmission line with a reactance slope of χ. The K inverters can be expressed as

$$K_{01} = \sqrt{\frac{R_g w \chi_1}{g_0 g_1}} \tag{3.138}$$

$$K_{j,j+1} = w \sqrt{\frac{\chi_j \chi_{j+1}}{g_j g_{j+1}}} \quad j = 1, 2, \ldots, n-1 \tag{3.139}$$

$$K_{n,n+1} = \sqrt{\frac{R_L w \chi_n}{g_n g_{n+1}}} \tag{3.140}$$

where χ_j is given by (3.137).

The K inverter may be realized as a quarter wavelength transmission line,[11] as a lumped shunt reactance[11] (Figure 3.30), as a disk in a coaxial line with nonzero line length[13] (Figure 3.31), or as an open circuit coupled transmission[12] (Figure 3.32).

IMPEDANCE TRANSFORMERS AND FILTERS 75

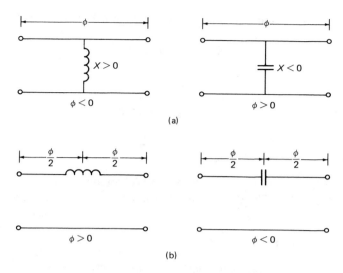

Figure 3.30. Lumped reactance (a) K-inverters and (b) J-inverters.

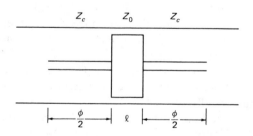

Figure 3.31. Coaxial disk realization of an impedance inverter.

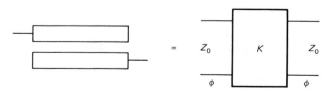

Figure 3.32. Coupled line equivalent to a K-inverter circuit.

For the lumped shunt reactances shown in Figure 3.30 Cohn found

$$K = Z_c \tan (\phi/2) \qquad (3.141)$$

where

$$\phi = -\arctan (2X/Z_c) \qquad (3.142)$$

$$X/Z_c = \frac{X/Z_c}{1 - (K/Z_c)^2}. \qquad (3.143)$$

Thus when K is found from (3.138) − (3.140), X and ϕ can be found from (3.142) and (3.143). The dual admittance inverter can be obtained in similar fashion.

3.5.1 The Coaxial Impedance Inverter

The lumped element approximation to a short section of low characteristic impedance line yields significant error in the design of coaxial bandpass filters having bandwidths $\leq$ 10% of the center frequency. This error increases as the bandwidth is reduced. A more accurate impedance inverter design takes account of the nonzero line length of the disk.

The coaxial impedance inverter shown in Figure 3.31 consists of a short length ℓ of line of low characteristic impedance Z_0, together with two lengths $\phi/2$ of line with characteristic impedance Z_c. The abrupt change in characteristic impedance between Z_c and Z_0 creates higher order modes that can be modeled by a discontinuity capacitance C_d. This capacitance is a function of the coaxial line diameters which can be evaluated from the chosen Z_c and Z_0. A formula for numerically evaluating C_d is given later in chapter 5. The equivalent circuit for the total impedance inverter including the discontinuity capacitances is found in Figure 3.33. By equating the ABCD matrix of Figure 3.33 with the ideal K inverter matrix in (3.144),

$$\begin{bmatrix} 0 & jK \\ j/K & 0 \end{bmatrix} \qquad (3.144)$$

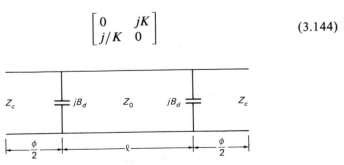

Figure 3.33. Equivalent circuit for the disk with discontinuity capacitance.

the desired length ℓ of the disk and the desired length ϕ of the transmission line on either side of the disk may be found[13,14]. The ABCD matrix of the disk impedance inverter is obtained by cascading the five individual sections in Figure 3.33.

$$A = D = -\left[B_d Z_0 \cos\phi + \frac{1}{2}\left(\frac{Z_c}{Z_0} + \frac{Z_0}{Z_c} - B_d^2 Z_0 Z_c\right)\sin\phi\right]$$
$$\times \sin\beta\ell + [\cos\phi - B_d Z_c \sin\phi]\cos\beta\ell \quad (3.145)$$

$$\frac{jB}{Z_c} = \left[B_d Z_0 \sin\frac{\phi}{2}\cos\frac{\phi}{2} - \frac{Z_0}{Z_c}\cos^2\frac{\phi}{2} - \left(B_d^2 Z_0 Z_c - \frac{Z_c}{Z_0}\right)\sin^2\frac{\phi}{2}\right]\sin\beta\ell$$
$$+ \left[-2\sin\frac{\phi}{2}\cos\frac{\phi}{2} + 2B_d Z_c \sin^2\frac{\phi}{2}\right]\cos\beta\ell \quad (3.146)$$

$$-jCZ_c = \left[-2B_d Z_0 \sin\frac{\phi}{2}\cos\frac{\phi}{2} - \frac{Z_0}{Z_c}\sin^2\frac{\phi}{2}\right.$$
$$\left. - \left(B_d^2 Z_0 Z_c - \frac{Z_c}{Z_0}\right)\cos^2\frac{\phi}{2}\right]\sin\beta\ell + \left[2\sin\frac{\phi}{2}\cos\frac{\phi}{2} + 2B_d Z_c \cos^2\frac{\phi}{2}\right]\cos\beta\ell$$

$$(3.147)$$

The sum (3.146) and (3.147) is

$$j\left(\frac{B}{Z_c} - CZ_c\right) = \left(\frac{Z_c}{Z_0} - \frac{Z_0}{Z_c} - B_d^2 Z_0 Z_c\right)\sin\beta\ell + 2B_d Z_c \cos\beta\ell. \quad (3.148)$$

Since the ABCD matrix for an ideal impedance inverter is given by (3.144), the left-hand side of (3.148) is

$$j\left(\frac{B}{Z_c} - CZ_c\right) = j\left(\frac{jK}{Z_c} - \frac{jZ_c}{K}\right). \quad (3.149)$$

Equating the left-hand sides of (3.148) and (3.149) yields an expression for the disk length ℓ in terms of the known K inverters and the chosen Z_c and Z_0.

$$\left(\frac{Z_c}{Z_0} - \frac{Z_0}{Z_c} - B_d^2 Z_0 Z_c\right)\sin\beta\ell + 2B_d Z_c \cos\beta\ell = \frac{Z_c}{K} - \frac{K}{Z_c} \quad (3.150)$$

This equation is in the form

$$a\sin\beta\ell + b\cos\beta\ell = c \quad (3.151)$$

where the coefficients are defined by (3.150) and are seen to be independent of $\beta\ell$. Rewriting this into the quadratic form

$$\tan^2 \beta\ell + \frac{2ab}{a^2 - c^2} \tan \beta\ell + \frac{b^2 - c^2}{a^2 - c^2} = 0. \tag{3.152}$$

Solution of this gives

$$\tan \beta\ell = \frac{-ab + c\sqrt{a^2 + b^2 - c^2}}{(a^2 - c^2)}, \tag{3.153}$$

which may be reduced to the following expression if the discontinuity capacitance is neglected.

$$\tan \beta\ell = \frac{\left(\frac{Z_c}{K} - \frac{K}{Z_c}\right)}{\sqrt{\left(\frac{Z_c}{Z_0}\right)^2 + \left(\frac{Z_0}{Z_c}\right)^2 - \left(\frac{Z_c}{K}\right)^2 - \left(\frac{K}{Z_c}\right)^2}} \tag{3.154}$$

The line length ϕ is obtained from (3.145), since $A = 0$ for an ideal impedance inverter.

$$\tan \phi = \frac{1 - B_d Z_0 \tan \beta\ell}{B_d Z_c + \frac{1}{2}(Z_0/Z_c + Z_c/Z_0 - B_d^2 Z_0 Z_c) \tan \beta\ell} \tag{3.155}$$

The equations (3.154) and (3.155) reduce to expressions for the lumped shunt reactance given by (3.141) − (3.143) when $\beta\ell \ll 1$.

3.6 COUPLED LINE FILTERS

The coupled line filter is widely used in stripline and microstrip circuits since it is economical to construct and provides a wide range of bandwidths. The original work on the open circuit coupled line filter was done by S. B. Cohn[11] who based his design on the intermediate equivalent K-inverter circuit. Because of the approximation he used, the maximum fractional bandwidth that could be obtained was approximately 20 percent. Subsequently, Matthaei[3,15] described an approach that would provide octave bandwidth filters. However, because of the tighter coupling requirements, the broader bandwidth designs needed an overlay construction. Cristal[16] improved Matthaei's design by using the entire immittance matrix rather than simply the image impedance and propagation constant in the derivation. The result of Cristal's approach is a closer correlation between his microwave design and the ideal response from a

IMPEDENCE TRANSFORMERS AND FILTERS

lumped element prototype circuit at the band edge. Minnis[17] was able to loosen the tight coupling requirements of the broadband design so that filters approaching 100 percent bandwidth could be fabricated as true edge-coupled lines on one side of a dielectric board.

The discussion in this section will apply Cristal's design equations and Minnis' modification to achieve edge-coupled narrowband and broadband filter designs applicable to microstrip and stripline construction. In the following section, therefore, an equivalent impedance inverter circuit for a highpass lumped element prototype will be derived. Then in section 3.6.2 an equivalence between a coupled line section and a line and stub equivalent circuit is found. In the following two sections an equivalence is made between the impedance inverter circuit and the line and stub equivalent circuit. This completes the transistion from the lumped prototype circuit → impedance inverter circuit → line and stub circuit → coupled line circuit. The design modification approach of Minnis is outlined in section 3.6.5 with two examples.

3.6.1 Highpass Impedance Inverter Circuit

Because the open circuit coupled line filter section transforms to a highpass type circuit, it is convenient to transform the lumped element prototype circuit to its highpass form as described in section 3.3. This together with the equivalent impedance inverter circuit is shown in Figure 3.34. As discussed in section 3.5, corresponding parts of each circuit are equated to obtain values for the $K_{i,i+1}$. In the present case, the response of two circuit elements taken together are made to be equivalent. The frequency variable is $s = j\omega$ and the

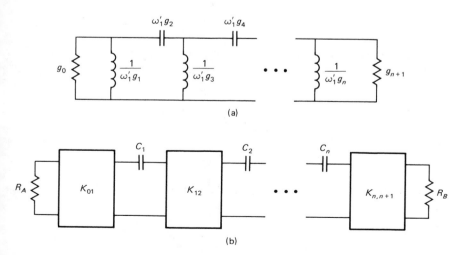

Figure 3.34. (a) The highpass lumped element prototype circuit and (b) the corresponding K-inverter circuit for odd n.

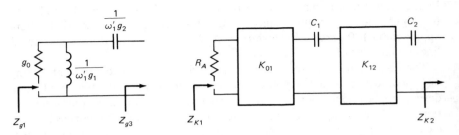

Figure 3.35. Equivalence of the first two sections of the lumped prototype and K-inverter circuits.

cutoff frequency for the lowpass prototype is $\omega_1' = 1$ rad/sec. For the first two sections shown in Figure 3.35, the impedance of the lumped circuit is

$$Z_{g1} = g_0 + \frac{\frac{g_2 \omega_1'}{s} + Z_{g3}}{1 + \frac{g_1 g_2 \omega_1'^2}{s^2} + \frac{Z_{g3} g_1 \omega_1'}{s}} \quad (3.156)$$

and the K-inverter circuit is

$$Z_{K1} = R_A + \frac{\left(\frac{K_{01}}{K_{12}}\right)^2 \left(\frac{1}{sC_2} + Z_{K3}\right)}{1 + \frac{1}{s^2 C_1 C_2 K_{12}^2} + \frac{Z_{K3}}{sC_1 K_{12}}}. \quad (3.157)$$

To force the equality of the input resistors, (3.157) is multiplied through by g_0/R_A.

$$\frac{g_0 Z_{K1}}{R_A} = g_0 + \frac{\frac{g_0}{R_A}\left(\frac{K_{01}}{K_{12}}\right)^2 \left(\frac{1}{sC_2} + Z_{K3}\right)}{1 + \frac{1}{s^2 C_1 C_2 K_{12}^2} + \frac{Z_{K3}}{sC_1 K_{12}}} \quad (3.158)$$

Equating like coefficients of s^{-2} in the denominators of (3.156) and (3.158) yields

$$K_{12} = \frac{1}{\omega_1' \sqrt{C_1 C_2 g_1 g_2}} \quad (3.159)$$

IMPEDENCE TRANSFORMERS AND FILTERS 81

and equating like coefficients of s^{-1} in the numerators provides an expression for the first impedance inverter.

$$K_{01} = \sqrt{\frac{R_A}{\omega_1' C_1 C_2 g_0 g_1}} \qquad (3.160)$$

Finally equating the frequency independent terms of the numerators yields

$$Z_{g3} = g_2 C_2 \omega_1' Z_{K3}. \qquad (3.161)$$

A similar procedure is used for the subsequent two sections shown in Figure 3.36. The input impedance for the lumped prototype circuit is

$$Z_{g3} = \frac{\frac{g_4}{s} + Z_{g5}}{1 + \frac{g_3 g_4 \omega_1'^2}{s^2} + \frac{Z_{g5} g_3 \omega_1'}{s}}, \qquad (3.162)$$

and the input impedance for the K-inverter circuit after multiplying through by $g_2 C_2 \omega_1'$ according to (3.161) is

$$g_2 C_2 \omega_1' Z_{K3} = \frac{g_2 C_2 \omega_1' \left(\frac{K_{23}}{K_{34}}\right)^2 \left(\frac{1}{sC_4} + Z_{K5}\right)}{1 + \frac{1}{s^2 C_3 C_4} + \frac{Z_{K5}}{sC_3 K_{34}}}. \qquad (3.163)$$

Again by equating like coefficients of s, expressions similar to (3.159) and (3.161) are obtained. This procedure is repeated where at each step $Z_{gi} = \omega_1' g_{i-1} C_{i-1} Z_{Ki}$.

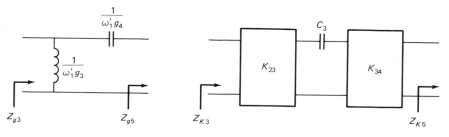

Figure 3.36. Equivalence of the third and fourth sections of the lumped prototype and K-inverter circuits.

82 MICROWAVE SEMICONDUCTOR CIRCUIT DESIGN

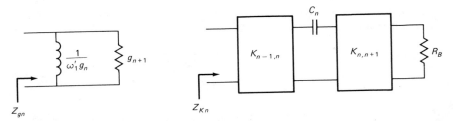

Figure 3.37. Equivalence of the end sections of the lumped prototype and K-inverter circuits.

For the final section pictured in Figure 3.37, the respective input impedances are

$$Z_{gn} = \frac{\dfrac{s}{\omega_1' g_n}}{1 + \dfrac{s}{g_n g_{n+1} \omega_1'}} \tag{3.164}$$

$$\omega_1' g_{n-1} C_{n-1} Z_{Kn} = \frac{\omega_1' g_{n-1} C_{n-1} s C_n K_{n-1,n}^2}{1 + \dfrac{s C_n K_{n,n+1}^2}{R_B}}. \tag{3.165}$$

Equating like coefficients of s provides expressions for the last two impedance inverters. In summary, these values are

$$K_{01} = \sqrt{\frac{R_A}{\omega_1' C_1 g_0 g_1}} \tag{3.166}$$

$$K_{n,n+1} = \sqrt{\frac{R_B}{\omega_1' C_n g_n g_{n+1}}} \tag{3.167}$$

$$K_{i,i+1} = \frac{1}{\omega_1' \sqrt{C_i C_{i+1} g_i g_{i+1}}} \quad i = 1, 2, \ldots n-1 \tag{3.168}$$

The values for the capacitances can be chosen arbitrarily without affecting the performance of the filter. Convenient values for C_i will be chosen in section 3.6.3 where the line and stub circuit is related to the K-inverter circuit.

3.6.2 The Line and Stub Equivalent Circuit

In section 3.4 the open circuit coupled line was represented by the equivalent circuit in Figure 3.38. In this section a symmetrical form is found that can be

IMPEDENCE TRANSFORMERS AND FILTERS 83

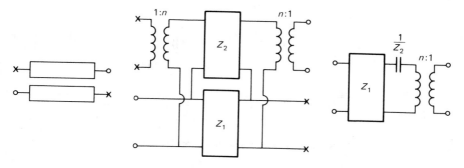

Figure 3.38. Model of the open circuit coupled line.

related to the impedance inverter circuit in Figure 3.34. The "capacitor" in Figure 3.38 represents a commensurate-length, open circuit stub of characteristic admittance $1/Z_2$. This stub may be split into two stubs and one of them brought out to the right-hand side of the ideal transformers as shown in Figure 3.39. The factor k is as yet undefined. The fourth Kuroda identity is applied to that part of the circuit enclosed by the dotted line. Since the Kuroda identity contains an ideal transformer with a turns ratio to the second power, let $m^2 = n$. The resulting circuit is shown in Figure 3.39b where

$$m^2 = 1 + \frac{Z_2(1-k)}{Z_1} \qquad (3.169)$$

Since from (3.122)

$$n^2 = 1 + \frac{Z_2}{Z_1} \qquad (3.170)$$

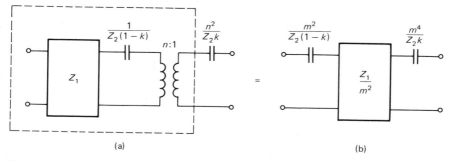

Figure 3.39. (a) The S-plane capacitor is split into two and (b) the circuit transformed by the fourth Kuroda identity.

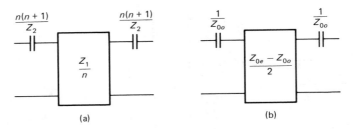

Figure 3.40. Symmetrical equivalent circuit for the open circuit coupled line.

the value for k is uniquely determined by

$$k = \frac{n}{n+1} \qquad (3.171)$$

if the application of the Kuroda identity is to be valid. Replacing m^2 with n, and k with (3.171) in Figure 3.39b, the coupled line is seen to have the symmetrical form shown in Figure 3.40a. Application of (3.116), (3.117) and (3.123) shows that Figure 3.40b is equivalent to Figure 3.40a. It is now clear why the highpass equivalent K-inverter circuit was used in section 3.6.1.

For the end section the coupled line equivalent circuit is shown in Figure 3.41. As a special constraint, the unit element Z_1 is chosen to have the value

$$R_A \stackrel{\Delta}{=} Z_1 = \tfrac{1}{2}(Z_{0e} + Z_{0o}). \qquad (3.172)$$

The open circuit stub impedance for the input section is found from (3.117) and (3.123)

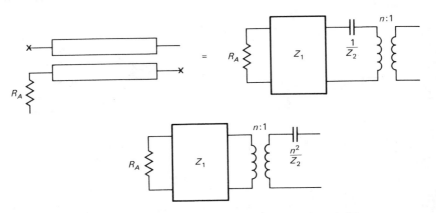

Figure 3.41. Equivalent circuit for the terminated coupled line.

IMPEDENCE TRANSFORMERS AND FILTERS 85

$$\frac{Z_2}{n^2} = \frac{Z_{0e}Z_{0o}}{R_A} \qquad (3.173)$$

and a similar expression is found for the output. The full equivalent circuit for the filter is found in Figure 3.42. The next step is to relate the K-inverter circuit with the desired design specifications to the line and stub equivalent circuit.

3.6.3 Design Equations for Coupled Line Filters

The equivalence between the K-inverter circuit and the line and stub equivalent circuit shown in Figure 3.42 is accomplished by equating the ABCD parameters of the corresponding sections.[16] For an interior section of the line and stub circuit, the ABCD parameter matrix is

$$\frac{1}{\sqrt{1-S^2}} \begin{bmatrix} 1 & \frac{Z_{0o}}{S} \\ 0 & 1 \end{bmatrix} \begin{bmatrix} 1 & SZ_u \\ \frac{S}{Z_u} & 1 \end{bmatrix} \begin{bmatrix} 1 & \frac{Z_{0o}}{S} \\ 0 & 1 \end{bmatrix}$$

$$= \frac{1}{\sqrt{1-S^2}} \begin{bmatrix} 1 + \frac{Z_{0o}}{Z_u} & \frac{Z_{0o}}{S}2 + \frac{Z_{0o}}{Z_u} + SZ_u \\ \frac{S}{Z_u} & 1 + \frac{Z_{0o}}{Z_u} \end{bmatrix} \qquad (3.174)$$

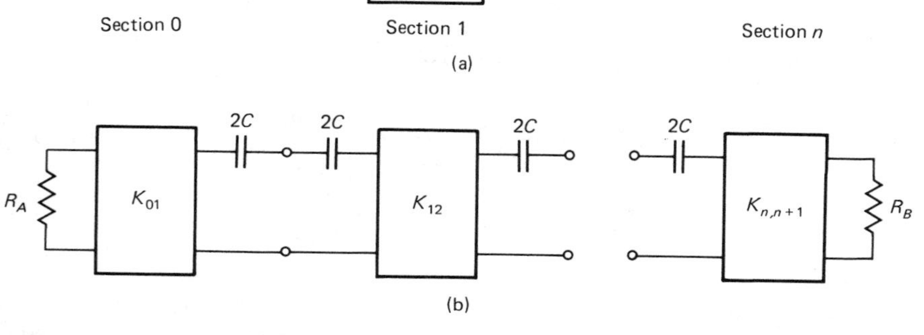

Figure 3.42. (a) Line and stub equivalent circuit where in general Z_{0e} and Z_{0o} are different for each section and (b) the K-inverter equivalent circuit where all the C's are equal.

where $Z_u \triangleq \frac{1}{2}(Z_{0e} - Z_{0o})$. The B element of this matrix may be reduced to

$$B_u = j\frac{(Z_{0e} - Z_{0o})^2 - (Z_{0e} + Z_{0o})^2 \cos^2\theta}{2\sin\theta\,(Z_{0e} - Z_{0o})} \tag{3.175}$$

where $S = j\tan\theta$. For the corresponding section of the impedance inverter (where subscripts on K are dropped) the ABCD matrix is

$$\begin{bmatrix} 1 & \dfrac{1}{2Cs} \\ 0 & 1 \end{bmatrix} \begin{bmatrix} 0 & jK \\ \dfrac{j}{K} & 0 \end{bmatrix} \begin{bmatrix} 1 & \dfrac{1}{2Cs} \\ 0 & 1 \end{bmatrix}$$

$$= \begin{bmatrix} \dfrac{j}{2CsK} & \dfrac{j}{4C^2s^2K} + jK \\ \dfrac{j}{K} & \dfrac{j}{2CsK} \end{bmatrix}. \tag{3.176}$$

The B element of this matrix is

$$B_u = K - \frac{1}{4C^2\omega^2 K}. \tag{3.177}$$

Equating separately the positive and negative parts of (3.175) and (3.177) at the band edge, expressions for the sum and difference of the even and odd mode impedances are found.

$$Z_{0e} - Z_{0o} = 2K\sin\theta_1 \tag{3.178}$$

$$Z_{0e} + Z_{0o} = \frac{\tan\theta_1}{\omega_1'C} \tag{3.179}$$

Thus for an interior section where $i = 1, 2, \ldots n - 1$,

$$(Z_{0e})_i = \frac{1}{\omega_1'C}\left[\frac{\tan\theta_1}{2} + \frac{\sin\theta_1}{\sqrt{g_i g_{i+1}}}\right] \tag{3.180}$$

$$(Z_{0o})_i = \frac{1}{\omega_1'C}\left[\frac{\tan\theta_1}{2} - \frac{\sin\theta_1}{\sqrt{g_i g_{i+1}}}\right] \tag{3.181}$$

The two end sections must be considered separately from the interior sections. Looking toward the load the input impedance for the line and stub circuit and the K-inverter circuit respectively are

IMPEDENCE TRANSFORMERS AND FILTERS

$$Z_{inu} = \frac{R_A}{n_a^2} + \frac{Z_{0e}Z_{0o}}{R_A S} \qquad (3.182)$$

$$Z_{inK} = \frac{K_{01}^2}{R_A} + \frac{1}{j2\omega_1' C} \qquad (3.183)$$

Equating the real parts of (3.182) and (3.183) provides an expression for the transformer turns ratio.

$$n_a = \frac{R_A}{K_{01}} \qquad (3.184)$$

Equating the imaginary parts of these two equations gives

$$Z_{0o} = \frac{R_A}{\omega_1' C} \frac{\tan \theta_1}{2} \frac{1}{Z_{0e}}. \qquad (3.185)$$

Substituting this into (3.172) yields an expression for the even mode impedance in terms of the lumped capacitor in the K-inverter model.

$$Z_{0e} = R_A \left[1 + \sqrt{1 - \frac{\tan \theta_1}{\omega_1' C R_A 2}} \right] \qquad (3.186)$$

Since the capacitance is arbitrary, it is convenient to choose it to be

$$\frac{1}{\omega_1' C} \stackrel{\Delta}{=} hR_A \stackrel{\Delta}{=} \frac{R_A}{\frac{K_{01}^2 C}{R_A} + \frac{\tan \theta_1}{2}} \qquad (3.187)$$

so the even mode impedance for the input section becomes

$$(Z_{0e})_0 = R_A \left[1 + \sqrt{\frac{h}{\omega_1' g_0 g_1}} \right]. \qquad (3.188)$$

The odd mode impedance is found in similar fashion.

$$(Z_{0o})_0 = R_A \left[1 - \sqrt{\frac{h}{\omega_1' g_0 g_1}} \right] \qquad (3.189)$$

The turns ratio given in (3.184) can now be expressed in terms of the above values for Z_{0e} and Z_{0o} using (3.123) and (3.172).

$$n_a = \frac{2R_A}{(Z_{0e})_0 - (Z_{0o})_0} = \sqrt{\frac{g_0 g_1}{h}} \qquad (3.190)$$

For the right-hand side end section, the equations for the characteristic impedances are derived in similar fashion. However, it should be noted for this section only that h' defined below is used rather than h.

$$(Z_{0e})_n = R_B \left[1 + \sqrt{\frac{h'}{\omega_1' g_n g_{n+1}}} \right] \qquad (3.191)$$

$$(Z_{0o})_n = R_B \left[1 - \sqrt{\frac{h'}{\omega_1' g_n g_{n+1}}} \right] \qquad (3.192)$$

$$n_b = \frac{2R_B}{(Z_{0e})_n - (Z_{0o})_n} \qquad (3.193)$$

$$h' \stackrel{\Delta}{=} \frac{1}{\dfrac{K_{n,n+1}^2 \, C}{R_B} + \dfrac{\tan \theta_1}{2}} \qquad (3.194)$$

3.6.4 Coplanar Line Realization

Having obtained the even and odd mode characteristic impedances in the previous section, equations given in chapter 5 may be used to find the line widths and spacings for microstrip and stripline. However, straightforward application of these formulas may result in unrealizable circuits. Minnis[17] has shown that by referring back to the line and stub equivalent circuit, modifications can be made that, except in extreme cases, do not significantly alter the electrical performance of the filter, yet make it realizable as an edge-coupled structure. The filter manipulation will be directed toward achieving coupling structures with a coupling factor

$$k = \frac{Z_{0e} - Z_{0o}}{Z_{0e} + Z_{0o}} = \frac{Z_u}{Z_u + Z_s} \qquad (3.195)$$

that lies in the approximate range from 0.17 to 0.56 and a mean impedance

$$\sqrt{Z_{0e} Z_{0o}} = \sqrt{Z_s(2Z_u + Z_s)} \qquad (3.196)$$

less than 100 ohms. The symbols Z_s and Z_u refer to the characteristic impedance of the series open circuit stub and the unit element between the two stubs respectively. The equivalent circuit shown in Figure 3.43 is found by removing

IMPEDENCE TRANSFORMERS AND FILTERS   89

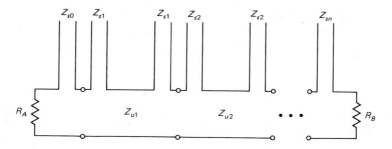

Figure 3.43. Simplified line and stub equivalent circuit.

the ideal transformers on each end of the circuit in Figure 3.42, and multiplying the internal impedances by the square of the turns ratio. When $g_i = g_{n+1-i}$, as occurs in Butterworth and odd order Chebyshev filters, the characteristic impedances of the end stubs are obtained from (3.173) and (3.188)–(3.190).

$$Z_{s0} = R_A \left[\frac{g_0 g_1}{h} - 1 \right] \quad (3.197)$$

$$Z_{sn} = R_B \left[\frac{g_n g_{n+1}}{h} - 1 \right] \quad (3.198)$$

For the interior sections the stub impedances are

$$Z_{si} = g_0 g_1 R_A \left[\frac{\tan \theta_1}{2} - \frac{\sin \theta_1}{\sqrt{g_i g_{i+1}}} \right] \quad i = 1, 2, \ldots n-1 \quad (3.199)$$

Table 3.4. Coupled Line Filter Design Parameters.

	$w = .20, N = 5$			$w = .67, N = 9$		
i	g_i	Z_{ui}	Z_{si}	g_i	Z_{ui}	Z_{si}
0	1		181.0212	1		51.4636
1	1.1468	45.1636	135.8576	1.1957	39.3623	12.1013
2	1.3712	34.4154	146.6058	1.4426	29.4602	22.0034
3	1.9503	34.4154	146.6058	2.1346	27.8286	23.6350
4	1.3712	45.1636	135.8576	1.6167	27.3781	24.0854
5	1.1468		181.0212	2.2054	27.3781	24.0854
6	1			1.6167	27.8286	23.6350
7				2.1346	29.4602	22.0034
8				1.4426	39.3623	12.1013
9				1.9567		51.4636
10				1		

and the characteristic impedances of the lines between the stubs are

$$Z_{ui} = g_0 g_1 R_A \frac{\sin \theta_1}{\sqrt{g_i g_{i+1}}}, \quad i = 1, 2, \ldots n - 1 \quad (3.200)$$

Thus all the interior sections are seen to be independent of the factor h.

Two examples will be used to illustrate design of bandpass filters using coupled lines: a 5 section, 20% bandwidth, 0.1 dB ripple Chebyshev filter and a 9 section, 67% bandwidth, 0.1 dB ripple Chebyshev filter. Table 3.4 summarizes these designs.

Considering first the 20% bandwidth design, by combining the end stub with its neighbor and introducing a redundant unit element, the circuit of Figure 3.44a is obtained. Next the fourth Kuroda identity is applied to some arbitrary portion of $Z_{s0} + Z_{s1}$ to give the circuit in Figure 3.44b. The second stub is then split into two. If it is desirable to have the end section be a symmetrical coupled line, then the two stubs on either side should be the same as indicated in Figure 3.44c. Finally, the ideal transformers are removed on each end by multiplying all impedances by $1/n^4$. This leaves the impedance of the first stub in the second section as

$$Z'_{s1} = \frac{Z_{s0} + Z_{s1} - (n^2 + 1)Z'_{s0}}{n^4} \quad (3.201)$$

$$= \frac{Z_{s0} + Z_{s1} - (n^4 - 1)Z_0}{n^4}.$$

Choosing a convenient value for $Z'_{s1} = 20$ gives $n^2 = 2.289$ and the numerical values for the filter in Figure 3.44d. For the asymmetrical section, Minnis suggests applying the equation twice for the strip width and spacing, once for each side of the section. The strip spacing is then taken to be the mean of the two values. This approximation seems to work reasonably well. The formulas for strip width and spacing are found in chapter 4.

For the octave bandwidth filter all the adjacent stubs are combined together. Part of the first stub is split off to provide a symmetrical coupled line end section. Furthermore, the fourth Kuroda identity is applied to the two central sections to give the result shown in Figure 3.45. The value for n^2 is

$$n^2 = 1 + \frac{24.0854}{27.3781} = 1.8797. \quad (3.202)$$

Now the ideal transformer is removed by dividing the impedances in the interior section by n^2. The resulting filter and its coupled line topology are shown

IMPEDENCE TRANSFORMERS AND FILTERS 91

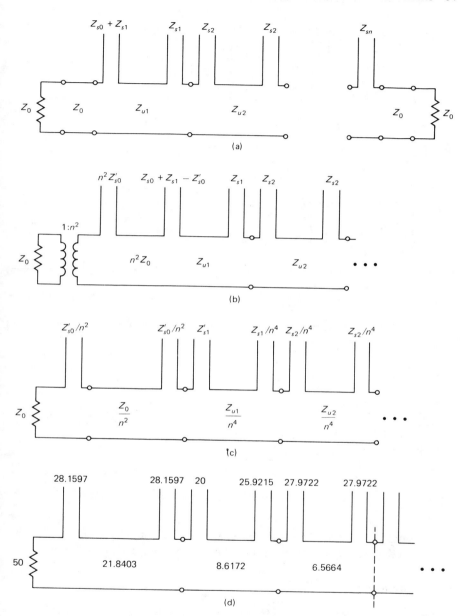

Figure 3.44. Realization of 20 percent bandwidth coupled line filter.

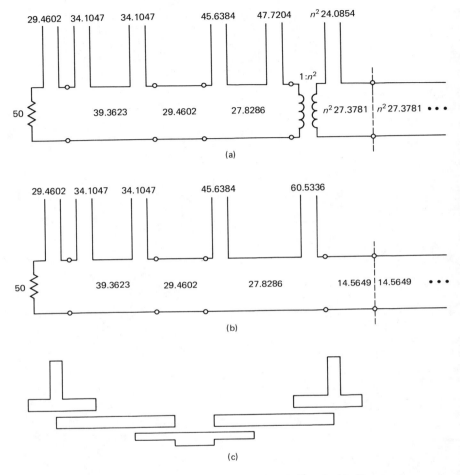

Figure 3.45. Development of octave bandwidth realizable filter.

in Figure 3.45b and c. It should be noted that the coupling coefficient for the first section has been reduced from 0.765 to 0.535 using this technique.

3.7 SUMMARY

The design of filters and matching networks encompasses a vast range of design and realization techniques that cannot be condensed into a single chapter. However, several references are available that go into greater depth. The now classic work of Matthaei, Young, and Jones[3] provides a solid foundation and extensive list of references. Later, Young[18] published a review of microwave filter designs up to 1965 that included tradeoffs between various filter types. More recently, Malherbe[10] incorporated many of the newer design techniques. The present chapter should provide a solid basis for filter and transformer designs that could be used in many applications.

REFERENCES

1. R. E. Collin, *Foundations for Microwave Engineering*. New York: McGraw Hill, 1966.
2. S. B. Cohn, "Optimum Design of Stepped Transmission Line Transformers," *IEEE Trans. on Microwave Theory and Techniques*, Vol. MTT-3, pp. 16–21, April 1955.
3. G. L. Matthaei, L. Young, and E. M. T. Jones, *Microwave Filters, Impedance-Matching Networks, and Coupling Structures*. New York: McGraw Hill, 1965.
4. L. F. Lind, "Synthesis of Equally Terminated Low-Pass Lumped and Distributed Filters of Even Order," *IEEE Trans. on Microwave Theory and Techniques*, Vol. MTT-17, pp. 43–45, January 1969.
5. P. I. Richard, "A Special Class of Functions with Positive Real Part in a Half-Plane," *Duke Math Journal*, Vol. 14, pp. 777–786, September 1947.
6. P. I. Richard, "Resistor-Transmission-Line Circuits," *Proc. of the IRE*, Vol. 36, pp. 217–220, February 1948.
7. E. A. Guillemin, *Synthesis of Passive Network*. New York: John Wiley, 1957.
8. H. Ozaki and J. Ishii, "Synthesis of a Class of Strip-Line Filters," *IRE Trans. on Circuit Theory*, Vol. CT-5, pp. 104–109, June 1958.
9. M. C. Horton and R. J. Wenzel, "General Theory and Design of Optimum Quarter-Wave TEM Filters," *IEEE Trans. on Microwave Theory and Techniques*, Vol. MTT-13, pp. 316–327, May 1965.
10. J. A. G. Malherbe, *Microwave Transmission Line Filters*. Dedham, Mass: Artech, 1979.
11. S. B. Cohn, "Direct-Coupled-Resonator Filters," *Proceedings of the IRE*, Vol 45, pp. 187–196, Feb. 1957.
12. S. B. Cohn, "Parallel-Coupled Transmission-Line-Resonator Filters," *IRE Trans. on Microwave Theory and Techniques*, Vol. MTT-6, pp. 223–231, April 1958.
13. W. A. Davis and P. J. Khan, "Coaxial Bandpass Filter Design," *IEEE Trans. on Microwave Theory and Techniques*, Vol. MTT-19, pp. 373–380, April 1971.
14. A. Giefing, "Coaxial Impedance Inverter Design," *IEEE Trans. on Microwave Theory and Techniques*, Vol. MTT-21, p. 135, March 1973.
15. G. L. Matthaei, "Design of Wide-Band (and Narrow-Band) Band-Pass Microwave Filters on the Insertion Loss Basis," *IRE Trans. on Microwave Theory and Techniques*, Vol. MTT-8, pp. 580–593, November 1960.
16. E. G. Cristal, "New Design Equations for a Class of Microwave Filters," *IEEE Trans. on Microwave Theory and Techniques*, Vol. MTT-19, pp. 486–593, November 1960.
17. B. J. Minnis, "Printed Circuit Coupled-Line Filters for Bandwidths Up to and Greater than an Octave," *IEEE Trans. on Microwave Theory and Techniques*, Vol. MTT-29, pp. 215–222, March 1981.
18. L. Young, "Microwave Filters—1965," *IEEE Trans. on Microwave Theory and Techniques*, Vol. MTT-13, pp. 489–508, September 1965.

Chapter 4
Broadband Directional Couplers

4.1 INTRODUCTION

Broadband directional couplers used for multioctave bandwidths must be designed in a fundamentally different way than the rat race and branch line directional couplers described in chapter 2. One of the most useful discoveries regarding broadband coupled line directional couplers is the analogy between these couplers and quarter wavelength impedance transformers. Because of this analogy, multisection directional coupler designs can be obtained directly from the stepped impedance transformer design given in chapter 3. However, the steps in the resulting coupling coefficients lead to nontheoretical interactions between the lines. These interactions become serious enough above 4 to 6 GHz to degrade the coupler directivity. The ultimate performance for directivity and bandwidth therefore requires smoothly tapered coupled lines. This chapter contains designs for two different forms of the tapered line directional coupler. The first of these is the asymmetric coupler that has a relative phase angle between the coupled and direct output ports of 0° or 180°. The second is a symmetric coupler that has a relative phase angle of 90° between the two output ports.

In the next section the relationship between the impedance transformer and the directional coupler is found. This is followed by a derivation of the relationship between the characteristic impedance distribution and the reflection coefficient for a tapered line. With this background, the design procedures for the asymmetric and symmetric tapered coupled line directional couplers are found in sections 4.4 and 4.5 respectively.

4.2 COUPLER SYNTHESIS

The broadband directional coupler has a horizontal line of symmetry as indicated in Figure 4.1. Because of this symmetry, the even and odd mode analysis described in chapter 2 also applies to couplers. This means that the analysis of the 4-port directional coupler circuit can be reduced to the analysis of two 2-port circuits. As was shown in Figure 2.5, all transverse lines can be open circuited for even mode excitation and short circuited for odd mode excitation without changing the response of the circuit. For the coupled line directional coupler there are no transverse lines. Consequently, a single section directional

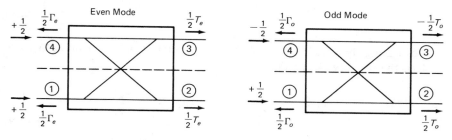

Figure 4.1. Even and odd mode excitation of a directional coupler.

coupler of arbitrary electrical length θ when excited by an even mode voltage can be represented by the ABCD matrix of a simple transmission line.

$$M_e = \begin{bmatrix} A & B \\ C & D \end{bmatrix} = \sqrt{1 + \Omega^2} \begin{bmatrix} 1 & j\Omega Z_{0e} \\ \dfrac{j\Omega}{Z_{0e}} & 1 \end{bmatrix} \quad (4.1)$$

$$\Omega = \tan \theta$$

The odd mode matrix is identical to (4.1) except the odd mode characteristic impedance Z_{0o} replaces the even mode impedance Z_{0e}.

The outputs for the four ports of the coupler were found in chapter 2 to be

$$S_{11} = \tfrac{1}{2}(\Gamma_e + \Gamma_o) \quad (4.2)$$
$$S_{21} = \tfrac{1}{2}(T_e + T_o) \quad (4.3)$$
$$S_{31} = \tfrac{1}{2}(T_e - T_o) \quad (4.4)$$
$$S_{41} = \tfrac{1}{2}(\Gamma_e - \Gamma_o). \quad (4.5)$$

From Table 1.1, the reflection and transmission coefficients for a circuit with equal input and output impedances are given in terms of the ABCD parameters as follows.

$$\Gamma_e = \frac{A + B/Z_0 - CZ_0 - D}{A + B/Z_0 + CZ_0 + D} \quad (4.6)$$

$$T_e = \frac{2}{A + B/Z_0 + CZ_0 + D} \quad (4.7)$$

The odd mode reflection and transmission coefficients have a similar form where Z_{0o} replaces Z_{0e}. The scattering parameters are completely determined from (4.6), (4.7) and (4.1). For the coupler to be matched, $S_{11} = 0$, so that from (4.2)

$$\Gamma_e = -\Gamma_o. \tag{4.8}$$

Substituting the values for the ABCD parameters from (4.1) into (4.6) and (4.7), the equality (4.8) yields

$$\frac{j\Omega\left(\dfrac{Z_{0e}}{Z_0} - \dfrac{Z_0}{Z_{0e}}\right)}{2 + j\Omega\left(\dfrac{Z_{0e}}{Z_0} + \dfrac{Z_0}{Z_{0e}}\right)} = \frac{-j\Omega\left(\dfrac{Z_{0o}}{Z_0} - \dfrac{Z_0}{Z_{0o}}\right)}{2 + j\Omega\left(\dfrac{Z_{0o}}{Z_0} + \dfrac{Z_0}{Z_{0o}}\right)}. \tag{4.9}$$

This imples that for any angle $\Omega = \tan\theta$,

$$Z_0 = \sqrt{Z_{0e}Z_{0o}}. \tag{4.10}$$

For this matched condition, (4.7) indicates that $T_e = T_o$. Consequently, (4.4) indicates that port 3 is the isolated port since $S_{31} = 0$ for any θ. The direct output at port 2 is

$$S_{21} = T_e = \frac{2\sqrt{1 + \Omega^2}}{2 + j\Omega\left(\dfrac{Z_{0e}}{Z_0} + \dfrac{Z_0}{Z_{0e}}\right)} \tag{4.11}$$

and the coupled output at port 4 is

$$S_{41} = \Gamma_e = \frac{j\Omega\left(\dfrac{Z_{0e}}{Z_0} - \dfrac{Z_0}{Z_{0e}}\right)}{2 + j\Omega\left(\dfrac{Z_{0e}}{Z_0} + \dfrac{Z_0}{Z_{0e}}\right)}. \tag{4.12}$$

The derivation of (4.11) and (4.12) was based on (4.1) which represents a transmission line with characteristic impedance Z_{0e} and terminated with a load Z_0. Therefore, the coupled output S_{41} of the directional coupler is the same as the reflection coefficient of an impedance transformer, and the direct output S_{21} of the directional coupler is the same as the transmission coefficient of an impedance transformer. A coupler design is found by simply interpreting the impedance steps of the transformer as the even mode impedances of the coupler. This very useful analogy between the coupler and impedance transformer was pointed out by Young.[1] It provides a method of using the theory developed for impedance transformers for the design of coupled line directional couplers.

The relative phase between the direct and coupled outputs of the single section coupler is 90° as can be seen by comparing (4.11) and (4.12). The insertion loss through the direct channel is

$$L = -10 \log \left|\frac{1}{S_{21}}\right|^2$$
$$= -10 \log \left[1 + \frac{1}{4} \sin^2 \theta \left(\frac{Z_{0e}}{Z_0} - \frac{Z_0}{Z_{0e}}\right)^2\right]. \quad (4.13)$$

The relative magnitude of the two outputs is

$$\left|\frac{S_{41}}{S_{21}}\right|^2 = \frac{1}{4} \sin^2 \theta \left(\frac{Z_{0e}}{Z_0} - \frac{Z_0}{Z_{0e}}\right)^2. \quad (4.14)$$

The coupling coefficient for a directional coupler is defined as

$$k = \frac{Z_{0e} - Z_{0o}}{Z_{0e} + Z_{0o}}. \quad (4.15)$$

Using (4.15) together with the matching condition (4.10), the even and odd mode impedances can be obtained in terms of the coupling coefficient.

$$Z_{0e} = Z_0 \sqrt{\frac{1+k}{1-k}} \quad (4.16)$$

$$Z_{0o} = Z_0 \sqrt{\frac{1-k}{1+k}} \quad (4.17)$$

4.3 TRANSMISSION LINE TAPER

Stepped quarter wavelength directional couplers can be synthesized directly from the impedance transformer design described in chapter 2. But, coupler directivity is better if it is synthesized from a smoothly varying transformer rather than a stepped transformer. Design of an impedance taper requires an expression that relates the input reflection coefficient of the taper with the characteristic impedance of the taper as a function of distance x along the taper.

As shown in Figure 4.2 the characteristic impedance $Z(x)$ is located at x on the transmission line taper. At position x the input impedance is Z_{in}, and at $x + dx$ the input impedance is $Z_{in} + dZ_{in}$. The relationship between these values is defined in the transmission line equation where $d\theta = (\omega/c)dx$.

$$Z_{in} = Z(x) \frac{Z_{in} + dZ_{in} + jZ(x) \tan d\theta}{Z(x) + j(Z_{in} + dZ_{in}) \tan d\theta} \quad (4.18)$$

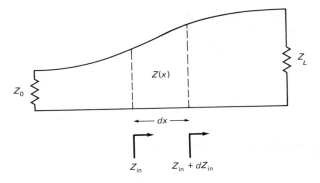

Figure 4.2. Tapered transmission line with characteristic impedance $Z(x)$.

For small angles, $\tan d\theta = d\theta$. Simplifying (4.18) by neglecting differentials to the power two or higher gives

$$Z_{in} \cong \frac{Z(x)\,[Z_{in} + dZ_{in} + jZ(x)\,d\theta]}{Z(x)\left[1 + j\dfrac{Z_{in}}{Z(x)}\,d\theta\right]} \tag{4.19}$$

$$\cong Z_{in} + dZ_{in} + jZ(x)\,d\theta - j\frac{Z_{in}^2}{Z(x)}\,d\theta. \tag{4.20}$$

The differential equation for the input impedance is therefore

$$\frac{dZ_{in}}{d\theta} = j\frac{Z_{in}^2}{Z(x)} - Z(x). \tag{4.21}$$

At any position on the taper the Z_{in} and $Z(x)$ are related by the reflection coefficient $\Gamma(x)$.

$$Z_{in} = Z(x)\,\frac{1 + \Gamma(x)}{1 - \Gamma(x)}. \tag{4.22}$$

This is differentiated to find a value for $dZ_{in}/d\theta$ in terms of $\Gamma(x)$.

$$\frac{dZ_{in}}{d\theta} = \frac{1 + \Gamma}{1 - \Gamma}\frac{dZ(x)}{d\theta} + \frac{2Z(x)}{(1 - \Gamma)^2}\frac{d\Gamma}{d\theta} \tag{4.23}$$

All explicit references to Z_{in} in (4.21) may now be removed by substituting in (4.22) and (4.23).

$$\frac{1 + \Gamma}{1 - \Gamma}\frac{dZ(x)}{d\theta} + \frac{2Z(x)}{(1 - \Gamma)^2}\frac{d\Gamma}{d\theta} = jZ(x)\left[\frac{(1 + \Gamma)^2}{(1 - \Gamma)^2} - 1\right] \tag{4.24}$$

Rearranging the terms in (4.24) gives

$$\frac{d\Gamma}{d\theta} = j2\Gamma - \frac{(1 - \Gamma^2)}{2Z(x)} \frac{dZ(x)}{d\theta}. \tag{4.25}$$

Using $dln(Z) = dZ/Z$ and $d\theta = \beta dx$, the differential equation for the reflection coefficient is obtained.

$$\frac{d\Gamma}{dx} = j2\beta\Gamma - \frac{1}{2}(1 - \Gamma^2) \frac{d\ln[Z(x)]}{dx} \tag{4.26}$$

This is known as the Riccati equation, and it is nonlinear because of the Γ^2 term. However, if $|\Gamma|^2 \ll 1$, then (4.26) reduces to a first order linear differential equation in Γ.

$$\frac{d\Gamma}{dx} = j2\beta\Gamma - \frac{1}{2} \frac{d\ln[Z(x)]}{dx} \tag{4.27}$$

Solution of this equation is found by multiplying it through with the integrating factor $\exp(-2j\int_0^x \beta \, du)$ to form an exact differential.

$$\frac{d}{dx}\left[\Gamma \exp\left(-j2 \int_0^x \beta \, du\right)\right] = -F(x) \exp\left[-j2 \int_0^x \beta \, du\right] \tag{4.28}$$

$$F(x) \stackrel{\Delta}{=} \frac{1}{2} \frac{d\ln[Z(x)]}{dx} \qquad 0 \leq x \leq L \tag{4.29}$$

If β is independent of distance and line dimensions, then

$$\Gamma(x) = \frac{1}{2} \int_x^L \frac{d\ln[Z(u)]}{du} e^{-j2\beta(u-x)} du \tag{4.30}$$

where $\Gamma(L) = 0$. This gives the reflection coefficient at any position x along the taper. If $x = 0$, then (4.30) gives the reflection coefficient at the input of the taper. The expression for $\Gamma(0)$ is the Fourier transform of the function $F(x)$ given by (4.29) where $F(x)$ is zero for $x < 0$ and $x > L$. The Fourier transform pair are

$$\Gamma(2\beta) = \frac{1}{2} \int_0^L \frac{d\ln[Z(x)]}{dx} e^{-j2\beta x} dx \tag{4.31}$$

$$\frac{1}{2}\frac{d\ln[Z(x)]}{dx} = \frac{1}{2\pi}\int_{-\infty}^{\infty}\Gamma(2\beta)e^{j2\beta x}\,d(2\beta) \qquad (4.32)$$

In principle this solves the synthesis problem for the impedance taper. For a given desired reflection coefficient, the required $Z(x)$ can be found from (4.32). The problem is complicated by the requirement that Γ be chosen so that $F(x)$ is zero for $x < 0$ and $x > L$. If the characteristic impedance $Z(x)$ is replaced with $Z_{0e}(x)$ and Γ with S_{41}, then (4.32) represents the solution of the tapered directional coupler.

4.4 ASYMMETRIC 180° COUPLER DESIGN

The asymmetric directional coupler consists of two coupled transmission lines. The two lines gradually become more tightly coupled until they reach the center of the coupler where they suddenly return to a pair of uncoupled lines (Figure 4.3). DuHamel and Armstrong[2] appear to be the first to have recognized that this coupler could provide a phase difference of 0° or 180° between the coupled and direct output ports over a wide frequency range. To obtain the scattering matrix for the asymmetric coupler, the expressions (4.2) to (4.5) must be augmented to account for the asymmetry between ports 1 and 4, and ports 2 and 3.

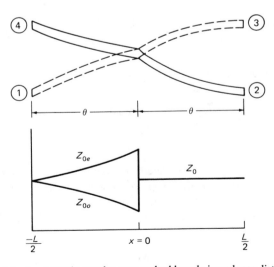

Figure 4.3. Asymmetric coupler even and odd mode impedance distribution.

$$S_{11} = S_{44} = \tfrac{1}{2}(\Gamma_{1e} + \Gamma_{1o}) \qquad (4.33)$$
$$S_{21} = S_{12} = S_{34} = S_{43} = \tfrac{1}{2}(T_{1e} + T_{1o}) \qquad (4.34)$$
$$S_{31} = S_{13} = S_{24} = S_{42} = \tfrac{1}{2}(T_{1e} - T_{1o}) \qquad (4.35)$$
$$S_{41} = S_{14} = \tfrac{1}{2}(\Gamma_{1e} - \Gamma_{1o}) \qquad (4.36)$$

The scattering parameters at port 2 and 3 are found by exciting these ports with a pair of even and odd mode generators just as was done for ports 1 and 4. These are obtained by analogy with (4.33) and (4.36).

$$S_{22} = S_{33} = \tfrac{1}{2}(\Gamma_{2e} + \Gamma_{2o}) \qquad (4.37)$$
$$S_{23} = S_{32} = \tfrac{1}{2}(\Gamma_{2e} - \Gamma_{2o}) \qquad (4.38)$$

The evaluation of (4.33) to (4.38) is facilitated by assuming the coupler is matched, so that

$$k' \triangleq \frac{Z_0}{Z_{0e}(0)} = \frac{Z_{0o}(0)}{Z_0}. \qquad (4.39)$$

As seen in Figure 4.3, the two transmission lines achieve their maximum coupling at $x = 0$. The transmission line beginning at port 1 is reflectionless until it reaches the center at $x = 0$. Here it meets the discontinuity where the impedance abruptly changes from Z_{0e} (or Z_{0o}) to Z_0. Since the reference plane for the reflection coefficients at the input to the coupler, the phase angle 2θ is used to account for the electrical distance the wave travels to and from the reflection. At port 1 the even and odd mode reflection coefficients are found.

$$\Gamma_{1e} = \frac{Z_0 - Z_{0e}(0)}{Z_0 + Z_{0e}(0)} e^{-j2\theta} \qquad (4.40)$$

$$= \frac{k' - 1}{k' + 1} e^{-j2\theta} \qquad (4.41)$$

$$\Gamma_{1o} = \frac{Z_0 - Z_{0o}(0)}{Z_0 + Z_{0o}(0)} e^{-j2\theta} \qquad (4.42)$$

$$= \frac{1 - k'}{1 + k'} e^{-j2\theta} = -\Gamma_{1e} \qquad (4.43)$$

The transmission line going from port 2 to the center of the coupler has a characteristic impedance Z_0 that does not vary with distance. This line is terminated at $x = 0$ with the impedance $Z_{0e}(0)$ (or $Z_{0o}(0)$ for the odd mode).

$$\Gamma_{2e} = \frac{Z_{0e}(0) - Z_0}{Z_{0e}(0) + Z_0} e^{-j2\theta} \tag{4.44}$$

$$= \frac{1 - k'}{1 + k'} e^{-j2\theta} \tag{4.45}$$

In a way similar to that used to obtain (4.43), it can be shown that $\Gamma_{2o} = -\Gamma_{2e}$. The magnitude of the transmission coefficient is obtained from the principle of energy conservation, and the insertion phase is obtained from the electrical length of the coupler.

$$T = \sqrt{1 - |\Gamma|^2}\, e^{-j2\theta} \tag{4.46}$$

$$T_{1e} = T_{1o} = \frac{2\sqrt{k'}}{1 + k'} e^{-j2\theta} \tag{4.47}$$

The six parameters $\Gamma_{1e}, \Gamma_{1o}, \Gamma_{2e}, \Gamma_{2o}, T_{1e}, T_{1o}$ completely determine the scattering matrix for the coupler. Substituting these values into (4.33) through (4.38), the scattering matrix is found.

$$[S] = \begin{bmatrix} 0 & \tau & 0 & -\gamma \\ \tau & 0 & \gamma & 0 \\ 0 & \gamma & 0 & \tau \\ -\gamma & 0 & \tau & 0 \end{bmatrix} \tag{4.48}$$

$$\gamma \stackrel{\Delta}{=} \frac{1 - k'}{1 + k'} e^{-j2\theta} \tag{4.49}$$

$$\tau \stackrel{\Delta}{=} \frac{2\sqrt{k'}}{1 + k'} e^{-j2\theta} \tag{4.50}$$

The relative amplitude and phase between the coupled and direct outputs are found from (4.48).

$$\frac{S_{41}}{S_{21}} = -\frac{\gamma}{\tau} \text{ (coupled side)} \tag{4.51}$$

$$\frac{S_{32}}{S_{12}} = \frac{\gamma}{\tau} \text{ (uncoupled side)} \tag{4.52}$$

The relative phase angles are 0° or 180°, and the relative magnitude of the two output signals may be readily synthesized by the appropriate choice of k'.

The design for the tapered asymmetric coupler can be obtained without using any optimization programs. Any one of several taper designs might be used, but the Klopfenstein taper[3,4] has been found to give excellent results. The

BROADBAND DIRECTIONAL COUPLERS 103

input reflection coefficient for this taper is found by increasing indefinitely the number of sections of a stepped Chebyshev impedance transformer while keeping the overall length L of the transformer constant. This taper provides the lowest possible minor sidelobe amplitudes consistent with the fixed length of the taper or the shortest possible length for a fixed minor sidelobe amplitude. The process of going from an N step quarter wavelength transformer to a continuous taper converts the bandpass characteristic of the stepped transformer to a highpass characteristic of the tapered transformer. The reason is that the first and each succeeding step in the transformer produces a secondary maximum in the reflection coefficient. As the step sizes become smaller, the frequency where this secondary maximum occurs increases until the bandwidth for the smooth taper becomes infinite. The characteristic Chebyshev ripples, however, remain. These are created by two small steps: one at the beginning of the taper and one at the end. The reflection coefficient for the taper, as obtained by Klopfenstein[3] and Collin,[4] is

$$\rho e^{j\beta L} = \rho_o \frac{\cos[\sqrt{(\beta L)^2 - A^2}]}{\cosh(A)} \quad (4.53)$$

where the lower edge of the passband is defined as

$$A = \beta_o L. \quad (4.54)$$

The reflection coefficient at very low frequencies is

$$\rho_o = \frac{Z_2 - Z_1}{Z_2 + Z_1}$$
$$\cong \frac{1}{2} \ln \frac{Z_2}{Z_1} \quad (4.55)$$

The theoretical reflection coefficient for the Klopfenstein taper is shown in Figure 4.4. An important quality factor for the taper is the ratio of the major to minor sidelobe levels. Investigation of (4.53) shows that the maximum $\rho = \rho_o$ and the ripple in the passband for $\beta L \geq A$ is

$$\rho_{\text{rip}} = \rho_o/\cosh(A). \quad (4.56)$$

Consequently the ratio of the minor to major sidelobes is

$$20 \log \frac{\rho_{\text{rip}}}{\rho_o} = 20 \log[\cosh(A)]. \quad (4.57)$$

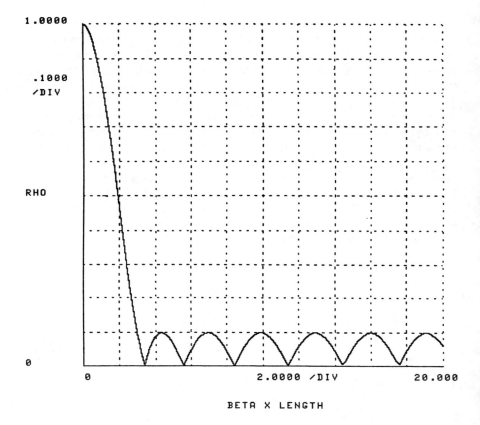

Figure 4.4. Reflection coefficient for a Klopfenstein taper with a 20 dB sidelobe level.

The impedance distribution for the Klopfenstein taper is obtained by substituting (4.53) into (4.32) and integrating the result. The answer as corrected by Collin[4] and Tresselt[5] is

$$\ln Z(x) = \frac{1}{2} \ln (Z_1 Z_2) + \frac{A^2 \ln (Z_2/Z_1)}{2 \cosh (A)} \Phi\left[\frac{2x}{L}, A\right] \quad (4.58)$$

where Z_1 and Z_2 are the source and load impedances matched by the transformer. The function Φ is

$$\Phi(x,A) = -\Phi(-x,A) = \int_0^x \frac{I_1(A\sqrt{1-y^2})}{A\sqrt{1-y^2}} dy \quad x \leq 1 \quad (4.59)$$

where I_1 is the modified Bessel function. Klopfenstein noted that this function may be evaluated in closed form at three points.

$$\Phi(0,A) = 0 \tag{4.60}$$
$$\Phi(x,0) = x/2 \tag{4.61}$$
$$\Phi(1,A) = \frac{\cosh A - 1}{A^2} \tag{4.62}$$

At the two end points of the taper

$$\ln Z(-L/2) = \ln Z_1 + \frac{\ln(Z_2/Z_1)}{2\cosh A} \tag{4.63}$$

and

$$\ln Z(L/2) = \ln Z_2 - \frac{\ln(Z_2/Z_1)}{2\cosh A}. \tag{4.64}$$

Although the step discontinuities at the ends are necessary to obtain the theoretical Chebyshev response, it is often argued that in practice these steps should be smoothed out.

For intermediate points, the evaluation of $\Phi(x,A)$ is most conveniently handled by using a recursion formula.[6]

$$\Phi(x,A) = \sum_{k=0}^{\infty} a_k b_k \tag{4.65}$$

$$a_0 = 1 \tag{4.66}$$

$$b_0 = x/2 \tag{4.67}$$

$$a_k = \frac{A^2}{4k(k+1)} a_{k-1} \tag{4.68}$$

$$b_k = \frac{\frac{x}{2}(1-x^2)^k + 2kb_{k-1}}{2k+1} \tag{4.69}$$

This relationship was derived by expanding the modified Bessel function into a power series and integrating term by term. Grossberg[6] also included a short FORTRAN program in his paper that used a maximum of 20 terms to evaluate (4.65).

Design of an asymmetrical coupler entails choosing two out of three parameters: the cutoff frequency β_o, the length L of the coupler, and the sidelobe level. The even mode impedance distribution is then obtained from (4.58) using the transformer-coupler analogy. Ordinarily $Z_1 = 1$ and Z_2 is related to the desired coupling frequency response $|C_\infty|$ at $\omega = \infty$.

$$Z_2 = \frac{Z_{0e}(0)}{Z_0} = \frac{1 + |C_\infty|}{1 - |C_\infty|} \qquad (4.70)$$

From (4.58) the even mode distribution for the left-hand side of the coupler is obtained. The right-hand side consists of two uncoupled lines with electrical length equal to that of the left-hand side. For designs requiring complete overlap of the two coupled lines, the dielectrical substrate dimensions and dielectric constant must be chosen to be consistent with $Z_{0e}(0)$.

4.5 SYMMETRIC 90° COUPLER DESIGN

The design of a continuously tapered directional coupler is determined from the choice of the desired coupling response, bandwidth, and overall length. The length must be long enough to obtain a sufficiently small passband ripple in the frequency response characteristic. The procedure followed by Kammler[7] was first to calculate a discrete $(2N - 1)$ step quarter wavelength directional coupler. Between each of these steps a smooth curve is fitted. For this purpose a spline curve[8] is used. The spline curve is the mathematical equivalent of drawing a smooth line through a set of points with a French curve as might be performed by a draftsman. This continuous coupler taper is broken up into several hundred small sections, each of which is described by an *ABCD* matrix as in (4.1). By cascading each of the elemental *ABCD* matrices together, the frequency response of both the amplitude and phase is obtained for a few hundred frequency points in the passband. The calculated coupling response is compared with the desired response, and the error is used to obtain a correction term to decrease the error. This basic technique has been used by Kammler to design a coupler with a bandwidth of 14 octaves and maximum ripple of 0.268 dB.

The optimization technique described in chapter 6 could in principle be used here in the design of the coupler. However, computer optimization of several hundred variables is impractical, so Kammler parameterized the frequency response error function. He simply chose $n \leq 20$ points to be optimized and assumed the spline curve would provide a good approximation to the optimum curve between the calculated optimized points. Since the phase difference of 90° between the direct and coupled output ports is automatically obtained for the symmetrical coupler, only the magnitude of the coupling coefficient $|C(\omega)|$ needs to be optimized.

The error function that is to be minimized over N_f frequencies is given by

$$\epsilon(\omega,\mathbf{p}) = 20 \log |C(\omega_i)| - 20 \log C_n \tag{4.71}$$

where C_n is the desired coupling frequency response. The vector

$$\mathbf{p} = [p_1, p_2, \ldots, p_n]^T \tag{4.72}$$

is the set of $(Z_{0e})_i$ that are chosen for optimization. Ordinarily these parameters are chosen to be equally spaced along the length of the coupler. They may also be chosen in nonuniform intervals in order to exert greater control of the optimization near the tight coupling region of the coupler. The error function given by (4.71) will oscillate back and forth creating m extrema of the error at frequencies $\omega = \Omega_1, \Omega_2, \ldots, \Omega_m$. At each of these extrema the parameter σ_i is defined as

$$\sigma_i = \begin{cases} +1 & \text{for a local maximum} \\ -1 & \text{for a local minimum} \end{cases} \tag{4.73}$$

and the average extremal deviation is defined as

$$d_{\text{avg}} = \frac{1}{m} \sum_{i=1}^{m} |\epsilon(\Omega_i, \mathbf{p}_i)|. \tag{4.74}$$

This latter function is used to level and compress the error curve when there are either too few or too many extrema. To obtain an improved error curve, the Jacobian matrix is found as shown below.

$$J_{ij} = \frac{\partial \epsilon(\Omega_i, \mathbf{p})}{\partial p_j} \tag{4.75}$$

This is found numerically by calculating $\Delta\epsilon(\Omega_i, \mathbf{p})$ when p_j is changed to $p_j + \Delta p_j$ so that

$$\Delta\epsilon = J \Delta\mathbf{p}. \tag{4.76}$$

The ripples in the passband should ideally follow a Chebyshev characteristic, $T_n(x) = \cos(n \arccos x)$, and thus have $n + 1$ ripples. In the exactly determined case where $m = n + 1$, the error would decrease if it were modified by

$$\Delta\epsilon_i = \sigma_i d - \epsilon(\Omega_i, \mathbf{p}) \quad i = 1, 2 \ldots m \tag{4.77}$$

By equating (4.76) and (4.77), a set of m equations with n unknown Δp values are obtained.

$$\begin{bmatrix} J_{11} & J_{12} & \ldots J_{1n} & -\sigma_1 \\ J_{21} & & & -\sigma_2 \\ \vdots & & & \vdots \\ J_{n1} & \ldots & J_{nn} & -\sigma_n \\ J_{m1} & \ldots & J_{mm} & -\sigma_m \end{bmatrix} \begin{bmatrix} \Delta p_1 \\ \Delta p_2 \\ \vdots \\ \Delta p_n \\ d \end{bmatrix} = \begin{bmatrix} -\epsilon_1 \\ -\epsilon_2 \\ \vdots \\ -\epsilon_n \\ -\epsilon_m \end{bmatrix} \quad (4.78)$$

The parameter d is a dummy variable, so it is discarded after (4.78) is solved for Δp. A new estimate for the parameter set is given by

$$p_j = p_j + \Delta p_j \quad (4.79)$$

For the underdetermined case where $m < n + 1$,

$$\Delta \epsilon_i = \tfrac{1}{2} \sigma_i \, d_{\text{avg}} - \epsilon(\Omega_i, \mathbf{p}) \qquad i = 1, 2 \ldots m \quad (4.80)$$

and the new estimate for $\Delta \mathbf{p}$ is obtained from the Penrose pseudo-inverse

$$\Delta \mathbf{p} = J^T (JJ^T)^{-1} \Delta \epsilon \quad (4.81)$$

where J^T is the transpose of the Jacobian matrix. When $m > n + 1$, (4.80) is again used to find $\Delta \epsilon$ and the least squares pseudo-inverse is used to find the new parameter set as indicated in (4.82).

$$\Delta \mathbf{p} = (J^T J)^{-1} J^T \Delta \epsilon \quad (4.82)$$

As an example of Kammler's procedure, a 3 dB coupler is designed by cascading two 8.343 dB couplers. Making use of the coupler-transformer analogy

$$C = \Gamma = 0.3827 \quad (4.83)$$

where C is the coupling coefficient of the coupler, and Γ is the reflection coefficient of the transformer (Figure 4.5). The transformer mismatch is obtained when the load Z_L is replaced by Z_0.

$$\frac{Z_L}{Z_0} = \frac{1 + |C|}{1 - |C|} = 2.23983 \quad (4.84)$$

For a 6-section transformer with a bandwidth from 2 GHz to 12 GHz the discrete step transformer has the following normalized impedances that correspond to the parameterization of the tapered coupler.

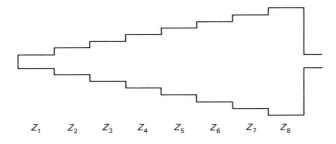

Figure 4.5. Stepped impedance transformer prototype for the symmetric coupler.

$$Z_1 = 1.00000$$
$$Z_2 = 1.09684$$
$$Z_3 = 1.21755$$
$$Z_4 = 1.39198$$
$$Z_5 = 1.60910$$
$$Z_6 = 1.83963$$
$$Z_7 = 2.04208$$
$$Z_8 = 2.23983$$

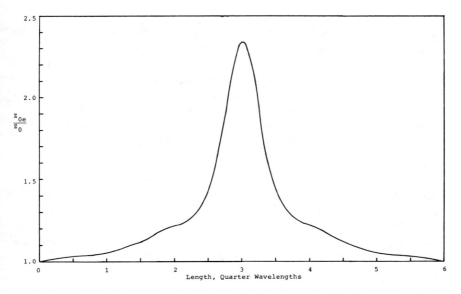

Figure 4.6. Even mode characteristic impedance distribution for the 8.343 dB symmetric directional coupler.

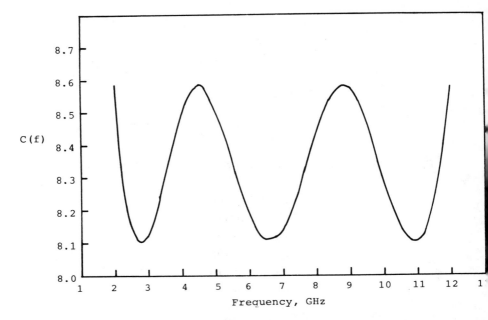

Figure 4.7. Frequency coupling response for the symmetric 8.343 dB directional coupler with a bandwidth of 2 to 12 GHz.

With these initial values, the procedure described above was used to design the coupler shown in Figure 4.6. It may be realized by using overlayed stripline construction as described by Shelton.[9] If complete crossover is desired at the center of the coupler, the design may be modified by constraining the tight coupling coefficient within a restricted range, changing the design bandwidth, or changing the coupler length. This is in addition to making the appropriate choice of dielectric material. Figure 4.7 shows the predicted frequency coupling response for this coupler. The maximum ripple in the passband is ± 0.25 dB.

4.6 CONCLUSIONS

The symmetric coupler provides 90° differential phase between the coupled and direct output ports while the asymmetric coupler provides either 0° or 180° differential phase. In addition Millican[10] has used nonsymmetrical couplers to provide a constant relative phase difference between the coupled and direct outputs such that they lie between 90° and 180°.

REFERENCES

1. L. Young, "The Analyical Equivalence of the TEM-Mode Directional Couplers and Transmission-Line Stepped Impedance Filters," *Proc. IEE,* Vol. 110, pp. 275–281, February 1963.

2. R. H. DuHamel and M. E. Armstrong, "The Tapered-Line Magic-T," Abstracts of 15th Annual Symposium of the USAF Antenna Research and Development Program, Monticello, Illinois, October 12–14, 1965. Also, reprinted in *Parallel Coupled Lines and Directional Couplers,* L. Young, ed. Dedham, Mass: Artech, 1972, pp. 207–233.
3. R. W. Klopfenstein, "A Transmission Line Taper of Improved Design," *Proc. IRE,* Vol. 44, pp. 31–35, January 1965.
4. R. E. Collin, "The Optimum Tapered Transmission Line Matching Section," *Proc. IRE,* Vol. 44, pp. 539–548, April 1965.
5. C. P. Tresselt, "Design and Computed Theoretical Performance of Three Classes of Equal-Ripple Nonuniform Line Couplers," *IEEE Trans. on Microwave Theory and Techniques,* Vol. MTT-17, pp. 218–230, April 1969.
6. M. A. Grossberg, "Extremely Rapid Computation of the Klopfenstein Impedance Taper," *Proc IEEE,* Vol. 56, pp. 1629–1630, September 1968.
7. D. W. Kammler, "The Design of Discrete N-Section and Continuously Tapered Symmetrical Microwave TEM Directional Couplers," *IEEE Trans. on Microwave Theory and Techniques,* Vol. MTT-17, pp. 577–590, August 1969.
8. T. E. N. Greville, "Spline Functions, Interpolation, and Numerical Quadrature," in *Mathematical Methods for Digital Computers,* A. Ralston and H. S. Wilf, eds. New York: Wiley, 1967, pp. 156–168.
9. J. P. Shelton, Jr., "Impedances of Offset Parallel-Coupled Strip Transmission Lines," *IRE Trans. on Microwave Theory and Techniques,* Vol. MTT-8, pp 7–15, January 1966.
10. G. L. Millican, "Practical Strip-Line Microwave Circuit Design," *IEEE Trans. on Microwave Theory and Techniques,* Vol. MTT-17, pp. 696–705, September 1969.

Chapter 5
Mechanical Realization of Selected Transmission Lines

5.1 TRANSMISSION LINE SYNTHESIS

Previous chapters have been concerned primarily with the circuit design problem. Little regard was given to the mechanical geometry of the required transmission lines or the effect of discontinuities on the circuit behavior. The present chapter contains a summary of formulas needed for synthesis of some of the most common transmission lines as well as an evaluation of a few common discontinuities. In contrast with other chapters in this book, no attempt has been made to derive these equations, but references are provided giving the source of these formulas.

Microwave energy can be carried on numerous types of transmission lines. Some of these are (1) rectangular waveguide, (2) two-wire line, (3) coaxial line, (4) stripline, (5) open microstrip, (6) slot line, (7) image line, (8) inverted microstrip, (9) trapped inverted microstrip, (10) coplanar waveguide, (11) suspended stripline, and (12) dielectric waveguide. Only the first five of these will be the subject of this chapter since they are the most widely used. Some discontinuity problems associated with each of the five specific transmission line types will be discussed in their respective sections. Knowing the numerical value for a discontinuity is not helpful unless the circuit can be modified to accommodate it and thereby cancel its deleterious effects. Consequently, the method of modifying, for example, a coaxial disk will be discussed in the sections on coaxial lines. In the final section, the general problem of modifying the line length because of a shunt or series discontinuity reactance will be discussed.

5.2 RECTANGULAR WAVEGUIDE

Rectangular waveguide is usually constructed with a height b that is approximately one-half the width a (Figure 5.1). Its primary advantages are its low loss and high power-handling capability. However, it is bulky, particularly at the lower microwave frequencies, dispersive, and does not propagate transverse electromagnetic (TEM) waves. It is bandlimited at the low frequency end by the waveguide cutoff frequency and at the high frequency end by the support

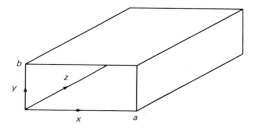

Figure 5.1. Rectangular waveguide.

of higher order propagating modes. In the following expressions, the assumed time dependence is $e^{j\omega t}$, so the sign of the propagation constant determines whether the wave is propagating in the forward or reverse direction. The fields are obtained by solution of Maxwell's equations with a rectangular boundary condition. Solutions are found in many references, including Collin.[1]

5.2.1 TE or H Mode

The following fields are found when the waveguide is excited by an H_z field with an amplitude of A_{mn}.

$$E_x = jA_{mn} \frac{\omega\mu_o n\pi}{k_c^2 b} \cos\frac{m\pi x}{a} \sin\frac{n\pi y}{b} e^{\mp j\beta z} \tag{5.1}$$

$$E_y = -jA_{mn} \frac{\omega\mu_o m\pi}{k_c^2 a} \sin\frac{m\pi x}{a} \cos\frac{n\pi y}{b} e^{\mp j\beta z} \tag{5.2}$$

$$E_z = 0 \tag{5.3}$$

$$H_x = \pm jA_{mn} \frac{\beta m\pi}{k_c^2 a} \sin\frac{m\pi x}{a} \cos\frac{n\pi y}{b} e^{\mp j\beta z} \tag{5.4}$$

$$H_y = \pm jA_{mn} \frac{\beta n\pi}{k_c^2 b} \cos\frac{m\pi x}{a} \sin\frac{n\pi y}{b} e^{\mp j\beta z} \tag{5.5}$$

$$H_z = A_{mn} \cos\frac{m\pi x}{a} \cos\frac{n\pi y}{b} e^{\mp j\beta z} \tag{5.6}$$

$$k_c^2 = \left(\frac{m\pi}{a}\right)^2 + \left(\frac{n\pi}{b}\right)^2 \tag{5.7}$$

$$\beta = \sqrt{(\omega/c)^2 - k_c^2} \tag{5.8}$$

The lowest propagating mode is the TE_{10} mode. Because of resistive losses in the waveguide walls, power decays as $P = P_o e^{-2\alpha z}$ where

$$\alpha = \frac{R_m k_c^2 \left[b + \frac{a}{2}(1 + \beta^2/k_c^2) \right]}{\beta \mu_o \omega ab} \quad (5.9)$$

$$R_m = \sqrt{\frac{\omega \mu_o}{2\sigma}} \quad (5.10)$$

σ = conductivity of the metal

Care should be taken in using the tabulated values for σ. The tables list values for pure copper, silver, gold, brass, etc. while the waveguide structure may use a rough, oxidized surface. The surface roughness may, in fact, be the most important factor in determining an effective value for σ.[2]

5.2.2 TM or E Mode

The following fields are found when the waveguide is excited by an E_z field with an amplitude of A_{mn}.

$$E_x = \mp j A_{mn} \frac{\beta m \pi}{k_c^2 a} \cos \frac{m \pi x}{a} \sin \frac{n \pi y}{b} e^{\mp j\beta z} \quad (5.11)$$

$$E_y = \mp j A_{mn} \frac{\beta n \pi}{k_c^2 b} \sin \frac{m \pi x}{a} \cos \frac{n \pi y}{b} e^{\mp j\beta z} \quad (5.12)$$

$$E_z = A_{mn} \sin \frac{m \pi x}{a} \sin \frac{n \pi y}{b} e^{\mp j\beta z} \quad (5.13)$$

$$H_x = j A_{mn} \frac{n \pi}{k_c^2 \omega \mu_o b} \sin \frac{m \pi x}{a} \cos \frac{n \pi y}{b} e^{\mp j\beta z} \quad (5.14)$$

$$H_y = -j A_{mn} \frac{m \pi}{k_c^2 \omega \mu_o a} \cos \frac{m \pi x}{a} \sin \frac{n \pi y}{b} e^{\mp j\beta z} \quad (5.15)$$

$$H_z = 0 \quad (5.16)$$

Values for k_c and β are of the same form as the TE case. The lowest order TM mode is the TM_{11} mode.

5.2.3 TE_{10} Mode Characteristic Impedance[3]

The rectangular waveguide does not have a simple relationship between the transverse electric and magnetic fields and the RF voltages and currents respectively. Consequently, there is no unique definition for Z_0. In this case the three parameters, power, voltage, and current are not all related to one another. A choice must be made for two out of the three of these in order to define Z_0. The power flow through the guide is

MECHANICAL REALIZATION OF SELECTED TRANSMISSION LINES

$$P = \tfrac{1}{2} Re \iint \mathbf{E} \times \mathbf{H}^* \cdot \hat{z}\, dS \tag{5.17}$$
$$= A_{10}^2 \frac{\eta k_o \beta a^2 b}{4\pi^2}$$

where k_o is the free space wave number ω/c and $\eta = \sqrt{\mu/\epsilon}$ is the intrinsic wave impedance. The RF voltage across the top and bottom walls of the waveguide at $x = a/2$ is

$$V = b E_y(a/2) \tag{5.18}$$
$$= -j A_{10} \frac{\eta k_o ab}{\pi}.$$

The RF longitudinal current is

$$I = \int_0^a H_x\, dx \tag{5.19}$$
$$= j A_{10} \frac{2\beta a^2}{\pi^2}.$$

The volt-ampere impedance definition is

$$Z_{vi} = -\frac{V}{I} \tag{5.20}$$
$$= \frac{b}{a} \frac{k_o \pi \eta}{2\beta},$$

the power-current definition is

$$Z_{pi} = \frac{2P}{|I|^2} \tag{5.21}$$
$$= \frac{b}{a} \frac{k_o \pi^2 \eta}{8\beta},$$

and the power-voltage definition is

$$Z_{pv} = \frac{|V|^2}{2P} \tag{5.22}$$
$$= \frac{b}{a} \frac{2 k_o \eta}{\beta}.$$

Among these three definitions, the power-voltage definition is commonly acknowledged as being the best of the three. In any case the characteristic impedance is proportional to the height to width ratio.

5.2.4 Waveguide Height Step Discontinuity

A step in the height of the waveguide produces higher order modes near the step. These may be modeled by a shunt lumped susceptance B as shown in Figure 5.2.[4]

$$B = \frac{2}{\pi}\left\{\ln\left[\frac{1-\alpha^2}{4\alpha}\left(\frac{1+\alpha}{1-\alpha}\right)^{1/2[\alpha+1/\alpha]}\right] + \frac{2}{A}\right\}/Z_{pv} \quad (5.23)$$

$$A = \left(\frac{1+\alpha}{1-\alpha}\right)^{2\alpha}\frac{1+\sqrt{1-(\beta b/\pi)^2}}{1-\sqrt{1-(\beta b/\pi)^2}} - \frac{1+3\alpha^2}{1-\alpha^2} \quad (5.24)$$

$$\alpha = \frac{b'}{b} \quad (5.25)$$

$$\beta = \sqrt{k_o^2 - \left(\frac{\pi}{a}\right)^2} \quad (5.26)$$

The error in this equation is said to be less than 3% when $\beta d/\pi < 0.7$.

5.3 TWO-WIRE LINE

The two-wire line and the transmission lines discussed in subsequent sections of this chapter are considered to be TEM lines. The losses that are usually small cause a longitudinal voltage, but this is considered to be insignificant. Consequently the characteristic impedance may be obtained from

$$Z_0 = \frac{1}{vC} \quad (5.27)$$

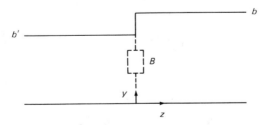

Figure 5.2. Waveguide step discontinuity.

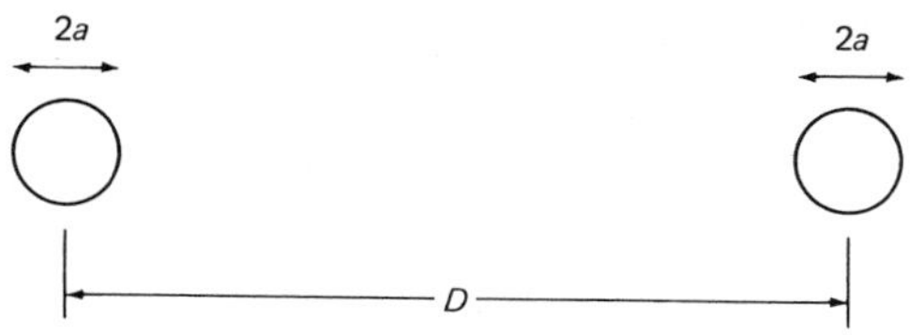

Figure 5.3. Two-wire transmission line.

where v is the propagation velocity of the wave and C is the static capacitance between the conductors. For the two-wire line and its equivalent one-wire above a ground plane,[5]

$$Z_0 = \frac{\eta}{\pi} \text{Arccosh} \frac{D}{2a} \tag{5.28}$$

where the dimensions are indicated in Figure 5.3. If the power decays as $e^{-2\alpha z}$ the attenuation constant is

$$\alpha = \alpha_c + \alpha_d \tag{5.29}$$

where the conductor and dielectric losses are

$$\alpha_c = \frac{1}{2a} \sqrt{\frac{\omega \mu}{2\sigma_c}} \frac{1}{\eta \, \text{Arccosh} \, (D/2a)} \tag{5.30}$$

$$\alpha_d = \frac{\sigma_d \eta}{2} \tag{5.31}$$

and σ_c and σ_d are the conductivities of the conductor and dielectric respectively. The two-wire line is inexpensive and widely used in UHF applications.

5.4 COAXIAL LINE

Coaxial transmission line is widely used and it comes in rigid, semirigid and flexible forms. It is readily available with an outer diameter of 7 mm that operates up to 18 GHz. More recently, coaxial lines with an outer diameter of 3.5 mm that operate up to 26 GHz are becoming more common. With dimensions shown in Figure 5.4 the characteristic impedance and loss is given as follows.

$$Z_0 = \frac{\eta}{2\pi} \ln \frac{b}{a} \tag{5.32}$$

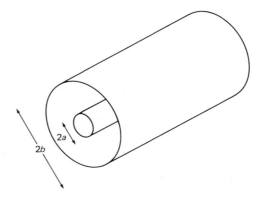

Figure 5.4. Coaxial line with outer diameter $2b$ and inner diameter $2a$.

If the dielectric constant is $\epsilon = \epsilon' - j\epsilon''$, then the loss in the dielectric contributes to the attenuation constant α as

$$\alpha_d = \frac{k_o \epsilon''}{2\sqrt{\epsilon'}}. \tag{5.33}$$

The loss in the conductor also contributes to the attenuation.

$$\alpha_c = \frac{R_m}{2\eta \ln(b/a)}\left(\frac{1}{a} + \frac{1}{b}\right) \tag{5.34}$$

$$R_m = \sqrt{\frac{\omega\mu}{2\sigma}} \tag{5.35}$$

As for the two-wire line, the total attenuation constant is

$$\alpha = \alpha_c + \alpha_d. \tag{5.36}$$

In practice, most of the loss comes from the conductor, and the loss gets smaller as the diameter increases.

5.4.1 Coaxial Line Discontinuities

A step in the inner conductor causes an effective discontinuity capacitance on the high impedance side and a step in the outer conductor causes an effective discontinuity capacitance on the low impedance side. These discontinuities are illustrated in Figure 5.5. The numerical values for the discontinuity capacitances are given by Somlo.[6]

For a step on the inner conductor, the capacitance is

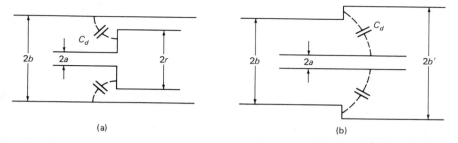

Figure 5.5. Coaxial step discontinuity caused by (a) step on the inner conductor and (b) step on the outer conductor.

$$C_d = 2b\epsilon \left[\frac{\alpha^2 + 1}{\alpha} \ln \frac{1 + \alpha}{1 - \alpha} - 2 \ln \frac{4\alpha}{1 - \alpha^2} \right]$$
$$+ \; 0.111(1 - \alpha^2)(\tau - 1)2\pi b \; 10^{-12} \text{ F}$$

ϵ = dielectric constant in Farads/meter
b = outer conductor radius, meters (5.37)
a = inner conductor radius, meters
r = stepped radius, meters

$$\alpha = \frac{b - r}{b - a}$$

$$\tau = \frac{b}{a}.$$

For a step on the outer conductor, the capacitance is

$$C_d = 2a\epsilon \left[\frac{\alpha^2 + 1}{\alpha} \ln \frac{1 + \alpha}{1 - \alpha} - 2 \ln \frac{4\alpha}{1 - \alpha^2} \right]$$
$$+ \; 0.412(1 - \alpha)(\tau - 1.4)2\pi a \; 10^{-12} \text{ F} \quad (5.38)$$

$$\alpha = \frac{b' - a}{b - a}$$

b' = stepped outer conductor radius, meters.

The remaining parameters are the same as the inner conductor step. For the inner conductor step, the error is stated as $\pm 0.03(2\pi b)$ pF for $0.01 \leq \alpha \leq 1.0$ and $1.0 \leq \tau \leq 6.0$. The error for the step on the outer conductor is $\pm 0.06(2\pi a)$ pF for $0.01 \leq \alpha \leq 0.7$ and $1.5 \leq \tau \leq 6.0$.

5.4.2 Disk Modification from a Discontinuity[7]

A disk in a coaxial line would ideally have a characteristic impedance Z_0 as shown in Figure 5.6a. Because of the discontinuity capacitance the circuit is actually that shown in Figure 5.6b. Cohn[8] showed that a short transmission

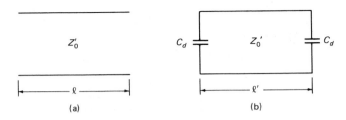

Figure 5.6. (a) Length of transmission line representing a disk, and (b) length of transmission line with discontinuity capacitance.

line may be represented by the π equivalent circuit in Figure 5.7. If $Z_0 = Z_0'$, ℓ' replaces ℓ where

$$\ell' = \frac{2}{\beta} \text{Arctan}\left(\tan \frac{\beta\ell}{2} - Z_0\omega C_d\right). \tag{5.39}$$

A more exact method would be to modify both Z_0 and ℓ. This requires an iterative solution as found by Levy and Rozzi.[9] If ℓ_0, Z_0, and $C_0 = C_d$ are the length, characteristic impedance, and discontinuity capacitance for the original disk, then the new length after $k + 1$ iterations is

$$\cos(\beta\ell_{k+1}) = \cos(\beta\ell_k) + \omega(C_k - C_{k-1})Z_k \sin(\beta\ell_k) \tag{5.40}$$

$$Z_{k+1} = Z_k \frac{\sin(\beta\ell_k)}{\sin(\beta\ell_{k+1})} \qquad k = 0,1,2 \ldots \tag{5.41}$$

where $C_{-1} = 0$. Normally 3 to 11 iterations are all that are needed to achieve satisfactory results.

5.5 STRIPLINE

Stripline consists of two ground plane conductors and one center conductor as shown in Figure 5.8. Stripline is favored over microstrip in passive circuit appli-

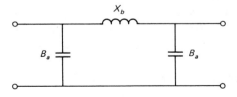

Figure 5.7. A π equivalent circuit for a transmission line where $X_b = Z_0 \sin(\beta\ell)$ and $B_a = Y_0 \tan(\beta\ell/2)$.

MECHANICAL REALIZATION OF SELECTED TRANSMISSION LINES 121

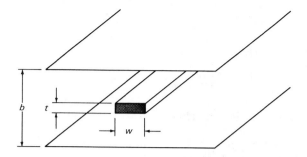

Figure 5.8. Stripline with center conductor of width w and thickness t.

cations where high precision, low dissipation and radiation loss, and small dispersion are needed. In the following paragraphs synthesis of the width for a given characteristic impedance, calculation of the open end effect and step discontinuities, and synthesis of the coupled stripline will be discussed. The following list of symbols will be pertinent to the stripline equations.

b = Stripline spacing between ground planes
w = Conductor width
s = Spacing between coupled lines
t = Conductor thickness
Z_{0e}, Z_{0o} = Even and odd mode characteristic impedance
$\epsilon = \epsilon_r \epsilon_o$ Dielectric constant of the substrate
c = Velocity of light in a vacuum

5.5.1 Stripline Characteristic Impedance and Loss

The synthesis equations for stripline with nonzero thickness are given by Wheeler[10] and Gupta et al.[11]

$$\frac{w}{b} = \frac{w_o}{b} - \frac{\Delta w}{b} \qquad (5.42)$$

where

$$\frac{w_o}{b} = \frac{8(1-\tau)}{\pi} \frac{\sqrt{e^A + 0.568}}{e^A - 1}$$
$$\frac{\Delta w}{b} = \frac{\tau}{\pi}\left(1 - \frac{1}{2}\ln\left[\left(\frac{\tau}{2-\tau}\right)^2 + \left(\frac{\tau/4\pi}{w_o/b - 0.26\tau}\right)^m\right]\right) \qquad (5.43)$$

$$A = \frac{4Z_0}{\eta} \tag{5.44}$$

$$\tau = t/b \tag{5.45}$$

$$m = 2\left[1 + \frac{2}{3}\frac{\tau}{1-\tau}\right]^{-1}. \tag{5.46}$$

The conductor loss is given by Howe[12] as

$$\alpha_c = \frac{2.659 R_m \sqrt{\epsilon_r}}{Z_0}\left(1 + \frac{2w_o}{b-t}\right) \tag{5.47}$$

$$-\frac{1}{\pi}\left[\frac{3\tau}{2-\tau} + \ln\frac{\tau}{2-\tau}\right]\bigg) D \times 10^{-3} \tag{5.48}$$

$$D = \frac{30e^{-A}}{w_o\sqrt{\epsilon_r}}\left[\frac{3.135}{Q} - \left(\frac{8}{\pi}\frac{b-t}{w_o}\right)^2(1+Q)\right] \tag{5.49}$$

$$Q = \sqrt{1 + 6.27\left(\frac{\pi}{8}\frac{w_o}{b-t}\right)^2}.$$

The dielectric loss and total attenuation constant is given in (5.33) and (5.36).

5.5.2 Stripline Open End Effect

In Figure 5.9 the electrical length is shown as extending beyond the mechanical length of the transmission line. This is because of the stray capacitance at the end of the line. This additional effective length is given by Altschuler and Oliner[13] as

$$\Delta\ell = c'\left(\frac{c' + 2w}{4c' + 2w}\right) \tag{5.50}$$

$$c' = \frac{b}{\pi}\ln 2. \tag{5.51}$$

5.5.3 Stripline Step Discontinuity

The equivalent circuit for the discontinuity associated with a step in the characteristic impedance of a stripline (Figure 5.10) consists of a series discontinuity reactance and two line lengths. The ideal transformer represents the change in impedance level.

$$\ell_1 = -\ell_2 = \frac{b\ln 2}{\pi} \tag{5.52}$$

$$X_d = \frac{f}{2c}(b-t)\eta \ln\left[\csc\left(\frac{\pi Z_1}{2Z_2}\right)\right] \quad Z_1 < Z_2 \tag{5.53}$$

MECHANICAL REALIZATION OF SELECTED TRANSMISSION LINES 123

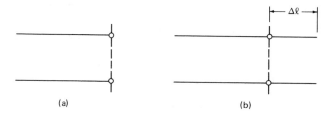

Figure 5.9. Open circuit end effect for the (a) mechanical line with (b) equivalent electrical circuit.

Since the discontinuity reactance is directly proportional to frequency, it can be modeled as a lumped inductance.

5.5.4 Coupled Striplines

The calculation of the line widths and spacing of two equal width coupled striplines shown in Figure 5.11 was given by Cohn.[14] When $s/t < 5$, a trial and error solution is required.

$$C_f(t/b) = \frac{\epsilon}{\pi} [2\tau \ln (\tau + 1) - (\tau - 1) \ln (\tau^2 - 1)] \quad (5.54)$$

$$\tau = \frac{1}{1 - t/b}$$

$$C_{fe}(0, s/b) = \frac{2\epsilon}{\pi} \ln \left[1 + \tanh \left(\frac{\pi s}{2b} \right) \right] \quad (5.55)$$

$$C_{fe}(t/b, s/b) = \frac{C_f(t/b)}{C_f(0)} C_{fe}(0, s/b) \quad \text{all } s/t \quad (5.56)$$

$$C_{fo}(0, s/b) = \frac{2\epsilon}{\pi} \ln \left[1 + \coth \left(\frac{\pi s}{2b} \right) \right] \quad (5.57)$$

$$C_{fo}(t/b, s/b) = \frac{C_f(t/b)}{C_f(0)} C_{fo}(0, s/b) \quad s/t \geq 5 \quad (5.58)$$

$$C_{fo}(t/b, s/b) = C_{fo}(0, s/b) + \epsilon t/s \quad s/t < 5 \quad (5.59)$$

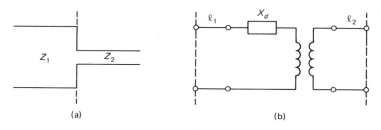

Figure 5.10. (a) Stripline step discontinuity and (b) equivalent circuit.

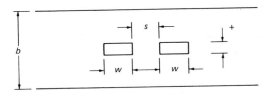

Figure 5.11. Coupled stripline.

To find s/b, calculate the following dimensionless parameter γ.

$$\gamma = \frac{\pi}{4}\eta\left(\frac{1}{Z_{0o}} - \frac{1}{Z_{0e}}\right)\frac{C_f(0)}{C_f(t/b)} \tag{5.60}$$

For $s/t \geqq 5$ the closed form solution for s is

$$s = \frac{2b}{\pi}\text{Arctanh}(e^{-\gamma}) \qquad s/t \geqq 5. \tag{5.61}$$

When $s/t < 5$, the value of s that satisfies (5.62) is found by trial and error.

$$\frac{1}{Z_{0o}} - \frac{1}{Z_{0e}} = \frac{2}{\sqrt{\mu_0\epsilon}}[C_{fo}(t/b,s/b) - C_{fe}(t/b,s/b)] \tag{5.62}$$

Once s is known, the line width is found.

$$\frac{w}{b} = (1 - t/b)\left[\frac{\eta}{4Z_{0e}} - \frac{1}{2\epsilon}\{C_f(t/b) + C_{fe}(t/b,s/b)\}\right] \tag{5.63}$$

5.5.5 Offset Parallel Coupled Lines

Shelton[15] developed a set of equations for the design of overlayed coupled striplines that are valid when $w/(1 - s) \geqq 0.35$ (Figure 5.12). For tight coupling when $w_c/s \geqq 0.7$, the synthesis equations are given below when s, Z_0, and $\rho = Z_{0e}/Z_{0o}$ are known.

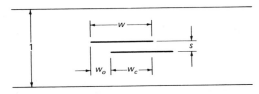

Figure 5.12. Offset coupled striplines (Shelton[15]).

MECHANICAL REALIZATION OF SELECTED TRANSMISSION LINES

$$A = \exp\left[\frac{\pi}{2} \sqrt{\frac{\mu_o}{\epsilon_o \epsilon_r} \frac{1}{Z_0}} \left(\frac{1 - \rho s}{\sqrt{\rho}}\right)\right] \tag{5.64}$$

$$B = \frac{A - 2 + \sqrt{A^2 - 4A}}{2} \tag{5.65}$$

$$p = \frac{(B-1)\left(\frac{1+s}{2}\right) + \sqrt{\left(\frac{1+s}{2}\right)^2 (B-1)^2 + 4sB}}{2} \tag{5.66}$$

$$r = \frac{sB}{p} \tag{5.67}$$

$$C_{fo} = \frac{1}{\pi}\left\{-\frac{2}{1-s}\ln s + \frac{1}{s}\ln\left[\frac{pr}{(p+s)(1+p)(r-s)(1-r)}\right]\right\} \tag{5.68}$$

$$C_o = \sqrt{\frac{\mu_o \rho}{\epsilon_o \epsilon_r} \frac{1}{Z_0}} \tag{5.69}$$

$$w = \frac{s(1-s)}{2}(c_o - C_{fo}) \tag{5.70}$$

$$w_o = \frac{1}{2\pi}\left\{(1+s)\ln\frac{p}{r} + (1-s)\ln\left[\frac{(1+p)(r-s)}{(s+p)(1-r)}\right]\right\} \tag{5.71}$$

For maximum coupling where $w_o = 0$, the maximum ratio of the even to odd mode impedances is found from

$$\frac{1 - \rho_{max} s}{\sqrt{\rho_{max}}} = \frac{2}{\pi}\sqrt{\frac{\epsilon_r \epsilon_o}{\mu_o}} Z_0 \ln 4. \tag{5.72}$$

For loose coupling when $2w_o/(1 + s) \geq 0.85$ the synthesis equations are the following.

$$\Delta C = \sqrt{\frac{\mu_o}{\epsilon_o \epsilon_r}} \frac{\rho - 1}{\sqrt{\rho}} \tag{5.73}$$

$$K = [\exp(\pi \Delta C/2) - 1]^{-1} \tag{5.74}$$

$$a = \sqrt{\left(\frac{s - K}{s + 1}\right)^2 + K} - \frac{s - K}{s + 1} \tag{5.75}$$

$$q = K/a \tag{5.76}$$

$$C_{fo} = \frac{2}{\pi} \left[\frac{1}{1+s} \ln \frac{1+a}{a(1-q)} - \frac{1}{1-s} \ln q \right] \quad (5.77)$$

$$w_c = \frac{1}{\pi} \left[s \ln \frac{q}{a} + (1-s) \ln \frac{1-q}{1+a} \right] \quad (5.78)$$

$$C_f = -\frac{2}{\pi} \left[\frac{1}{1+s} \ln \left(\frac{1-s}{2} \right) + \frac{1}{1-s} \ln \left(\frac{1+s}{2} \right) \right] \quad (5.79)$$

$$w = \frac{1-s^2}{4} [C_o - C_{fo} - C_f] \quad (5.80)$$

In this final expression, C_o is obtained from (5.69).

5.6 MICROSTRIP

Microstrip (Figure 5.13) is used in circuits where discrete devices are bonded to the circuit, where easy access is needed for tuning, and where a compact design is needed. Since the electromagnetic fields lie partly in air and partly in the dielectric, obtaining solutions for the characteristic impedance and effective dielectric constant is more complicated than it is for stripline. Furthermore, microstrip is only approximately a TEM transmission line, but unless the circuit is to be used for very broad bandwidth applications or it is physically many wavelengths long, dispersion will not be a problem. Thus the TEM approximation will give useful results. The following paragraphs will consider the same topics as were covered for stripline, i.e., (1) characteristic impedance and loss, (2) open end effect, (3) step discontinuity, and (4) coupled lines.

5.6.1 Microstrip Characteristic Impedance and Loss

A simple synthesis formula for a thin microstrip was given by Wheeler.[16]

$$\frac{w}{h} = 8 \frac{\sqrt{\frac{A}{11}(7 + 4/\epsilon_r) + \frac{1}{0.81}(1 + 1/\epsilon_r)}}{A} \quad (5.81)$$

$$A = \exp\left(\frac{Z_0 \sqrt{\epsilon_r + 1}}{42.4}\right) - 1 \quad (5.82)$$

This initial value for the width may be used in the following analysis formulas given by Hammerstad and Jensen[17] to obtain a very accurate value for the width and the effective dielectric constant for a conductor with a nonzero thickness.

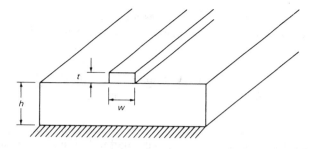

Figure 5.13. Microstrip line.

$$\Delta u_a = \frac{t/h}{\pi} \ln\left[1 + \frac{4\exp(1)}{t/h\ \coth^2\sqrt{6.517 w/h}}\right] \quad (5.83)$$

$$\Delta u_r = \tfrac{1}{2}[1 + 1/\cosh\sqrt{\epsilon_r - 1}]\,\Delta u_a \quad (5.84)$$

$$u_a = w/h + \Delta u_a \quad (5.85)$$

$$u_r = w/h + \Delta u_r \quad (5.86)$$

$$Z_{0a}(x) = \frac{\eta}{2\pi}\ln\left[\frac{f(x)}{x} + \sqrt{1 + \left(\frac{2}{x}\right)^2}\right] \quad (5.87)$$

$$f(x) = 6 + (2\pi - 6)\exp[-(30.666/x)^{.7528}] \quad (5.88)$$

$$\epsilon_e(x,\epsilon_r) = \frac{\epsilon_r + 1}{2} + \frac{\epsilon_r - 1}{2}\left(1 + \frac{10}{x}\right)^{-a(x)b(\epsilon_r)} \quad (5.89)$$

$$a(x) = 1 + \frac{1}{49}\ln\left[\frac{x^4 + (x/52)^2}{x^4 + 0.432}\right]$$
$$+ \frac{1}{18.7}\ln\left[1 + \left(\frac{x}{18.1}\right)^3\right] \quad (5.90)$$

$$b(\epsilon_r) = 0.564\left[\frac{\epsilon_r - 0.9}{\epsilon_r + 3}\right]^{0.053} \quad (5.91)$$

From the given value of t and trial solution w/h, (5.83) to (5.86) give unique values for u_a and u_r. The characteristic impedance and effective dielectric constant are obtained below using the functions (5.87) through (5.91).

$$Z_0(w/h, t, \epsilon_r) = \frac{Z_{0a}(u_r)}{\sqrt{\epsilon_e(u_r, \epsilon_r)}} \quad (5.92)$$

$$\epsilon_{\text{eff}}(w/h, t, \epsilon_r) = \epsilon_e(u_r, \epsilon_r)\left[\frac{Z_{0a}(u_a)}{Z_{0a}(u_r)}\right]^2 \quad (5.93)$$

Since w/h increases when Z_0 decreases and vice versa, one very simple and effective method for finding the new approximation for w/h is by using the ratio in (5.94).

$$\left(\frac{w}{h}\right)_{i+1} = \left(\frac{w}{h}\right)_i \frac{\text{Calculated } Z_0 \text{ from (5.92)}}{\text{Desired } Z_0} \quad (5.94)$$

The conduction losses are given by Pucel, Masse', and Hartwig[18,19] and put in the present form by Gupta, Garg and Chadha[11] as shown below.

$$\alpha_c = \begin{cases} 0.159 A \dfrac{R_m[32 - u_r^2]}{hZ_0[32 + u_r^2]} & w/h \leq 1 \quad (5.95a) \\[2ex] 7.02 \; 10^{-6} A \dfrac{R_m Z_0 \epsilon_{\text{eff}}}{h} \left[u_r + \dfrac{0.667 u_r}{u_r + 1.444}\right] & w/h \geq 1 \quad (5.95b) \end{cases}$$

$$A = 1 + u_r \left[1 + \frac{1}{\pi} \ln (2B/t)\right] \quad (5.96)$$

$$B = \begin{cases} h & w/h \geq \tfrac{1}{2}\pi \quad (5.97a) \\ 2\pi w & w/h \leq \tfrac{1}{2}\pi \quad (5.97b) \end{cases}$$

The value for R_m is given by (5.35).

The loss caused by the dielectric is

$$\alpha_d = \frac{\epsilon_r}{2\sqrt{\epsilon_{\text{eff}}}} \frac{\epsilon_{\text{eff}} - 1}{\epsilon_r - 1} \frac{k_o \epsilon''}{\epsilon'} \quad (5.98)$$

where $\epsilon = \epsilon' - j\epsilon'' = \epsilon_0 \epsilon_r - j\epsilon''$.

5.6.2 Microstrip End Effect

The added electrical length beyond the mechanical length of an open microstrip line is given by Hammerstad and Bekkadal[20] and modified by Garg and Bahl[21] as follows.

$$\frac{\Delta \ell}{h} = 0.412 \left[\frac{\epsilon_{\text{eff}} + 0.3}{\epsilon_{\text{eff}} - 0.258}\right] \left[\frac{w/h + 0.262}{w/h + 0.813}\right] \quad (5.99)$$

5.6.3 Microstrip Step Discontinuity

The formula for a microstrip discontinuity is given by Gupta, Garg, and Chadha.[11] The T equivalent circuit for the junction is shown in Figure 5.14.

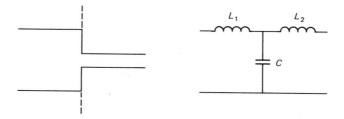

Figure 5.14. Approximate T equivalent circuit for step discontinuity.

$$\frac{C}{\sqrt{w_1 w_2}} = [10.1 \log (\epsilon_r) + 2.33] \frac{w_1}{w_2} - 12.6 \log (\epsilon_r) - 3.17 \quad (5.100)$$

This expression is in error by less than 10% when $\epsilon_r \leq 10$ and $1.5 \leq w_1/w_2 \leq 3.5$. A more accurate expression for alumina substrates with $\epsilon_r = 9.6$ is

$$\frac{C}{\sqrt{w_1 w_2}} = 130 \log \frac{w_1}{w_2} - 44 \quad (5.101)$$

where the error is less than 0.5% for $3.5 \leq w_1/w_2 \leq 10$. The values for the inductances in the equivalent circuit are obtained from the following relations.

$$\frac{L}{h} = 40.5 \left[\frac{w_1}{w_2} - 1.0 \right] - 75 \log \frac{w_1}{w_2} + 0.2 \left[\frac{w_1}{w_2} - 1.0 \right]^2 \quad (5.102)$$

The inductance per unit length of a microstrip is

$$L_{w1} = \frac{Z_0(w_1) \sqrt{\epsilon_{\text{eff}}(w_1)}}{c} \quad (5.103)$$

$$L_{w2} = \frac{Z_0(w_2) \sqrt{\epsilon_{\text{eff}}(w_2)}}{c} \quad (5.104)$$

so that

$$L_1 = \frac{L_{w1}}{L_{w1} + L_{w2}} L \quad (5.105)$$

$$L_2 = \frac{L_{w2}}{L_{w1} + L_{w2}} L. \quad (5.106)$$

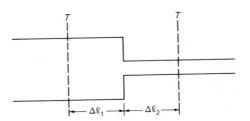

Figure 5.15. Approximate compensation for step discontinuity.

The effect of the inductance may be accommodated in the step junction by shortening the line lengths by $\Delta \ell_1$ and $\Delta \ell_2$ given in (5.107) as illustrated in Figure 5.15.

$$\Delta \ell_1 = \Delta \ell_2 = \frac{L}{(L_{w1} + L_{w2})h} \qquad (5.107)$$

An alternative method for accommodating the discontinuity is discussed in section 5.7.

5.6.4 Microstrip Coupled Lines

Solution for the strip width w and spacing s for a coupled microstrip line requires a trial and error procedure. However a closed form approximate solution given by Akhtarzad, Rowbotham, and Johns[22] provides an initial trial solution for the more accurate analysis formula of Garg and Bahl.[23] The approximate solution[22] follows.

$$w_e = w(Z_{0e}/2, \epsilon_{\text{eff}}) \qquad (5.108)$$
$$w_o = w(Z_{0o}/2, \epsilon_{\text{eff}}) \qquad (5.109)$$

The widths w are calculated from (5.81) to (5.94).

$$Q_e = \cosh \frac{\pi w_e}{2h} \qquad (5.110)$$

$$Q_o = \cosh \frac{\pi w_o}{2h} \qquad (5.111)$$

$$g = \frac{Q_e + Q_o - 2}{Q_o - Q_e} \qquad (5.112)$$

$$H = \tfrac{1}{2}[(g + 1)Q_e + g - 1] \qquad (5.113)$$

$$s = \frac{2h}{\pi} \operatorname{Arccosh} g \qquad (5.114)$$

$$w = \frac{h}{\pi} \operatorname{Arccosh} (Hg - \sqrt{H^2 g^2 - H^2 - g^2 + 1}) \qquad (5.115)$$

Refinement in this value of w may be made by varying g and H in such a way that $(2H - g + 1)/(g + 1)$ remains constant until

$$\frac{w_o}{h} = \frac{2}{\pi} \text{Arccosh}\left(\frac{2H - g - 1}{g - 1}\right) + \frac{4}{\pi(1 + E/2)} \text{Arccosh}\,(1 + 2w/s)$$

(5.116)

where

$$E = \begin{cases} \epsilon_r & \epsilon_r \leq 6 \\ 6 & \epsilon_r > 6. \end{cases}$$

The values for s and w obtained from (5.114) and (5.115) can be in error by as much as 10%. However, they may be used as the initial trial solution for the iterative procedure based on the following analysis equations.[23]

$$C_e = C_p + C_f + C_f'$$ (5.117)

$$C_o = C_p + C_f + C_{gd} + C_{ga} + \frac{2\epsilon_o t}{s}$$ (5.118)

$$C_p = \frac{\epsilon_w}{h}$$ (5.119)

$$C_f = \tfrac{1}{2}[\sqrt{\epsilon_{\text{eff}}}/(cZ_0) - C_p]$$ (5.120)

The single strip equations (5.83) through (5.93) are used in (5.120) to find Z_0 and ϵ_{eff}.

$$C_f' = \frac{C_f}{1 + (Ah/s)\tanh(10s/h)}\sqrt{\frac{\epsilon_r}{\epsilon_{\text{eff}}}}$$ (5.121)

$$A = \exp[-0.1\exp(2.33 - 2.53w/h)]$$ (5.122)

$$C_{ga} = \epsilon_o \frac{K(k')}{K(k)} \cong \begin{cases} \dfrac{\epsilon_o}{\pi} \ln \dfrac{2(1 + \sqrt{k'})}{1 - \sqrt{k'}} & 0 < k^2 < 0.5 \\ \epsilon_o \pi \left[\ln \dfrac{2(1 + \sqrt{k})}{1 - \sqrt{k}}\right]^{-1} & 0.5 < k^2 < 1 \end{cases}$$ (5.123)

The function $K(k)$ is the elliptic function for which the above approximations have been made.

$$k = \frac{s/h}{s/h + 2w/h}$$ (5.124)

$$k'^2 = 1 - k^2$$ (5.125)

$$C_{gd} = \frac{\epsilon}{\pi} \ln \coth \frac{\pi s}{4h} + 0.65 C_f \left[\frac{0.02\sqrt{\epsilon_r}}{s/h} + (1 - \epsilon_r^{-2}) \right] \quad (5.126)$$

From these calculated capacitance values, Z_{0e} and Z_{0o} are found from

$$Z_{0e} = \frac{1}{c\sqrt{C_{ea}C_e}} \quad (5.127)$$

$$Z_{0o} = \frac{1}{c\sqrt{C_{oa}C_o}} \quad (5.128)$$

where C_{ea} and C_{oa} are the even and odd mode capacitances calculated with an air dielectric. The effective dielectric constants for the even and odd modes are

$$\epsilon_{\text{effe}} = \frac{C_e}{C_{ea}} \quad (5.129)$$

$$\epsilon_{\text{effo}} = \frac{C_o}{C_{oa}}. \quad (5.130)$$

5.7 ACCOMMODATION OF DISCONTINUITIES

The shunt discontinuity susceptance in Figure 5.16a or the series discontinuity reactance in Figure 5.16b modifies the reflection and transmission coefficients of the incident wave. Cohn[24] found for small discontinuities, the major effect caused by the discontinuity is a change in phase of the propagating wave. Simply by changing the line lengths to accommodate this phase change, the effect of the discontinuity may be largely eliminated. For a shunt B_k as shown in Figure 5.16a, the angles of the reflection and transmission coefficients are

$$\angle \rho_k = - \text{Arctan} \frac{B_k/Y_{k+1}}{Y_k/Y_{k+1} - 1} - \text{Arctan} \frac{B_k/Y_{k+1}}{Y_k/Y_{k+1} + 1} \quad (5.131)$$

$$\angle t_k = - \text{Arctan} \frac{B_k/Y_{k+1}}{Y_k/Y_{k+1} + 1}. \quad (5.132)$$

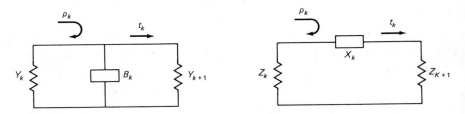

Figure 5.16. (a) Shunt and (b) series discontinuities.

MECHANICAL REALIZATION OF SELECTED TRANSMISSION LINES 133

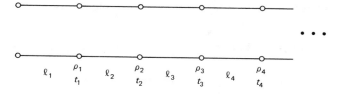

Figure 5.17. Multiple reflection coefficients ρ_k and transmission coefficients t_k from discontinuities on a transmission line.

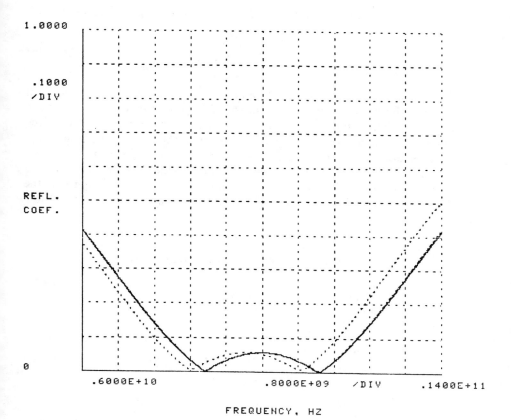

Figure 5.18. Ideal 2-section Chebyshev impedance transformer (——), response with discontinuity capicitances (——), response with discontinuity capacitances compensated by changes in line lengths (overlay ---).

For the series reactance X_k in Figure 5.15b, the angles are

$$\angle \rho_k = \text{Arctan} \frac{X_k/Z_{k+1}}{1 - Z_k/Z_{k+1}} - \text{Arctan} \frac{X_k/Z_{k+1}}{1 + Z_k/Z_{k+1}} \quad (5.133)$$

$$\angle t_k = \text{Arctan} \frac{X_k}{Z_{k+1}} - \text{Arctan} \frac{X_k/Z_{k+1}}{1 + Z_k/Z_{k+1}}. \quad (5.134)$$

The phase angles at each of the discontinuities caused by the multiple reflections shown in Figure 5.17 are given by

$$\phi_k = -\angle \rho_k - 2 \sum_{i=1}^{k-1} \angle t_i \quad (5.135)$$

and the new line length ℓ_k is found from the original line length ℓ_{ko} by (5.136).

$$\ell_k = \ell_{ko} - \frac{1}{\beta}(\phi_k - \phi_{k-1}) \quad (5.136)$$

In section 3.2 a design example for an impedance transformer was given. If this were built in coaxial line, the measured response would be that shown in the dashed line in Figure 5.18. The discontinuity capacitances have, in effect, lowered the center frequency of the transformer. When the line lengths are modified according to (5.136), the measured curve corresponds with the original desired response shown by the solid curve within the resolution of the graph. The design parameters for the original transformer design and the design after accounting for the discontinuity capacitances are shown in Table 5.1.

Table 5.1. Coaxial Impedance Transformer.

Bandwidth	40%
Center frequency	10 GHz
Input impedance	10 Ω
Output impedance	100 Ω
Coaxial outer diameter	7 mm
Number of sections	2

IMPEDANCE Z_i	ORIGINAL LENGTH ℓ_{ko}	MODIFIED LENGTH ℓ_k
10		
18.3035	0.2951	0.2826
54.6344	0.2951	0.2850
100		

REFERENCES

1. R. E. Collin, *Foundations for Microwave Engineering*. New York: McGraw-Hill, 1966.
2. R. D. Lending, "New Criteria for Microwave Component Surfaces," *Proc. of the National Electronics Conference*, Vol. 11, pp. 391–401, 1955.
3. C. G. Montgomery, R. H. Dicke, and E. M. Purcell, eds., *Principles of Microwave Circuits*. New York: McGraw-Hill, 1948.
4. J. Schwinger and D. S. Saxon, *Discontinuities in Waveguides*. New York: Gordon and Breach, p. 123, 1968.
5. S. R. Seshadri, *Fundamentals of Transmission Lines and Electromagnetic Fields*. Reading, Mass.: Addison-Wesley, 1971.
6. P. I. Somlo, "The Computation of Coaxial Line Step Capacitances," *IEEE Transactions on Microwave Theory and Techniques*, Vol. MTT-15, pp. 48–53, January 1967.
7. W. A. Davis and P. J. Khan, "Coaxial Bandpass Filter Design," *IEEE Transactions on Microwave Theory and Techniques*, Vol MTT-19, pp. 373–380, April 1971.
8. S. B. Cohn, "Principles of Transmission Line Filter Design," in *Very High Frequency Techniques*, Vol 2. New York: McGraw-Hill, Chapter 26, 1947.
9. R. Levy and T. E. Rozzi, "Precise Design of Coaxial Low-Pass Filters," *IEEE Transactions on Microwave Theory and Techniques*, Vol. MTT-16, pp. 142–147, March 1968.
10. H. A. Wheeler, "Transmission Line Properties of a Stripline Between Parallel Planes, *IEEE Trans. on Microwave Theory and Techniques*, Vol. MTT-26, pp. 866–876, November 1978.
11. K. C. Gupta, R. Garg, and R. Chadha, *Computer Aided Design of Microwave Circuits*. Dedham, Mass.: Artech House, 1981.
12. H. Howe Jr., *Stripline Circuit Design*. Dedham, Mass.: Artech House, 1974.
13. H. M. Altschuler and A. A. Oliner, "Discontinuities in the Center Conductor of Symmetric Strip Transmission Line," *IRE Trans. on Microwave Theory and Techniques*, Vol. MTT-8, pp. 328–339, May 1960.
14. S. B. Cohn, "Shielded Coupled-Strip Transmission Line," *IRE Trans. on Microwave Theory and Techniques*, Vol. MTT-3, pp. 29–38, October 1955.
15. J. P. Shelton, "Impedances of Offset Parallel-Coupled Strip Transmission Lines," *IRE Trans. on Microwave Theory and Techniques*, Vol. MTT-14, pp. 7–15, January 1966.
16. H. A. Wheeler, "Transmission-Line Properties of a Strip on a Dielectric Sheet on a Plane," *IEEE Trans. on Microwave Theory and Techniques*, Vol. MTT-25, pp. 631–647, August 1977.
17. E. Hammerstad and O. Jensen, "Accurate Models for Microstrip Computer-Aided Design," *1980 IEEE MTT-S International Microwave Symposium Digest*, pp. 407–409, May 1980.
18. R. A. Pucel, D. J. Masse', and C. P. Hartwig, "Losses in Microstrip," *IEEE Transactions on Microwave Theory and Techniques*, Vol. MTT-16, pp. 342–350, June 1968.
19. ———, "Corrections to 'Losses in Microstrip'," *IEEE Transactions on Microwave Theory and Techniques*, Vol. MTT-16, p. 1064, December 1968.
20. E. O. Hammerstad and F. Bekkadal, *A Microstrip Handbook*. ELAB report STF 44 A74169, The University of Trondheim, The Norwegian Institute of Technology, 1975.
21. R. Garg and I. J. Bahl, "Microstrip Discontinuities," *International Journal of Electronics*, Vol. 45, pp. 81–87, July 1978.
22. S. Akhtarzad, T. R. Rowbotham, P. B. Johns, "The Design of Coupled Microstrip Lines," *IEEE Transactions on Microwave Theory and Techniques*, Vol. MTT-23, pp. 486–492, June 1975.
23. R. Garg and I. J. Bahl, "Characteristics of Coupled Microstrips," *IEEE Transactions on Microwave Theory and Techniques*, Vol. MTT-27, pp. 700–705, July 1979.
24. S. B. Cohn, "Optimum Design of Stepped Transmission Line Transformers," *IRE Transactions on Microwave Theory and Techniques*, Vol MTT-3, pp. 16–21, April 1955.

Chapter 6
Computer Aided Design, Manufacturing, and Test

6.1 INTRODUCTION

A new technology is often introduced as a better way of performing an existing job. As the new technology becomes widespread, and its advantages more clearly recognized, it stimulates further growth and development. Eventually, the new technological convenience comes to be viewed as an essential need. The automobile is an example of this process.

When the first motor vehicle was introduced, it allowed a person to move a little faster than he could before. It was an invention society found convenient, but not necessary. With the passage of time, however, the automobile opened new possibilities that would never have occurred without it. Today, most people consider the automobile to be an unquestionable necessity. More recently, industry, education, and government have found computers convenient for performing calculations and controlling processes. In much of the world, they are now becoming a necessity. This technology has important ramifications for the field of microwave engineering. It is no longer necessary, for example, to have assembly lines of people doing a small portion of a long calculation to reduce the chance of error. It is no longer necessary for authors of scientific papers to reduce many of their mathematical results to an asymptotic formula or some other approximation in order to obtain numerical results. Furthermore, our very way of thinking about problems may be changing. It is possible that we may start thinking less in terms of analytical functions and more in terms of mathematical procedures, i.e., algorithms.

Although the application of computers is relatively new, and it is certain that many more developments will be coming in the future, it seems appropriate to include a chapter on the impact of the computer on microwave engineering. This chapter is divided into three major sections corresponding to the three areas in which the microwave engineer has been most heavily influenced. The first of these is computer aided design (CAD), where a circuit may be designed and optimized to meet a set of specifications. The second is computer aided manufacturing (CAM) where the design is laid out on a graphics terminal connected to a computer. When the engineer is satisfied with the picture on the screen, the image is transferred to a machine that makes a photographic mask of the circuit. After assembling the circuit, the third technique employs computer aided testing (CAT) to verify the design. These three topics are discussed in sections 6.2, 6.3, and 6.4 respectively.

6.2 COMPUTER AIDED DESIGN

Application of computer aided design (CAD) involves numerically evaluating analytical expressions, optimizing linear circuits, and performing nonlinear analysis of active circuits. Subsection 6.2.1 will outline a design approach for a linear circuit, while subsection 6.2.2 will provide an approach to analysis of a nonlinear circuit.

6.2.1 Linear Circuit Design

The initial effort in the design of a microwave circuit should focus on closed form solutions such as those presented earlier in the book. Even if a satisfactory design does not come from this effort, an initial topology and some initial circuit parameters are obtained. These can be a great benefit in the optimization process. The commercially available program Super Compact® includes a synthesis section that may be helpful in finding an appropriate circuit topology with initial circuit parameters. The next step in the design process is to provide the computer with the S parameters of any nonlinear device included in the circuit. These S parameters may be in the form of (1) experimental data points taken either by the user or provided by the manufacturer, (2) a set of polynomials that approximate the S parameters, or (3) a linear circuit where the S parameters of the circuit model have been optimized to closely match the device data. Together with controlled voltage and current sources, this can model a wide range of diode and transistor amplifiers.

To optimize all the parameters in a large circuit over the entire frequency range of interest is often uneconomical. Hence, the designer often chooses a few parameters that most affect the circuit response. He must also choose the range of variation of these parameters. At this point some judgment and experience is very helpful. The advantage of the CAD approach is that this experience can be gained quickly.

The error criteria must next be chosen to stop the program when the computer calculations are acceptably close to the exact solution. The part of the frequency range that is most important and needs more weighting in the optimization must also be chosen.

After the program has provided an optimized circuit design, it is good practice to test the circuit in an analysis program to insure that the anticipated shape of the circuit response matches the desired shape. The optimization program has an analysis section in it that may be adequate. Sometimes the application requires special considerations not contained in the optimization program, such as parasitic reactances, radiation from a microstrip bend, or a non-ideal connector. In such cases a custom analysis program may be needed to verify and possibly fine tune the design.

Optimization programs generally use either cascade analysis or nodal analysis. In the first approach, two-port circuits are cascaded together. The program Compact® uses this approach. Although this program could consider four-port circuits, it requires some special considerations by the user. In the nodal approach, circuit elements are attached to specific numbered nodes. This approach allows for more flexible circuit topologies than the cascade approach and is the one now being used by Super Compact®.

For many applications it is more convenient to use a special purpose optimization program rather than a general purpose program. Sometimes a simple linear approach, such as that used in (5.94), may be sufficient. A more powerful technique is the Newton-Raphson method. This method formulates a solution error function, ϵ, that becomes zero when the solution is found. If x_1 is the initial guess for the solution and x_2 is the solution, then a Taylor series expansion of the error function is

$$\epsilon(x_2) = 0 = \epsilon(x_1) - \epsilon'(x_1)(x_2 - x_1). \tag{6.1}$$

This may be solved for x_2 to obtain an improved guess for the solution. In general, if x_i is the last guess, then the new estimate is

$$x_{i+1} = x_i - \frac{\epsilon(x_i)}{\epsilon'(x_i)}. \tag{6.2}$$

When a solution must be obtained for several parameters simultaneously, a generalized Newton-Raphson formula may be used.[1] Two restrictions must be followed in using this method. First, the objective function U that is to be minimized must be represented as a sum of squares of functions.

$$U = \sum_{i=1}^{N} |\epsilon_i|^2 \tag{6.3}$$

For example if $\{p\}$ represents a set of N circuit parameters that are varied to meet a specified condition (e.g., minimum reflection coefficient), then the generalized form of (6.1) is

$$0 = \epsilon_i(p_1^k, p_2^k, \ldots p_N^k) + \sum_{j=1}^{N} \frac{\partial \epsilon_i}{\partial p_j}(p_j^{k+1} - p_j^k); \quad i = 1, 2, \ldots N \tag{6.4}$$

In matrix notation the solution for the $(k + 1)$th trial solution is

$$\mathbf{p}^{k+1} = \mathbf{p}^k - \mathbf{J}^{-1}\epsilon(\mathbf{p}^k) \tag{6.5}$$

The second restriction is that the Jacobian matrix

$$\mathbf{J} = \begin{bmatrix} \dfrac{\partial \epsilon_1}{\partial p_1} & \dfrac{\partial \epsilon_1}{\partial p_2} & \cdots & \dfrac{\partial \epsilon_1}{\partial p_N} \\ \vdots & & & \\ \dfrac{\partial \epsilon_N}{\partial p_1} & & \cdots & \dfrac{\partial \epsilon_N}{\partial p_N} \end{bmatrix} \tag{6.6}$$

is a square matrix.[1]

6.2.2 Nonlinear Circuit Analysis

Analysis of nonlinear circuits has historically taken two approaches: (1) time domain integration of a differential equation that yields the transient solution of the circuit and (2) the frequency domain harmonic balance technique that provides a steady-state solution directly. A steady-state solution can be obtained from method 1 by numerical integration of the time domain equations over many periods, but this is usually a very time-consuming method. Aprille and Trick[2,3] have shortened this process somewhat by predicting the steady-state condition using Newton's method. The disadvantage of the time domain method is that each time a new circuit is calculated in a network optimization problem, a new steady-state analysis must be performed.

In the second approach, each voltage and current are represented by a Fourier series that satisfies the circuit constraints at all harmonic frequencies of the applied voltage (or current). Both Bailey[4] and Lindenlaub[5] express the dynamic equations of the network in terms of a set of simultaneous first order differential equations. A time domain error function of the form

$$U = \int_0^T |\epsilon|^2 \, dt \tag{6.7}$$

is minimized to test for a solution with a prescribed error. Each component of ϵ_i is

$$\epsilon_i \cong \frac{dx_i}{dt} - f_i(x_1, x_2, \ldots x_n, t) \tag{6.8}$$

where each dynamic variable x_i is expanded in a Fourier series containing N frequencies. An optimization technique is used to adjust the Fourier coefficients to obtain minimum error. Nakhla and Vlach[6] improve the efficiency of this method by reducing the large number of variables to be optimized in any stage. This is done by breaking the network into a number of easily analyzed subnetworks. The frequency domain method provides a fast method of obtaining new circuit solutions in an optimization problem, but it has the disadvantage of requiring $2N + 1$ variables for N frequencies.

In recent years, a third basic formulation of the nonlinear problem has been done with the aid of the Volterra series. This is a power series with memory in which the output of the nonlinear system is expressed in terms of powers of the input. Solution of the system is accomplished using an iterative procedure. At each step higher order nonlinear transfer functions are obtained from lower order nonlinear terms. Increasing the order of the nonlinearity rapidly increases the computational complexity of the problem. In practice, the Volterra series is not used beyond the second or third order of nonlinearity. The Volterra series method, however, allows complicated nonlinear problems to be treated in a systematic way.

The harmonic balance technique seems most appropriate when analyzing microwave circuits where the transient solution is not needed. Since microwave circuits cannot be easily modeled with frequency independent elements, this formulation has no time dependent voltages, currents, or circuit elements. Rather, this method depends on readily measurable frequency domain circuit impedances and the nonlinear device characteristics for a number of harmonic frequencies.

The response of a nonlinear element to a DC excitation can be modeled as a single polynomial.

$$V(i) = a_o + a_1 i + a_2 i^2 + a_2 i^3 + \cdots \quad (6.9)$$

When an AC current is impressed on the nonlinearity, harmonic frequencies are generated that complicate the problem considerably. For the linear problem, the linear circuit impedance is

$$Z = \begin{cases} R_0 \text{ at DC} \\ |Z_1| \angle \psi_1 \text{ at } \omega \end{cases} \quad (6.10)$$

where $\psi_1 = \arctan(X_1/R_1)$ is the angle of the impedance at radian frequency ω. When a current with arbitrary amplitude and phase

$$i = i_0 + i_{c1} \cos \omega t + i_{s1} \sin \omega t \quad (6.11)$$

is impressed on the linear impedance, the resulting voltage is

COMPUTER AIDED DESIGN, MANUFACTURING, AND TEST   141

$$v = v_0 + v_{c1} \cos \omega t + v_{s1} \sin \omega t. \quad (6.12)$$

From Ohm's law, this voltage can also be expressed in terms of the current and the impedance.

$$v = i_0 R_0 + |Z_1|[i_{c1} \cos(\omega t + \psi_1) + i_{s1} \sin(\omega t + \psi_1)] \quad (6.13)$$

The individual voltage components are therefore

$$v_0 = i_0 R_0 \quad (6.14)$$
$$v_{c1} = i_{c1} R_1 + i_{s1} X_1 \quad (6.15)$$
$$v_{s1} = -i_{c1} X_1 + i_{s1} R_1. \quad (6.16)$$

If a set of harmonically related currents excite the linear circuit, the currents and voltages may be represented either as a Fourier series or as column matrix of Fourier coefficients.

$$i = \sum_{k=0}^{N} i_{ck} \cos k\omega t + i_{sk} \sin k\omega t \quad (6.17)$$

$$\mathbf{I} = [i_{c0}, i_{c1} \ldots i_{cN}, i_{s1} \ldots i_{sN}]^T \quad (6.18)$$

The superscript "T" designates the transpose of the matrix. The multifrequency expression for Ohm's law is now given by

$$\mathbf{V} = \mathbf{ZI} \quad (6.19)$$

$$\begin{bmatrix} v_0 \\ v_{c1} \\ v_{c2} \\ \vdots \\ v_{cN} \\ v_{s1} \\ v_{s2} \\ \vdots \\ v_{sN} \end{bmatrix} = \begin{bmatrix} R_0 & 0 & 0 & \ldots & 0 & 0 & 0 & \ldots & 0 \\ 0 & R_1 & 0 & \ldots & 0 & X_1 & 0 & \ldots & 0 \\ 0 & 0 & R_2 & \ldots & 0 & 0 & X_2 & \ldots & 0 \\ \vdots & \vdots & \vdots & & \vdots & \vdots & \vdots & & \vdots \\ 0 & 0 & 0 & \ldots & R_N & 0 & 0 & \ldots & X_N \\ 0 & -X_1 & 0 & \ldots & 0 & R_1 & 0 & \ldots & 0 \\ 0 & 0 & -X_2 & \ldots & 0 & 0 & R_2 & \ldots & 0 \\ \vdots & \vdots & \vdots & & \vdots & \vdots & \vdots & & \vdots \\ 0 & 0 & 0 & \ldots -X_N & 0 & 0 & \ldots & R_N \end{bmatrix} \begin{bmatrix} i_0 \\ i_{c1} \\ i_{c2} \\ \vdots \\ i_{cN} \\ i_{s1} \\ i_{s2} \\ \vdots \\ i_{sN} \end{bmatrix}$$

(6.20)

The cosine and sine current components at a given frequency are multiplied by the appropriate matrix elements to obtain the voltage at the same frequency.

A nonlinear device will provide power at harmonic frequencies, and these may be obtained by multiplying two Fourier series together. An equivalent procedure is to multiply a $(2N + 1) \times (2N + 1)$ matrix by a column matrix consisting of the Fourier coefficients of the second series. The elements of the square matrix are found by carrying out the product of the two Fourier series.

$$P = \left[\sum_{n=0}^{\infty} d_{cn} \cos n\omega t + d_{sn} \sin n\omega t\right]\left[\sum_{m=0}^{\infty} b_{cm} \cos m\omega t + b_{sm} \sin m\omega t\right] \quad (6.21)$$

$$\begin{aligned} P &= \sum_{n=0}^{\infty} d_{cn} \cos n\omega t \sum_{m=0}^{\infty} b_{cm} \cos m\omega t \\ &+ \sum_{n=0}^{\infty} d_{sn} \sin n\omega t \sum_{m=0}^{\infty} b_{sm} \sin m\omega t \\ &+ \sum_{n=0}^{\infty} d_{cn} \cos n\omega t \sum_{m=0}^{\infty} b_{sm} \sin m\omega t \\ &+ \sum_{n=0}^{\infty} d_{sn} \sin n\omega t \sum_{m=0}^{\infty} b_{cm} \cos m\omega t \end{aligned} \quad (6.22)$$

Each of the four terms in (6.22) can be put into the form of a double sum by using the formula for the Cauchy product of two power series.[7]

$$\sum_{n=0}^{\infty} c_n \sum_{m=0}^{\infty} e_m = \sum_{n=0}^{\infty} \sum_{m=0}^{n} c_n e_{n-m} \quad (6.23)$$

When this is applied to (6.22) and the standard identities are used for the product of two trigonometric functions, the product of the Fourier series is given in (6.24).

$$\begin{aligned} P &= \sum_{n=0}^{\infty} \sum_{m=0}^{n} \tfrac{1}{2} d_{c(n-m)} b_{cm} [\cos n\omega t + \cos(n-2m)\omega t] \\ &+ \sum_{n=0}^{\infty} \sum_{m=0}^{n} \tfrac{1}{2} d_{s(n-m)} b_{sm} [-\cos n\omega t + \cos(n-2m)\omega t] \\ &+ \sum_{n=0}^{\infty} \sum_{m=0}^{n} \tfrac{1}{2} d_{s(n-m)} b_{cm} [\sin n\omega t + \sin(n-2m)\omega t] \\ &+ \sum_{n=0}^{\infty} \sum_{m=0}^{n} \tfrac{1}{2} d_{c(n-m)} b_{sm} [\sin n\omega t - \sin(n-2m)\omega t] \end{aligned} \quad (6.24)$$

If the series is truncated to N frequencies, a square matrix **H** of dimension $2N + 1$ can be found. This is made up of combinations of the d coefficients from the first Fourier series.

$$\mathbf{P} = \mathbf{HB}$$

or

$$\begin{bmatrix} p_{c0} \\ p_{c1} \\ \vdots \\ p_{cN} \\ p_{s1} \\ \vdots \\ p_{sN} \end{bmatrix} = \begin{bmatrix} h_{0,0} & h_{0,1} & \cdots & h_{0,N} & h_{0,N+1} & \cdots & h_{0,2N} \\ h_{1,0} & h_{1,1} & \cdots & h_{1,N} & h_{1,N+1} & \cdots & h_{1,2N} \\ \vdots & \vdots & \vdots & \vdots & \vdots & & \vdots \\ h_{N,0} & \cdots & & & & & \\ h_{N+1,0} & \cdots & & & & & \\ \vdots & & & & & & \\ h_{2N,0} & \cdots & & & & & h_{2N,2N} \end{bmatrix} \begin{bmatrix} b_{c0} \\ b_{c1} \\ \vdots \\ b_{cN} \\ b_{s1} \\ \vdots \\ b_{sN} \end{bmatrix}$$

(6.25)

The equivalent matrix representation of the product of the two Fourier series is a square matrix **H** made up of coefficients from the first series only, and a column matrix made up of coefficients from the second series only. The first double sum in (6.24) corresponds to the upper left quadrant of **H**, the second sum to the upper right quadrant, the third sum to the lower left quadrant, and the last sum to the lower right quadrant. In the first quadrant, the element h_{rm} relates the coefficient b_{cm} with the $\cos r\omega t$. For the first cosine term in (6.24) $r = n$, so the d coefficient is simply $d_{c(r-m)}$. For the second cosine term let

$$\pm r = n - 2m$$

or

$$n - m = m \pm r \qquad (6.26)$$

so the d coefficient will therefore be $d_{c(m \pm r)}$. The terms in the remaining three quadrants of **H** are obtained similarly. By carefully accounting for all the Fourier series terms, the rules for defining the **H** matrix can be obtained.

$$\begin{cases} h_{rm} = \tfrac{1}{2}[d_{c(r+m)} + d_{c(r-m)} + (1 - \delta_{r0})d_{c(m-r)}] \\ \text{subscripts} \geq 0 \\ 0 \leq r \leq N \\ 0 \leq m \leq N \end{cases} \qquad (6.27)$$

$$h_{r,m+N} = \tfrac{1}{2}[-d_{s(r-m)} + d_{s(r+m)} + (1 - \delta_{r0})d_{s(m-r)}]$$

$$\text{subscripts} > 0$$
$$0 \leq r \leq N$$
$$0 \leq m \leq N$$
(6.28)

$$h_{r+N,m} = \tfrac{1}{2}[d_{s(r-m)} + d_{s(m+r)} - d_{s(m-r)}]$$

$$\text{subscripts} > 0$$
$$1 \leq r \leq N$$
$$0 \leq m \leq N$$
(6.29)

$$h_{r+N,m+N} = \tfrac{1}{2}[d_{c(r-m)} - d_{c(m+r)} + d_{c(m-r)}]$$

$$\text{subscripts} \geq 0$$
$$1 \leq r \leq N$$
$$1 \leq m \leq N$$
(6.30)

The Kronecker delta function used above is defined as

$$\delta_{r0} = \begin{cases} 1 & r = 0 \\ 0 & r \neq 0 \end{cases}$$
(6.31)

The matrix representation provides an alternative method of multiplying two or more Fourier series together if they are truncated to the same number of terms. In general, a product P of a set of M Fourier series F_i, each with N frequency terms can be multiplied directly

$$P = F_1 F_2 F_3 \cdots F_M$$
(6.32)

or transformed into matrices and evaluated by the recursion formula

$$\mathbf{P}_1 = \mathbf{H}_1$$
$$= \mathbf{P}_{k-1}\mathbf{F}_k \quad k = 2, 3, \ldots, M.$$
(6.33)

Here $\mathbf{H}_1$ is given by (6.25) as obtained from F_1, and $\mathbf{F}_k$ is the column matrix of the coefficients of F_k.

As long as the nonlinearity is not too strong, this technique may be used in the evaluation of the response of a nonlinear device in a linear circuit. A straightforward power series expansion of a nonlinear function will provide the response only to a DC excitation. However, if each term of the power series represents ascending powers of a DC plus sinusoidal excitation, then the linear and nonlinear terms must be expressed as ascending powers of a Fourier series. Each term in the Fourier series represents an harmonic generated by the nonlinearity. The amplitudes of the various harmonics are constrained by both the device and the linear circuit.

As an example, the capacitance of the varactor diode in Figure 6.1 may be expanded into a power series

$$\frac{C_0}{(1 - v/\phi)^\gamma} = C_0 \left[1 + \sum_{n=1}^{\infty} \frac{(\gamma)(\gamma + 1) \cdots (\gamma + n - 1)}{n!} \left(\frac{v}{\phi}\right)^n \right] \quad (6.34)$$

in which each v^n is the Fourier series product given by (6.32) or (6.33). A forcing current i_f with potentially several frequency components produces a voltage across the linear admittance Y_c and the nonlinear element. The impressed voltage on the nonlinear device can produce currents at the forcing frequencies as well as harmonics of these frequencies. This in turn will modify the voltage across the diode that will again modify the currents. When Kirchoff's current law is satisfied at the node for each frequency (Figure 6.1), a final solution of the phase and amplitude of the voltage components **V** across the device and the current components **I** through the device is obtained. The solution error that is to be minimized by the generalized Newton-Raphson method given in (6.5) is

$$\epsilon = \mathbf{I}_f - Y_c \mathbf{V} - \mathbf{I}. \quad (6.35)$$

The solution of the varactor diode problem in a given circuit with given current excitation is summarized in Table 6.1. The admittance across the diode, defined as the ratio of the current to voltage at the fundamental frequency, is shown in Figure 6.2.

6.3 COMPUTER AIDED MANUFACTURING

Recently developed computer programs[8,9] have been designed to aid in the layout and generation of microstrip and stripline circuit masks. While these pro-

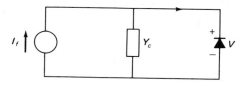

Figure 6.1. Linear circuit with a nonlinear device.

Table 6.1. Example Solution of the Nonlinear Varactor Circuit.

Varactor
Input
Current
Fundamental
Frequency

$$C = \frac{10^{-12}/\pi}{(1 - v/1.1)^{1/3}}$$

$i = 0.5 + 0.5 \cos(2\pi ft)$

$f = 5$ GHz

HARMONIC	CIRCUIT ADMITTANCE	DEVICE VOLTAGE	DEVICE CURRENT
0	0.2	2.5	0
1	$0.1 + j0.1$	$0.16 + j0.23$	$1.2\ 10^{-2} - j6.9\ 10^{-3}$
2	$0.2 + j0.3$	$-3.9\ 10^{-4} - j6.2\ 10^{-3}$	$1.9\ 10^{-3} + j1.1\ 10^{-3}$
3	$0.2 + j0.2$	$1.2\ 10^{-4} + j1.4\ 10^{-4}$	$-5.1\ 10^{-5} - j4.4\ 10^{-6}$
4	$0.1 - j0.4$	$8.3\ 10^{-6} + j5.5\ 10^{-5}$	$2.1\ 10^{-5} - j8.8\ 10^{-6}$
5	$0.4 + j0.2$	$1.1\ 10^{-6} - j5.4\ 10^{-6}$	$6.3\ 10^{-7} + j2.4\ 10^{-6}$

grams are not widely available today, the economic advantages they provide make their future widespread use almost inevitable. A CAM program is used after the basic design has been completed.

The layout of the design is begun by choosing the basic building blocks available from the program and placing them in their correct location on the video screen. These building blocks may be as simple as circles or rectangles, or as complex as a complete rat race coupler with a specified power split or coupled line directional coupler. The program is then called upon to interconnect the various components. When two ports are not colinear, they are automatically connected with a curved line that minimizes mismatch. When transmission lines are defined, their lengths can be easily calculated by the computer whether they are straight or curved. In this way phase matching of two or more lines can be accomplished easily. Finally, when the layout is satisfactory as presented on the graphics display, all the computer generated circuit coordinates are transferred to an automatic pattern generation machine. Thus, a former two-week circuit design and manual-cutting task has been reduced to less than a day. Furthermore, errors that might have slipped through the preview stage on the graphics terminal are fixed by a few lines of coding so that a completely new design layout is unnecessary. The immediate impact of CAM is to reduce turnaround time and increase engineering productivity. However, it appears that this technique may make a qualitative change in the way circuits are designed.

6.4 COMPUTER AIDED TEST

Computer aided testing of a circuit necessarily requires an interface between the digital domain of the computer and the analog domain of the physical cir-

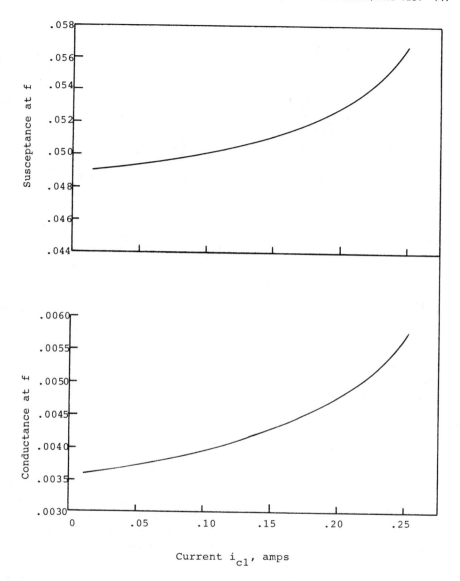

Figure 6.2. Effective admittance of the nonlinear varactor in the circuit described in Table 6.1.

cuit and test instruments. Obviously, analog to digital and digital to analog converters are needed. In addition a method must be found to coordinate all the instruments, receive and transmit data in a standard format, and provide timing of data transmission to avoid ambigious results. These functions are handled by the I/O bus and the interface (Figure 6.3).

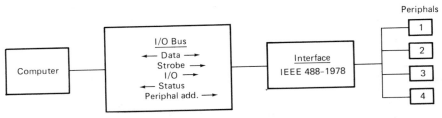

Figure 6.3. Computer control of measurement instruments.

The I/O bus is composed of several lines that link the computer with the peripheral instruments. There are a set of lines for the actual data transmission between the computer and the peripheral device, a set of lines for the address of the peripheral device, a strobe line to signify that data from the computer is available, an I/O line to signify the direction of the data flow to or from the computer, a flag line that tells the computer to wait for the peripheral device to complete its operation, and a status line that tells the computer that an addressed peripheral device is actually present.

The interface between the I/O bus and the peripheral devices can take on several forms. One recent form is called the IEEE 488-1978 bus. The pin usage, voltage and current levels, signal timing, and connector design are all specified by this standard. This allows manufacturers of a wide range of equipment following this standard to interface with one another without any special modifications. The automatic network analyzer is but one, though very useful, set of instruments now tied together with a version of the IEEE 488-1978 interface bus.

6.4.1 Frequency Domain Measurement

Automatic network analyzers are presently capable of measuring the scattering parameters of a circuit from 10 MHz to near 40 GHz. This instrument steps through the desired frequency range, measuring the reflection and transmission coefficients at each frequency much faster than could be done manually. However, its most important attribute is the ability to correct measurements for systematic instrumentation errors. These include system insertion loss, reflection, leakage, and directivity errors. By measuring a set of known standards, these systematic instrument errors may be found and later removed when measuring the unknown circuit. Calibration is typically accomplished by measuring (1) a well-matched 50-ohm load, (2) a short circuit, (3) either an open circuit, or a set of offset short circuits displaced a known distance from the measurement reference plane, (4) the insertion loss of the system, and (5) the cross talk. A sliding 50-ohm load is often used to provide the first standard since residual reflections can be averaged out by varying the position of the

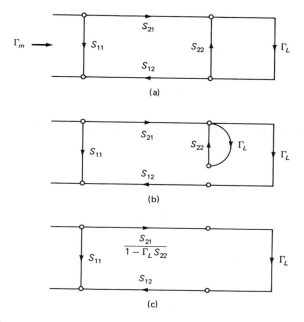

Figure 6.4. Error model for reflection measurement instrument.

sliding load. The cross talk is measured by terminating each of the two ports of the test set and measuring the resulting insertion loss. Several error models have been developed with various degrees of complexity, but a simple example for a reflection measurement will clarify the procedure.

The error model shown in Figure 6.4 is reduced by the flow graph technique described in chapter 1. The measured reflection coefficient Γ_m, when the instrument is terminated with a load with reflection coefficient Γ_L, is

$$\Gamma_m = S_{11} + \frac{S_{12}S_{21}\Gamma_L}{1 - \Gamma_L S_{22}}. \tag{6.36}$$

The three unknowns S_{11}, S_{22}, and $S_{12}S_{21}$ are found by replacing Γ_L with three known standards: a short ($\Gamma_L = -1$), a 50-ohm load ($\Gamma_L = 0$), and either an open circuit with a known fringing capacitance or an offset short of known length ($\Gamma_L = e^{-j\phi}$). For the short

$$\Gamma_{m1} = S_{11} - \frac{S_{12}S_{21}}{1 + S_{22}}, \tag{6.37}$$

for the 50-ohm load

$$\Gamma_{m2} = S_{11} \tag{6.38}$$

and for the open or offset short

$$\Gamma_{m3} = S_{11} + \frac{S_{12}S_{21}e^{-j\phi}}{1 - S_{22}e^{-j\phi}}. \tag{6.39}$$

This provides three equations for the three unknowns. Subtraction of (6.37) from (6.39) eliminates S_{11}, leaving

$$\Gamma_{m3} - \Gamma_{m1} = \frac{S_{12}S_{21}}{1 + S_{22}} \left[\frac{(1 + S_{22})e^{-j\phi}}{1 - S_{22}e^{-j\phi}} - 1 \right] \tag{6.40}$$

$$\triangleq \frac{S_{12}S_{21}}{1 + S_{22}} [U - 1] \tag{6.41}$$

However, from (6.37) and (6.38)

$$\Gamma_{m2} - \Gamma_{m1} = \frac{S_{12}S_{21}}{1 + S_{22}}, \tag{6.42}$$

which is the coefficient of (6.41). Solving (6.41) for U and substituting (6.42) gives

$$U = 1 + \frac{\Gamma_{m3} - \Gamma_{m1}}{\Gamma_{m2} - \Gamma_{m1}}. \tag{6.43}$$

This is a known, measured quantity. By comparing (6.40) and (6.41), the value for S_{22} is found

$$S_{22} = \frac{Ue^{j\phi} - 1}{U + 1}, \tag{6.44}$$

and from (6.42) $S_{12}S_{21}$ is found.

$$S_{12}S_{21} = (\Gamma_{m2} - \Gamma_{m1}) \frac{U(1 + e^{j\phi})}{U + 1} \tag{6.45}$$

When the reflection coefficient of an unknown circuit is measured, (6.36) can be used to extract the actual reflection coefficient Γ_L from the measured value Γ_m with the help of (6.38), (6.44) and (6.45). Using error correction techniques such as this, the residual SWR can be reduced to below 1.002 for frequencies up to 18 GHz.

COMPUTER AIDED DESIGN, MANUFACTURING, AND TEST   151

6.4.2 Time Domain Measurement[10]

Microwave circuit synthesis, measurement, and analysis is typically done in the frequency domain, and the results are the circuit-scattering parameters. However, the time domain method has been found useful in evaluating a wide range of microwave circuits. Frequency domain measurements and time domain measurements are best suited for two different types of problems. Frequency domain data will clearly distinguish two or more resonances if they are sufficiently far apart in frequency, even if the resonances are caused by elements physically close together. Similarly, time domain analysis will clearly distinguish two or more discontinuities if the discontinuities are sufficiently far apart, even if the discontinuities cause rapid variations in the circuit-scattering parameters over a relatively narrow range of frequencies.

Frequency domain analysis is most useful in deducing the equivalent circuit of a diode package where there are several lumped elements located in one position of the circuit or in evaluating narrowband filters. The time domain method is most useful in analyzing broadband circuits such as impedance transformers and connectors where the circuit discontinuities are small and spatially separated from one another. The time domain display gives the location, the characteristic type, and the magnitude of discontinuities within a test fixture, uncluttered by extraneous reflections that might be caused by resonant modes within the fixture or connectors. Furthermore, the effects of the connectors may be mathematically removed in the time domain and the result reconverted into the frequency domain. This impedance becomes the value that would have been measured through ideal reflectionless connectors.

The impulse responses of a 2-port circuit for simple series and shunt elements are shown in Figure 6.5. The "o" on the diagrams indicates the position

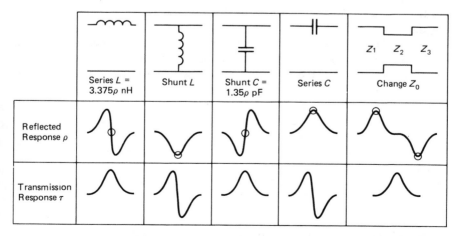

Figure 6.5. Impulse response of simple discontinuities *(Davis and Stinehelfer[10] © 1979 IEEE.)*

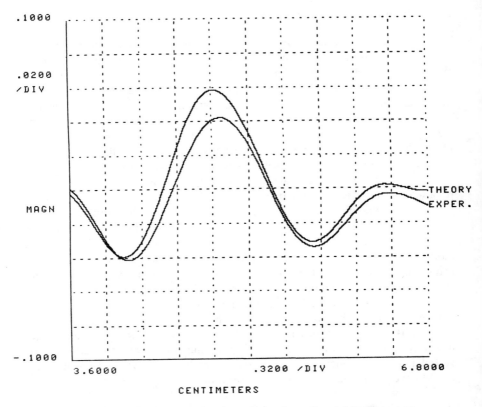

Figure 6.6. Microstrip launcher theoretical and experimental impulse response.

of the discontinuity, while the maximum amplitude is related to the magnitude of the discontinuity.

As an example, the time domain method may be used advantageously to find the equivalent circuit for the SMA coaxial line to microstrip launcher. The microstrip circuit is on 0.025 inch alumina substrate material, and the launcher

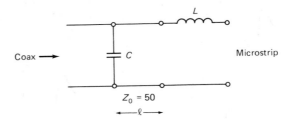

Figure 6.7. Equivalent circuit of a coaxial to microstrip launcher presented to a wave entering the microstrip line. $C = 0.054$, pF, $L = 0.11$ nH and $l = 0.959/\sqrt{\epsilon_{\text{eff}}}$ cm.

is a threaded bulkhead mount type.* The measured time domain response in the vicinity of the launcher and the response to the theoretical model are shown in Figure 6.6. The equivalent circuit model (Figure 6.7) consists of a shunt capacitor to ground, a short length of transmission line, and a series inductor. The shunt capacitance represents the effect of the sudden change in transmission line media at the interface of the launcher. The wave travels down the microstrip for a short distance before the electromagnetic fields are established in the microstrip mode. This distance is represented here by a line length and series inductance. This might possibly be more accurately modeled as a short transmission line with impedance greater than 50 ohms. The time domain method is very useful in helping to synthesize equivalent circuit models.

REFERENCES

1. J. W. Bandler, "Computer-Aided Circuit Optimization," in *Modern Filter Theory and Design*, G. C. Temes and S. K. Mitra, eds. New York: John Wiley, 1973.
2. J. Aprille and T. N. Trick, "A Computer Algorithm to Determine the Steady-State Response of Nonlinear Oscillators," *IEEE Trans. on Circuit Theory*, Vol. CT-19, pp. 354–360, July 1972.
3. J. Aprille and T. N. Trick, "Steady-State Analysis of Nonlinear Circuits with Periodic Inputs," *Proc. IEEE*, Vol. 60, pp. 107–114.
4. E. M. Baily, "Steady-State Harmonic Analysis of Nonlinear Networks," Ph.D. dissertation, Stanford University, Stanford, California, 1968.
5. J. C. Lindenlaub, "An Approach for Finding the Sinusoidal Steady-State Response of Nonlinear Systems," *Proc. 7th Annual Allerton Conference Circuit and System Theory*, pp. 323–329, 1969.
6. M. S. Nakhla and J. Vlach, "A Piecewise Harmonic Balance Technique for Determination of Periodic Response of Nonlinear Systems," *IEEE Trans. on Circuits and Systems*, Vol. CAS-23, pp. 85–91, February 1976.
7. E. D. Rainville, *Intermediate Differential Equations*, 2nd. Ed. New York: MacMillan, pp. 56–57, 1964.
8. T. Dowling, J. Birch, S. Temple, S. Monaghan, H. E. Stinehelfer, N. Cavallaro, and A. Davis, "A Novel Approach to Computer Automated Microwave Circuit Mask Design," *1982 IEEE MTT-S International Microwave Symposium Digest*, pp. 465–467, June 15–17, 1982.
9. W. H. Childs and J. McGregor, "Automatic Artwork Generation for Microwave Integrated Circuits," *1982 IEEE MTT-S International Microwave Symposium Digest*, pp. 468–470, June 15–17, 1982.
10. W. Alan Davis and H. E. Stinehelfer, "Applications of the Time Domain Method to Network Synalysis," *Computer-Aided Microwave Network and Device Characterization*, Session 4, Electro/82, Boston, Mass., May 25–27, 1982.

---

*Cablewave Systems part no. 705971-038

Chapter 7
Characteristics of Amplifiers and Oscillators

7.1 INTRODUCTION

Amplifiers and oscillators are an integral part of many microwave circuit designs. However, an amplifier design task can be hazardous to the aspiring engineer. It has often been said that an oscillator is "what you get when you design an amplifier." Although this version of Murphy's law may not be a rigorous definition of an oscillator or an amplifier, it does illustrate that there is a close relationship between the two. A linear amplifier provides to the load an enlarged replica of the amplitude of the input signal with a linear phase delay. An ideal linear amplifier is never achieved in practice, and in many cases a highly distorted output is quite acceptable. The basic parts of an amplifier must include an input signal source, an output load, an external source of power, and the amplifying device itself. An oscillator is made by feeding back to the input some of the output signal.

In this chapter a discussion of some general characteristics that apply to all microwave amplifiers and oscillators is presented. These characteristics are illustrated by looking at their limitations, their operation in multistage circuits, and the source of power required to produce the RF output.

7.2 ONE-PORT AND TWO-PORT DEVICES

There are two classes of microwave amplifiers: two-port and one-port. The two-port amplifiers, such as the bipolar and the field effect transistors, act as valves. They control a large amount of power being delivered to the output load by a DC power supply with a small input signal. The input and output ports present a positive resistance to the circuit. In the one-port amplifier, the incident power is reflected off a negative resistance. Since the reflection coefficient of a negative resistance is greater than 1, the reflected wave is an amplified version of the incident wave. The input and output signals can be separated either by a circulator or, when two devices are operated in parallel, by a 90° hybrid coupler.

Both amplifiers and oscillators have been successfully built from one-port and two-port devices. It seems, however, most natural to build amplifiers from two-port devices where the input and output are clearly separated and build oscillators from one-port devices where the input and output are present

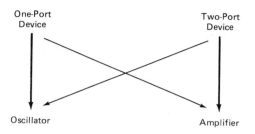

Figure 7.1. Preferred device types for oscillators and amplifiers designated by the dark lines.

together. This "rule" is illustrated by the dark lines in Figure 7.1. Amplifiers, however, are often built with one-port devices; and oscillators, from two-port devices, but this requires some extra effort. For example, FET oscillators require an additional feedback circuit, and one-port IMPATT diode amplifiers require the addition of either a circulator or hybrid coupler. Furthermore, the most successful high power IMPATT amplifiers are actually locked oscillators. As such they cannot amplify an amplitude modulated signal.

This "rule" is often broken, as illustrated by the light lines in Figure 7.1, to take advantage of certain device characteristics. Examples of these are the low noise quality of the reflection parametric amplifier, the high power capability of the IMPATT diode amplifier, or the ability to integrate the FET oscillator on a single substrate in order to build a monolithic circuit.

7.3 AMPLIFIER GAIN

The gain of an amplifier can be expressed in at least five different ways. It is therefore helpful to have a clear definition of the gain that is used in evaluating an amplifier. The amplifier to be considered is excited by a signal from a voltage source with internal impedance Z_g, and the amplified output is dissipated in a load impedance Z_L. These impedances play a key role in the following definitions of gain.

1. The *power gain* is defined as the ratio of the power dissipated in the load Z_L to the power delivered to the input of the amplifier. This gain definition is independent of Z_g, but certain amplifiers such as the negative resistance types depend strongly on Z_g. Consequently, this definition is limited in its usefulness.
2. The *available gain* is the ratio of the amplifier output power to the source available power. This definition depends on the value of Z_g, but is independent of Z_L. Again, since the characteristics of certain amplifiers depend on Z_L, this gain definition does not in general adequately describe amplifier gain.

3. The *exchangeable gain* is the ratio of the output exchangeable power to the input exchangeable power. The exchangeable power of the source is

$$P = \frac{|V|^2}{4\,Re(Z_g)}, \quad Re(Z_g) \neq 0.$$

This can be less than zero for negative resistance devices.[1] The exchangeable gain is an extension of the available gain definition. It is independent of Z_L and therefore suffers the same limitation as the available gain definition.

4. The *insertion gain* is the ratio of the output power to the power that would be dissipated in the load if the amplifier were not present. Exactly what replaces the amplifier in this definition is left unspecified. This becomes a major problem in frequency conversion circuits such as parametric upconverters. Therefore, this definition for gain is too vague to be generally useful.

5. The *transducer power gain* is the ratio of the power delivered to the load to the available power from the source. This definition depends on both Z_g and Z_L, an advantage the previous definitions lacked. Furthermore, in contrast to the exchangeable gain, the present definition gives positive gain for negative resistance amplifiers.

It is the transducer power gain definition that has been found to be the most useful one. Indeed, it has already been used in chapter 3 to describe the gain of a passive filter.

7.4 AMPLIFIER LIMITATIONS

An amplifier circuit must be designed within the limits imposed by fundamental physics and available technology. The amplifier specifications described in this section are limited by the following: (1) the operating frequency and bandwidth must be designed to the limits imposed by the application and volume constraints, (2) the output power must be consistent with device efficiency and available paths to a thermal heat sink, and (3) the dynamic range must lie within the limits bounded by the distortion at the high level signals and the noise at the low level signals. Each of these aspects of amplifier design will be described in the following paragraphs.

7.4.1 Amplifier Frequency and Bandwidth

The operating frequency of an amplifying device may be limited by the response times of the electrons and holes in the semiconductor, or simply the limitations on how small the device can be made. At millimeter wave frequencies it is common practice to mount the unpackaged semiconductor device directly into a waveguide circuit to reduce as far as possible the parasitic reactances of the package. At lower microwave frequencies the device is often packaged in an hermetically sealed can that has been carefully chosen to minimize the parasitic reactances. These package parasitic reactances may dramatically reduce the center frequency and bandwidth capability of the semiconductor device. Since these parasitic elements are physically close to the diode, they have a much stronger effect on the device operation than the external circuit does. Thus, the degradation caused by the package cannot be completely corrected by the external circuit, so the designer must either design the package itself or simply live within its limitations. However, if an equivalent circuit for the package can be found, then the package itself can be incorporated into the design in order to maximize the available bandwidth.

Methods for characterizing a diode package have been the subject of numerous papers. In the 1950s and 1960s it represented a particularly challenging measurement problem. With the advent of automated measurement systems and error correction programs, the task of obtaining an equivalent circuit for a diode package has become much easier.

The diode package is represented by the lumped, lowpass, equivalent circuit shown in Figure 7.2. The equivalent circuit for the package can be obtained only by measuring the package in a circuit similar to the one in which the packaged device will eventually be used. The lumped equivalent circuit represents a field structure that changes with the environment in which the package is placed. Therefore, the package cannot be measured in a coaxial fixture and used later in a waveguide or microstrip circuit. Nevertheless, as shown in Figure 7.2, it is conceptually helpful to make the following associations: (1) the inductance L_1 with the inductance of the strip between the top hat and the device, (2) the capacitance C_2 with the capacitance between the top hat and

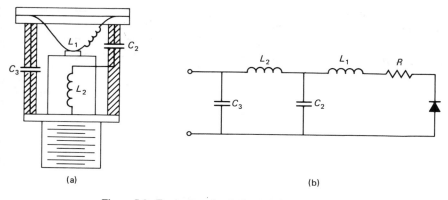

Figure 7.2. Equivalent circuit for a packaged diode.

the pedestal, (3) the inductance L_2 with the pedestal, and (4) the capacitance C_3 with the external shunt capacitance. In present day miniature packages the capacitances might typically lie between 0.2 and 1.2 pF, the strap inductance L_1 between 0.02 and 1 nH, and the pedestal inductance L_2 between 0.1 and 0.3 nH.

7.4.1.1 Device in the Package.
The first of two measurement methods discussed here was developed by Roberts.[2] This method was originally used to find the equivalent circuit for a high Q varactor diode. Its success is based on the low loss of the varactor diode. The diode itself is represented by a resistance R_s and a voltage dependent capacitance as shown in Figure 7.3. Impedance measurements are made of the packaged device placed at the end of a shorted transmission line over a broad range of frequencies at a constant bias voltage. This data, when plotted on a Smith chart, forms a circle near the outside of the chart. Roberts chose three pairs of frequency/reactance data points: the series resonant frequency f_a, $f_a/2$, and the parallel resonant frequency f_b. The corresponding reactances are $X(f_a) = 0$, $X(f_a/2)$, and $X(f_b) = \infty$. These three widely separated data points allow solution for C_1, L_1, and C_2.

$$C_2 = -\frac{3}{\pi f_a X(f_a/2)} \left[4\left(\frac{f_b}{f_a}\right)^2 - 1 \right]^{-1} \qquad (7.1)$$

$$C_1(V) = C_2 \left[\left(\frac{f_b}{f_a}\right)^2 - 1 \right] \qquad (7.2)$$

$$L_1 = \frac{1}{(2\pi f_a)^2 C_1} \qquad (7.3)$$

If sufficient data were taken to get the parallel resonant frequency, rederivation of (7.1) to (7.3) with an alternative frequency/reactance data pair would be needed.

The series resistance R_s can be found by simply measuring the real part of the input impedance at f_a. This assumes that R_s is independent of frequency. If this resistance measurement is made manually, losses in the instrumentation may be partially offset by using the equation (7.4) given by Bandler.[3]

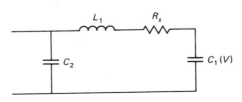

Figure 7.3. Equivalent circuit for a packaged varactor diode.

CHARACTERISTICS OF AMPLIFIERS AND OSCILLATORS

$$\frac{1}{S_m} = \frac{1}{S_L} + \frac{1}{S_{\max}} \qquad (7.4)$$

In this expression, S_m is the measured standing wave ratio (SWR), S_L is the desired SWR of the load R_s, and $S_{\max}$ is the maximum SWR when the packaged diode is replaced by a good short. The series resistance is determined from S_L rather than S_m.

7.4.1.2 Empty Package.
Owens and Cawsey[4] describe a method in which empty packages similar to the one being used with the semiconductor device are measured. In one of these packages the strap is left hanging, thereby representing an open circuited package. In the second package, the strap is shorted to the pedestal. The open circuited package allows solving for the three outer elements, C_2, L_2, and C_3. Measurement of the shorted package allows finding L_1 in addition.

Further refinements in this technique are now available with the aid of time domain analysis[5] and computer optimization. First, measurements of the empty package are made at harmonically related frequencies on an automatic network analyzer. This data is then transformed into the time domain where all extraneous discontinuities caused by the connectors and transitions are removed. The remaining time domain data of the package alone is converted back to the frequency domain. This data now represents the impedance of the package as if it were measured with perfect connectors.

From this data, 10 or 11 equally spaced data points are chosen for optimization. A convenient error function that the computer seeks to minimize is

$$e = \frac{1}{N} \sum_{i=1}^{N} \left| 1 - \frac{S_{11ti}}{S_{11mi}} \right|^2. \qquad (7.5)$$

This expression is the sum over N frequency points of the ratio of the complex reflection coefficients for the theoretical circuit S_{11t} and the measured data S_{11m}. After the optimization process has reduced the error in (7.5) to an acceptable value, plots of the experimental data may be compared with the response of the theoretical circuit.

This procedure assumes that the measured package is similar to the one containing the semiconductor device. In practice, many packages should be measured since manufacturing tolerances are sufficiently loose to cause significant variation between different packages.

7.4.2 Maximum Output Power Limitation

The maximum output power capability of an amplifier or oscillator is usually limited by the device itself rather than the circuit power-handling capability.

The maximum is therefore constrained by practical requirements that affect its ability to conduct heat away from the device, the thermal limitations of the device itself, and the device operating efficiency.

The device may be made to operate at high power levels by increasing the cross section diameter of the semiconductor so as to reduce the current density. However, there are limits to how large the device diameter can be made. Increasing the diameter of a varactor diode introduces unwanted additional capacitance, thereby reducing its maximum operating frequency. For the IMPATT or Gunn diode, increasing the area increases the device power-handling capability but also reduces the required circuit impedance. This makes it more difficult to match the device. Eventually, more loss is incurred in the circuit transformation than is gained by the increased power capability of the diode.

The thermal properties of a semiconductor are related to the Boltzmann factor $\exp(\mathcal{E}/kT)$ where $\mathcal{E}$ is the energy in electron volts, k is the Boltzmann constant, and T is the absolute temperature. The semiconductor must be kept within a safe operating temperature by controlling the ambient temperature and providing an adequate heat sink.

Heat transfer occurs by radiation, convection, and conduction. Only the latter mechanism is important in most semiconductor devices. The rate of heat conduction in units of energy/unit of time, or power, from a hot object to a cool one is proportional to the temperature difference ΔT. The proportionality constant is the thermal resistance R_T. This depends on the thermal resistivity of the materials involved and the shape of the conduction path. An equation in form similar to Ohm's law describes the average heat flow for constant or CW sources.

$$P = \frac{\Delta T}{R_T} \qquad (7.6)$$

Determining the shape factor for R_T may require extensive computer analysis using finite difference techniques. However, if the problem can be reduced to a two-dimentional geometrical shape, fairly accurate results can be quickly obtained by using the graphical technique of field mapping.[6] It is usually a good assumption to consider the thermal resistivity and the heat capacity to be independent of temperature, even for GaAs material.[7]

For certain semiconductor devices, such as the IMPATT diode, the average output power is larger when the device is operating in a pulsed mode rather than a continuous or CW mode. To prevent device burnout, the temperature must be limited to a given temperature rise ΔT above the ambient. A conservative criterion for reliability is to measure the temperature on the axis near the center of the device at the end of the pulse. If the diode is operating in a pulsed mode with a duty cycle D and pulse width τ (Figure 7.4), the allowable peak output power would be

CHARACTERISTICS OF AMPLIFIERS AND OSCILLATORS   161

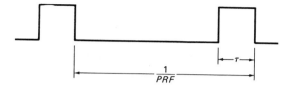

Figure 7.4. Pulse signal with pulse width τ, pulse repetition frequency *PRF*, and duty cycle *D* = *PRF* × τ.

$$P_{peak} = \frac{\eta}{(1-\eta)} \frac{\Delta T}{DR_T(\tau,D)}. \quad (7.7)$$

In this expression, η is the diode efficiency so that the output power is $\eta/(1-\eta)$ × the dissipated power. The thermal resistance $R_T(\tau,D)$ is defined by Holway[7] to be dependent on pulse width and duty cycle. The value for $R_T(\tau,D)$ is calculated after the device has been operating sufficiently long to have settled to a steady-state value. Holway calculated the specific values for $R_T(\tau,D)$ for a quadramesa GaAs double drift IMPATT diode which is shown in Figure 7.5. The peak power plotted in Figure 7.6 is obtained from (7.7) and

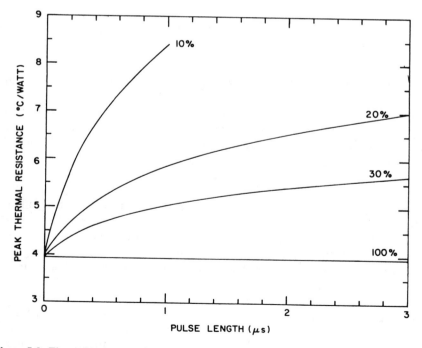

Figure 7.5. Thermal resistance of a pulsed diode as a function of pulse length and duty factor.[7] (Holway[7] © 1980.)

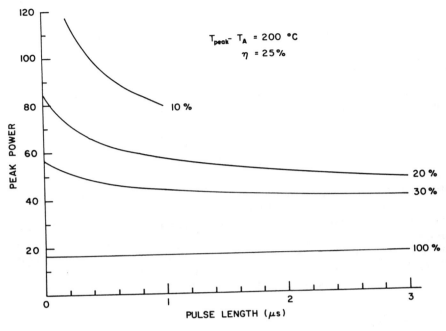

Figure 7.6. Peak output power from a quadramesa IMPATT diode as a function of pulse length and duty factor.[7] *(Holway*[7] © *1980.)*

the curve in Figure 7.5. It is apparent from these two figures that $R_T(\tau,D)$ approaches the CW value defined in (7.6) when $D \to \infty$ or when $\tau \to 0$.

In addition to these thermal limitations, the efficiency of two-port amplifiers may be controlled by appropriate choice of the operating point defined by the bias voltage. These amplifiers may be grouped into separate classes designated as class A, B, AB, and C.

A *class A* amplifier provides linear gain for all 360° of an input sine wave signal. The bias voltage is set near the center of the operating range so that the output signal produces approximately equal excursions in both the positive and negative polarites. The FET amplifier is often used in this mode up to about 20 GHz to provide low noise, high gain or linear amplification. Bipolar transistors are also used in class A operation to achieve high gain or linear power amplification.

A *class B* amplifier is biased so that only half of the sinusoidal signal is amplified; the other half is below cutoff. Recovery of the full 360° of the sine wave requires a push-pull circuit in which each half of the amplifier alternately supplies half the output signal. This form of circuit is often used in a transistor amplifier when DC power consumption is to be minimized.

A *class AB* amplifier is intermediate between a class A and class B amplifier. A small bias voltage is applied so that the amplifier is not completely cutoff when switching between one amplifier and its complement near zero level. This operation is often used to minimize crossover distortion in a push-pull amplifier.

A *class C* amplifier is one that is biased somewhat below cutoff so that less than half the input signal conducts.

A class B or AB push-pull amplifier provides greater output than the class A amplifier because larger input voltage swings are permissible for the class B type. Furthermore there is no power drain when there is no input signal. Recent experimental work at Westinghouse by Cohn, Degenford, and Freitag[8] showed that a class B operated FET amplifier in the 9 to 10 GHz frequency range has (1) power dissipation that is 48% less than the class A FET, and (2) power-added efficiency that is 60% greater than the class A FET. This was obtained with a phase behavior comparable to that expected from a class A operated FET. It also had no crossover distortion.

7.4.3 Dynamic Range

Amplifiers are often designed to operate in as close to a linear mode as possible where the output is directly proportional to the input. The distortion that occurs at high power levels can be defined in any one of several ways. At a given power level, the signal amplitude and phase may be distorted by both the linear circuit and the device over the operating bandwidth of frequencies. This type of distortion was discussed in section 7.3.2. In a nonlinear circuit at a given frequency, the signal will become distorted at high power levels. This distortion may be characterized by the signal level that produces 1 dB of gain compression, by the intercept point when two signals within the amplifier bandwidth excite the amplifier, and by cross modulation where the modulation of one signal crosses over and modulates another nearby signal. These different methods of measuring amplifier distortion have arisen from differences in the intended application of the amplifier. Although all three methods provide a measure of the nonlinearity of the amplifier, the complexity of the nonlinear problem precludes a mathematical relationship among these three quantities.

The 1 dB compression point is measured by simply increasing the input power level until the gain drops 1 dB. This is probably the simplest test that can be made for high level distortion. This method does not give a direct measure of the amplifier quality when it must amplify two or more signals simultaneously. In this case the third-order intermodulation intercept point is measured by exciting the amplifier with two equal amplitude signals at ω_1 and ω_2. The amplifier output contains among others, signals at ω_1 and a third order mixing term $2\omega_1 - \omega_2$. At low levels, a log–log plot of the output power versus the input power shows that the output at ω_1 has a slope of approximately 1 and the output at $2\omega_1 - \omega_2$ has a slope of 3 at low signal levels. The point where the low level slopes of these curves are extrapolated and intercept one another is called the intercept point. The higher the intercept point, the more linear is the amplifier.

Still a third method is sometimes used to directly measure the ability of an amplifier to handle a modulated signal. In this case two signals are fed into the amplifier, one of which is amplitude modulated with a modulation index of m.

$$v = (1 + m \cos \omega_m t) \cos \omega_1 t + \cos \omega_2 t \tag{7.8}$$

Because of the nonlinear mixing process, the output signal at ω_2 will also be modulated by frequency ω_m with a measurable modulation index m'. The cross modulation factor m'/m should be small for a nearly linear amplifier.

The distortion limiting the dynamic range at the low level is the noise power. At the microscopic level, noise is generated randomly by thermal fluctuations of electrons in resistors and semiconductors. In addition active devices with current flowing through them exhibit what is known as shot noise. Shot noise arises from random fluctuations of charge carriers in a DC electric current. Modern measurement techniques can sometimes extract a signal from below the noise level if the signal is repetitive. In this case the instrument can add the coherent signal plus random noise over many cycles. The random noise cancels out, and the coherent signal builds up and comes into view. However, no instrument can recover a nonrecurring signal that is less than the noise level. Chapter 8 will cover the origin and measurement of noise in more detail.

7.5 MULTISTAGE AMPLIFIERS

Multistage amplifiers are most easily analyzed when each stage is assumed to be unilateral, i.e., $S_{12} = 0$. In this case the total gain is the product of the gains of each stage and the output power is that of the final stage. If the amplifiers are nonunilateral, the calculation of parameters such as gain and bandwidth is much more complicated. Since most quality amplifiers have $S_{21} \gg S_{12}$, it is to assume that in most cases amplifiers are unilateral.

Usually the first stage of a multistage amplifier is designed to have low noise with reasonably high gain; and the last stage, to have high power and high efficiency. Amplifier stages in between might be designed to have medium power capability and linear gain. In the amplifier chain shown in Figure 7.7, each stage has a gain G_i, noise figure F_i, and a power added efficiency η_i. The actual noise figure defined by Kurokawa[9] is

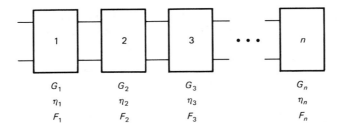

Figure 7.7. Gain, efficiency, and noise figure of an n stage amplifier chain.

$$F = \frac{\text{Actual noise output power}}{\text{Available noise input power} \times G} \tag{7.9}$$

where G is the transducer power gain defined in section 7.3. This definition of noise figure accounts for the noise generated in the load and source resistances. The noise power generated in a resistor R is given by the Nyquist formula

$$N = kT \Delta f \tag{7.10}$$

T is the absolute temperature, k is Boltzmann's constant, and Δf is the bandwidth. This formula will be derived later in chapter 8. For the n section amplifier chain in Figure 7.7, the total noise figure F_{Tn} is

$$F_{Tn} = \frac{N_{Tn}}{kT_o \Delta f G_1 G_2 G_3 \cdots G_n}, \tag{7.11}$$

where T_o is the standard temperature of 290°K, and N_{Tn} is the total noise delivered to the load. This quantity is the sum of the noise generated in the last stage and the total noise delivered from the previous stage times the gain of the last stage.

$$N_{Tn} = N_n + G_n N_{T(n-1)} \tag{7.12}$$

If the nth stage were removed and its noise figure measured alone, then its noise figure would be

$$F_n = \frac{kT_o \Delta f G_n + N_n}{kT_o \Delta f G_n} \tag{7.13}$$

or

$$F_n - 1 = \frac{N_n}{kT_o \Delta f G_n}. \tag{7.14}$$

Upon substituting (7.14) and (7.12) into (7.11), the total noise figure can be expressed in terms of the noise figure of the nth stage as

$$F_{Tn} = \frac{N_{T(n-1)}}{kT_o \Delta f G_1 G_2 \cdots G_{n-1}} + \frac{F_n - 1}{G_1 G_2 \cdots G_{n-1}} \qquad (7.15)$$

Repeating this process for all n stages, the total noise figure for a cascaded amplifier is

$$F_{Tn} = F_1 + \frac{F_2 - 1}{G_2} + \frac{F_3 - 1}{G_1 G_2} + \cdots + \frac{F_n - 1}{GG \cdots G_{n-1}} \qquad (7.16)$$

If the gain of the first stage is large, the noise figure of first stage is of primary importance in determining the overall noise F_{Tn} of the system. As long as the first stage is chosen appropriately, a receiver may have a relatively low noise figure even if its second stage is a mixer.

In a power amplifier chain the efficiency of the last stage will have the major effect on the efficiency of the total amplifier. For the kth stage in the chain, the power added efficiency η_k is

$$\eta_k = \frac{P_{ok} - P_{ik}}{P_{dk}} \qquad (7.17)$$

where for the kth stage P_o is the output power, P_i is the input power and P_d is the input from the source of power. Assuming that each amplifier is unilateral

$$P_{ik} = P_{i1} G_1 G_2 G_3 \cdots G_{k-1} \qquad (7.18)$$
$$P_{ok} = P_{ik} G_k. \qquad (7.19)$$

Substituting (7.18) and (7.19) into (7.17) gives the source of power needed for each stage.

$$P_{dk} = \frac{G_1 G_2 \cdots G_{k-1}(G_k - 1)}{\eta_k} P_{i1} \qquad (7.20)$$

Since the total added power for the chain is

$$P_{oN} - P_{i1} = P_{i1}(G_1 G_2 G_3 \cdots G_N - 1), \qquad (7.21)$$

then the total power added efficiency may be found by (7.21).

$$\eta_T = \frac{P_{oN} - P_{i1}}{\sum_{k=1}^{N} P_{dk}} \quad (7.22)$$

$$= \frac{(G_1 G_2 G_3 \cdots G_N - 1)}{\sum_{k=1}^{N} \frac{G_2 G_2 G_3 \cdots G_{k-1}(G_k - 1)}{\eta_k}} \quad (7.23)$$

To gain some insight into the implications of this expression, assume that each $G_k \gg 1$, so that

$$\eta_T \cong \frac{G_1 G_2 \cdots G_N}{\frac{G_1}{\eta_1} + \frac{G_1 G_2}{\eta_2} + \cdots + \frac{G_1 G_2 \cdots G_N}{\eta_N}} \quad (7.24)$$

$$= \frac{1}{\frac{1}{G_2 G_3 \cdots G_N \eta_1} + \cdots + \frac{1}{G_N \eta_{N-1}} + \frac{1}{\eta_N}}. \quad (7.25)$$

If the gain of the final amplifier is large, the efficiency of the total amplifier is primarily dependent on the efficiency of the last stage. The efficiency of the first stages make little impact on the overall efficiency since the amounts of power being dissipated in these early stages is small. If, for example, each amplifier of a three-stage amplifier has a gain of 10 dB, and the efficiencies of the amplifiers are $\eta_1 = 1\%$, $\eta_2 = 5\%$ and $\eta_3 = 10\%$, then the overall efficiency is 8.54%.

7.6 THE ENERGY SOURCE

The output power from an oscillator or the added output power from an amplifier clearly must come from an external DC or AC power supply. For certain two-port devices such as the transistor, the device acts as a valve in which a small input signal controls the output of a large DC power supply. In the traveling wave tube (TWT) amplifier, the signal is amplified by drawing power away from an electron beam that has been provided by a DC power supply. While in this case the energy source is a DC supply, the TWT is not regulating the DC output in the same way as the transistor.

For the one-port negative resistance devices, some use DC and others use AC external power sources. Gunn, IMPATT, TRAPATT, BARITT and tunnel diodes provide a negative resistance by converting power from a DC power supply to the microwave frequency region. The actual mechanism by which this is accomplished for several of these devices will be covered in subsequent

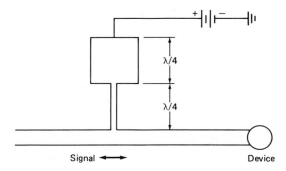

Figure 7.8. Microstrip bias circuit.

chapters. It is also possible to convert power from an AC pump source at f_p to provide negative resistance at the desired microwave signal frequency f_s. The MASER, the parametric relfection amplifier, and upconverters work on this principle. However, energy conservation demands that $f_p > f_s$ in order to achieve gain or oscillations at f_s.

Whatever the power source, equivalent circuits for an amplifier or oscillator often leave out this source. However, it is just at this point where an otherwise well-designed microwave circuit often succeeds or fails. To avoid distorting the output, the DC, pulsed, or high frequency pump source must be isolated from the signal frequency to avoid unwanted interactions. A high impedance line followed by a low impedance quarter wavelength line is one simple and often used method of providing DC current to a device on a microstrip circuit (Figure 7.8). Often some experimentation is still needed to find the optimum position of the bias line.

Providing an AC pump to a parametric amplifier requires the use of carefully placed filters in each of the pump, signal, and idler (difference frequency) circuits. These filters not only isolate the three frequencies from one another, but provide the proper impedance to the diode. Each device type has its own out of band characteristics. For this reason, the problem of providing the energy source must be largely handled individually for each device type.

REFERENCES

1. H. A. Haus and R. B. Adler, "An Extension of the Noise Figure Definition," *Proc. of the IRE,* Vol. 56, pp. 690–691, January 1968.
2. D. A. E. Roberts, "Measurements of Varactor Diode Impedance," *IEEE Trans. on Microwave Theory and Techniques,* Vol MTT-12, pp. 872–873, November 1965.
3. J. W. Bandler, "Precision Microwave Measurement of the Internal Parasitics of Tunnel-Diodes," *IEEE Trans. on Electron Devices,* Vol. ED-15, pp. 275–282, May 1968.
4. R. P. Owens and D. Cawsey, "Microwave Equivalent Circuit Parameters of Gunn-Effect-Device Packages," *IEEE Trans. on Microwave Theory and Techniques,* Vol. MTT-18, pp. 790–798, November 1970.

5. M. E. Hines and H. E. Stinehelfer, "Time Domain Oscillographic Network Analysis," *IEEE Trans. on Microwave Theory and Techniques,* Vol MTT-22, pp. 276–282, March 1974.
6. A. D. Moore, *Heat Transfer Notes for Electrical Engineering.* Ann Arbor: George Wahr Publishing Co., 1963.
7. L. H. Holway, "Transient Temperature Behaviour in Pulsed Double Drift IMPATT Diodes," *IEEE Trans. on Electron Devices,* Vol. ED-27, pp 433–442, February 1980.
8. M. Cohn, J. E. Degenford, and R. G. Freitag, "Class B Operation of Microwave FETS for Array Module Applications," *1982 IEEE MTT-S International Microwave Symposium Digest,* pp. 169–171, June 15–17, 1982.
9. K. Kurokawa, "Actual Noise Measure of Linear Amplifiers," *Proc. of the IRE,* Vol. 49, pp. 1391–1397, Sept. 1961.

Chapter 8
Noise

8.1 INTRODUCTION

Electrical noise is usually the factor that limits the sensitivity of a receiver. As a consequence, the theory of noise has played a fundamental role in microwave circuit design. The two primary sources of noise in electrical circuits are thermal noise and shot noise. The former is usually associated with random motion of charged carriers in a resistor; the latter with the random motions of electrons in a DC current. These two sources are not completely unreleated. In fact, shot noise is sometimes represented as an equivalent resistor.

This chapter provides a review of thermal and shot noise theory, and how it is measured. Since the noise associated with a linear amplifier is treated much differently than that in an oscillator, these two types of circuits are treated in separate sections.

8.2 THERMAL NOISE THEORY

Electrical noise arises in a resistor because of the random motion of its free electrons. As the temperature rises, the electron velocities as well as the number of collisions per unit time increase. The accelerating electrons produce an electromagnetic field that is measured as a noise voltage. Since the average noise voltage is zero, it has been the custom to express this quantity as an auto correlation function.

$$\langle v(t)^2 \rangle = \lim_{\tau \to \infty} \frac{1}{\tau} \int_{-\tau}^{\tau} v(t)^2 \, dt \qquad (8.1)$$

There have been a variety of methods used to find the noise voltage, all coming to the same conclusion. Nyquist first solved this using a transmission line model. Others have used such models as a lumped equivalent circuit, the random motions in a metal conductor, or the radiation from a black body. Since these are thermodynamic models, they all provide the same expression for the noise at frequencies and temperatures at which most microwave circuits operate. The exception to this is in cryogenically cooled devices such as masers and in some cases parametric amplifiers. To incorporate this latter condition into the noise formula, the black body radiation approach will be used here. It is based on quantum mechanics and is therefore valid at both low and high temperatures.

8.2.1 Black Body Radiation

A direct application of classical arguments that are based on the continuity of energy states led to the famous ultraviolet catastrophe. In that calculation the energy density of the black body increased without limit as the radiation frequency increased. This clearly unphysical circumstance was corrected by Planck who postulated that the energy states were not continuous, but possessed total energies of discrete values. These energy values for the harmonic oscillator are obtained by solving the Schrödinger equation. This derivation is found in most introductory texts on quantum mechanics.[1]

$$\mathcal{E} = (\tfrac{1}{2} + n)hf \qquad n = 0,1,2,\ldots \tag{8.2}$$

In this equation n is an integer quantum number and h is Planck's constant. The average energy cannot be obtained by the integral

$$\overline{\mathcal{E}} = \frac{\int_0^\infty \mathcal{E} P(\mathcal{E})\, d\mathcal{E}}{\int_0^\infty P(\mathcal{E})\, d\mathcal{E}} \tag{8.3}$$

because the energy levels are in reality discrete. The expression, $P(\mathcal{E}) = C \exp(-\mathcal{E}/kT)$, is called the Boltzmann factor and is derived in chapter 9. Planck substituted summations for the integrals in (8.3) to solve for the average energy.

$$\overline{\mathcal{E}} = \frac{\sum_{n=0}^{\infty} \mathcal{E} P(\mathcal{E})}{\sum_{n=0}^{\infty} P(\mathcal{E})} \tag{8.4}$$

$$= \frac{\sum_{n=0}^{\infty} (n + \tfrac{1}{2})hf e^{-\beta(n+1/2)hf}}{\sum_{n=0}^{\infty} e^{-\beta(n+1/2)hf}} ; \beta = \frac{1}{kT} \tag{8.5}$$

It may be easily shown by differentiation that

$$\frac{d}{d\beta} \ln \sum e^{-\beta(n+1/2)hf} = -\overline{\mathcal{E}}. \tag{8.6}$$

The argument of the logarithm is the geometric series whose sum is

$$e^{-\beta hf/2} \sum_{n=0}^{\infty} e^{-n\beta hf} = \frac{e^{-\beta hf/2}}{1 - e^{-\beta hf}}. \tag{8.7}$$

Substituting (8.7) into (8.6) gives the sought for average energy.

$$\begin{aligned}\overline{\mathcal{E}} &= \frac{d}{d\beta} [\ln (1 - e^{-\beta hf}) - \ln e^{-\beta hf/2}] \\ &= \frac{hfe^{-\beta hf}}{1 - e^{-\beta hf}} + \frac{1}{2} hf\end{aligned} \tag{8.8}$$

or

$$\overline{\mathcal{E}} = \frac{hf}{e^{hf/kT} - 1} + \frac{1}{2} hf \tag{8.9}$$

The energy distribution function given by (8.9) is a special case of the Bose-Einstein distribution derived in chapter 9.

8.2.2 The Nyquist Theorem

The thermal noise power for a given bandwidth Δf is obtained directly from (8.9).

$$N_T = \frac{hf \Delta f}{e^{hf/kT} - 1} + \frac{1}{2} hf \Delta f \tag{8.10}$$

It is applicable for very low temperatures and high frequencies. However, for electronic circuits operating at room temperature and at microwave frequencies, $hf/kT \ll 1$, so that a series expansion of (8.10) yields

$$N_T = kT \Delta f. \tag{8.11}$$

While the zero point power term, ½$hf \Delta f$, plays no role in the room temperature expression for noise, it does aid somewhat in the series conversion to (8.11). Furthermore, Siegman[2] has pointed out that this zero energy term may play an essential role in the minimum noise for a maser, and probably for other cryogenically cooled amplifiers.

The available noise power given by (8.11) is just the value obtained by Nyquist.[3] The voltage caused by the noise fluctuations generated in a resistor R is

$$\langle v^2 \rangle = 4RN_T \qquad (8.12)$$
$$= 4RkT\,\Delta f$$

and the expression for the noise current fluctuations is

$$\langle i^2 \rangle = 4GkT\,\Delta f. \qquad (8.13)$$

In applying these expressions to a circuit with several noise sources, it should be remembered that noise voltages are additive when the sources are in series

$$\langle v^2 \rangle = \sum_j \langle v^2 \rangle_j \qquad \text{series} \qquad (8.14)$$

and the noise currents are additive when the noise sources are in parallel.

$$\langle i^2 \rangle = \sum_j \langle i^2 \rangle_j \qquad \text{parallel} \qquad (8.15)$$

8.3 SHOT NOISE THEORY

Shot noise arises from random variations of a DC current. It has long been known to be the primary noise contribution in vacuum tubes operating under temperature limited conditions. The full shot noise is most apparent in a current source with zero shunt source admittance. The current source produces a stream of electrons that leave the source at random times, but on the average they provide a DC current I_o to the load (Figure 8.1). However, if an instrument could measure the current produced by each separate electron as it arrived at the load, the instrument would see a series of current impulses arriving at random time intervals. The charge q of each pulse is that of one electron.

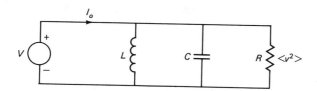

Figure 8.1. Equivalent circuit for shot noise.

If n is the average number of electrons emitted by the source in a given time interval Δt, then the DC current is

$$I_o = \frac{qn}{\Delta t}. \qquad (8.16)$$

Because of the shunt inductance in the circuit, the DC output power is zero. However, the AC power is not zero. Each current pulse provides an energy pulse to the capacitor with the value of

$$\mathcal{E} = \frac{q^2}{2C}. \qquad (8.17)$$

The average shot noise power delivered to the load is then

$$N_s = \frac{n\mathcal{E}}{\Delta t} \qquad (8.18)$$

which, upon substitution of (8.16) and (8.17), becomes

$$\begin{aligned} N_s &= \frac{nq^2}{2C\,\Delta t} \\ &= \frac{qI_o}{2C} \end{aligned} \qquad (8.19)$$

The equipartition theorem derived in chapter 9 states that the average energy of a system of uniform temperature is equally divided among the degrees of freedom of the system. A system with N degrees of freedom can be described uniquely by N variables. Furthermore, the equipartition theorem states that the average energy is

$$\overline{\mathcal{E}} = \frac{N}{2}kT \qquad (8.20)$$

where k is Boltzmann's constant. The circuit in Figure 8.1 has two energy storage elements, each one containing an average energy of $kT/2$. In particular for the capacitor, this is

$$\tfrac{1}{2}C\langle v^2 \rangle = \tfrac{1}{2}kT. \qquad (8.21)$$

But from the Nyquist formula for thermal noise given by (8.12), the value for the capacitance can be found.

$$C = \frac{1}{4R\,\Delta f} \qquad (8.22)$$

The shot noise is then obtained by substituting (8.22) into (8.19).

$$N_s = 2qRI_o\,\Delta f \qquad (8.23)$$

Division by R gives the mean square current fluctuation as first obtained by Schottky.

$$\langle i^2 \rangle = 2qI_o\,\Delta f \qquad (8.24)$$

8.4 AMPLIFIER NOISE CHARACTERIZATION

A very useful quantitative measure of the noise performance of an amplifier is the "actual noise figure" defined by Kurokawa[4] as

$$F = \frac{\text{Actual noise output power}}{\text{Available noise input power}} \times \frac{1}{G}. \qquad (8.25)$$

From the Nyquist formula,

$$F = \frac{1}{kT_o\,\Delta f G} \qquad (8.26)$$

where the noise temperature of the source is $T_o = 290°K$ and the center frequency is set at f. This is sometimes called the spot noise figure. The transducer power gain G used above is defined by

$$G = \frac{\text{Actual output power delivered to the load}}{\text{Available power from the source}} \qquad (8.27)$$

Since this definition of noise figure depends on both the source and load impedances, it can be especially helpful in design of negative resistance amplifiers. In such cases the load and source impedance play an important role in the amplifier operation.

Sometimes the noise figure is defined as the measure of how much an amplifier has degraded an incoming signal. The noise figure in this case is the signal to noise ratio at the amplifier input divided by the signal to noise ratio at the output. If the signal power is P, then

$$F = \frac{P/(kT\,\Delta f)}{GP/N}. \tag{8.28}$$

This equals the definition given in (8.26) only if the noise source temperature is 290°K. Thus this definition is not the preferred one, but may be conceptually helpful.

Another useful parameter is the actual noise measure[4] which is defined in terms of (8.25) and (8.27).

$$M = \frac{F-1}{1-\dfrac{1}{G}} \tag{8.29}$$

This parameter is particularly useful in assessing the noise properties of a low gain amplifier. The noise measure parameter aids in the selection between a low gain, low noise amplifier or a high gain, higher noise amplifier for the front end of a system. Because of its high resolution for very low noise amplifiers, it is more convenient to express the noise in terms of an equivalent noise temperature.

$$T = T_o(F-1) \tag{8.30}$$

Methods for measuring noise figure was the subject of a special IRE standards committee report that was published in 1960.[5] Of the three methods that were discussed in this report, the one called the "dispersed-signal-source method" was found to be the most practical. This method is often called the Y factor method.[6] As its name implies, the amplifier is excited by a wideband noise source. The amplifier response is measured both with the source on and with it off. From the two amplifier responses, the noise figure is found. The experimental result is not the spot noise figure as defined in (8.26), but the average noise figure over the frequency range of the amplifier bandwidth, i.e.,

$$\overline{F} = \frac{\int F(f)G(f)\,df}{\int G(f)\,df}. \tag{8.31}$$

It is often convenient to define a constant G_o at some reference frequency f_o so that

$$\int G(f)\,df = G_o\,\Delta f_o \tag{8.32}$$

where Δf_o is the equivalent signal bandwidth. The average noise figure can then be expressed as

$$\overline{F} = \frac{N_r}{kT_oG_o\Delta f_o} \qquad (8.33)$$

where N_r is the total output noise power delivered to the load from the noise source.

The noise figure is a measure of the signal degradation caused by the amplifier. The noise power added to the output by the amplifier is

$$N_a = N - kT_oG\,\Delta f \qquad (8.34)$$
$$= (F - 1)kT_oG\,\Delta f.$$

Returning now to the Y factor measurement method, let N_1 be the output noise power when the noise source is at T_o and N_2 the output noise power when the source is at temperature T_2. Consequently from (8.34)

$$N_1 = kT_oG_o\,\Delta f_o + kT_oG_o\,\Delta f_o(\overline{F} - 1) \qquad (8.35)$$
$$N_2 = kT_og_o\,\Delta f_o + kT_oG_o\,\Delta f_o(\overline{F} - 1) + k(T_2 - T_o)G_o\,\Delta f_o \qquad (8.36)$$

The first two terms in (8.35) and (8.36) represent respectively the thermal noise from the source resistance at T_o and the noise added by the amplifier itself. The last term in (8.36) is the excess noise added by the turned on noise source at the amplifier input. The ratio of N_2/N_1, called the Y factor, is given by

$$Y = \frac{N_2}{N_1} = \frac{\overline{F}T_o + T_2 - T_o}{\overline{F}T_o}. \qquad (8.37)$$

From this $\overline{F}$ may be found in terms of the known parameters.

$$\overline{F} = \frac{T_2 - T_o}{T_o(Y - 1)} \qquad (8.38)$$

The accuracy of this method depends on the ambient temperature being 290° K, the excess noise source being matched to the amplifier to give a correct value for T_2, the calibration accuracy of the noise source that again affects the accuracy of T_2, and the image responses of the amplifier itself. A mismatch does not affect the Y ratio N_2/N_1, since it affects both N_2 and N_1 in the same way. Because of the random phase of the noise, errors in T_2 cannot be corrected; only a maximum error range can be assigned. The spurious passbands and image responses of the amplifier may give a lower noise figure measurement than what it would actually be when the amplifier is used for its intended application. If the amplifier passes power from the input to the output over many frequency bands, then the output noise power is

$$N = k(T_o + T_a)[\Delta f_o G_o + \Delta f_1 G_1 + \cdots + \Delta f_n G_n] \quad (8.39)$$

where T_a is the amplifier equivalent noise temperature. However, if the noise figure is measured with a white noise source, the input noise power will not be $kT_o G_o \Delta f_o$ as required by the definition; instead, it will become the noise that passes through all the transmission bands of the amplifier.

$$kT_o(G_0 \Delta f_0 + G_1 \Delta f_1 + \cdots + G_n \Delta f_n). \quad (8.40)$$

The ratio of these two values will not give (8.38). However, if the measured noise figure is multiplied by the factor

$$r = 1 + \frac{\Delta f_1 G_1}{\Delta f_0 G_0} + \cdots + \frac{\Delta f_n G_n}{\Delta f_0 G_0}, \quad (8.41)$$

then the correct value for $\overline{F}$ is regained. For example a receiver with a signal and image response of equal bandwidth and gain is measured on a noise figure test set. Then the average measured noise figure is

$$\overline{F} = \frac{T_o + T_a}{T_o}. \quad (8.42)$$

In actual operation, only one sideband is used. This measured value is 3 dB too optimistic. It must be multiplied by the r factor that, in this case, is 2. In general the average noise figure is

$$\overline{F} = \frac{N}{kT_o G_o \Delta f_o}\left[1 + \frac{\Delta f_1 G_1}{\Delta f_0 G_0} + \cdots + \frac{\Delta f_n G_n}{\Delta f_0 G_0}\right]. \quad (8.43)$$

The method of measuring the noise figure given here[7] involves de-embedding the amplifier from three stages: the amplifier itself, an RF attenuator, and a mixer/IF amplifier. The noise figure for a cascade of three stages (Figure 8.2) was found in (7.16) to be

$$\overline{F} = \overline{F}_1 + \frac{\overline{F}_2 - 1}{G_1} + \frac{(\overline{F}_3 - 1)L_2}{G_1} \quad (8.44)$$

where G_1 is the transducer power gain of the amplifier, and L_2 is the loss ratio of the attenuator in stage 2. If the attenuator is at temperature $T = T_o$, then $L_2 = F_2$ and the total noise figure reduces to

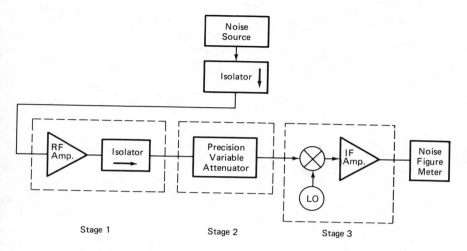

Figure 8.2. Noise figure measurement circuit.

$$\overline{F} = \overline{F}_1 - \frac{1}{G_1} + \frac{L_2 \overline{F}_3}{G_1}. \qquad (8.45)$$

The value for $\overline{F}_3$ is known since it is found by an independent measurement. The noise figure $\overline{F}$ is measured as a function of L_2 which gives a straight line with a slope of $\overline{F}_3/G_1$ and an intercept point at $\overline{F} = \overline{F}_1 - 1/G_1$ (Figure 8.3). Since $\overline{F}_3$ is known, the slope of the curve gives G_1. This value represents the amplitude response of the amplifier to a noise signal. The intercept point gives $\overline{F}_1$, which is the noise figure of the amplifier/isolator combination. This method depends on several measurements, and it therefore allows averaging individual measurement errors.

As an example, consider the measurement of a low noise negative resistance amplifier that has been set at three gain levels. The noise figure of stage 3 was measured as 10. The straight line plots in Figure 8.3 of noise figure $\overline{F}$ vs. L_2 results in a calculation for $\overline{F}_1$ (Table 8.1).

8.5 OSCILLATOR NOISE

The quality of noise and measurement techniques for noise in an oscillator differs considerably from that in a linear amplifier. Noise in an amplifier may be characterized by a single parameter such as noise figure or noise measure, a quantity derivable from noise figure and amplifier gain. As shown in Figure 8.4, noise is displayed as a distribution of noise sidebands on either side of the carrier signal. That portion of the noise above the carrier is referred to as the upper sideband, and that below the carrier as the lower sideband. The quality factor for an oscillator is usually given in terms of the ratio of the carrier power

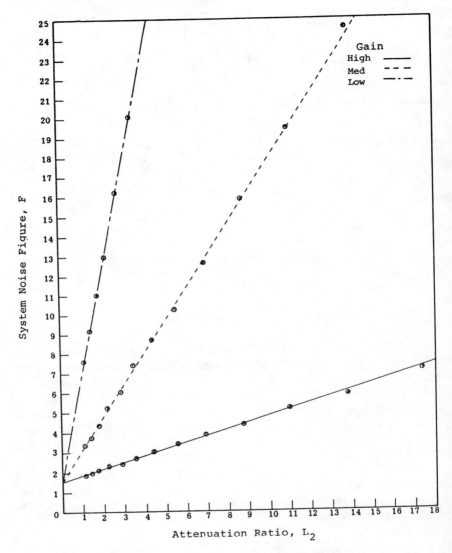

Figure 8.3. Experimental determination of noise figure.

Table 8.1. Noise Figure from Figure 8.3.

SLOPE F_3/G_1	NOISE GAIN G_1, RATIO	NOISE FIGURE F_1, RATIO	NOISE TEMP. T, °K
0.386	14.87	1.58	168°
1.632	7.87	1.71	206°
5.362	2.71	2.09	316°

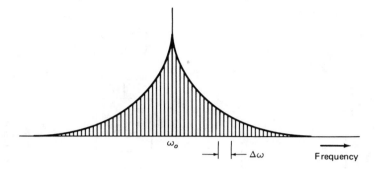

Figure 8.4. Noise power spectrum of a microwave oscillator.

to the noise power in the sidebands. This definition is complicated by the multiple sources of oscillator noise. The most important of these are the background thermal noise (analogous to the amplifier noise), amplitude modulation (AM) noise, and frequency modulation (FM) noise. These latter two noise contributions, expressed by Middleton[8] in mathematical form, are expressed as the noise voltage

$$v(t) = A_o Re\{[1 + mv_n(t - \tau)] \exp [j\omega t - j\phi_n(t)]\} \quad (8.46)$$

where

$$\phi_n(t) = D_o \int^t v_n(t') \, dt'. \quad (8.47)$$

In these expressions, D_o is the FM modulation index and m is the AM modulation index. The modulation noise components arise from modulation of the carrier signal by noise generated at lower frequencies. Usually noise very close to the carrier is predominately FM while noise further away from the carrier has equal components of AM and FM noise. Specification of the oscillator noise entails defining which of these three noise components are being measured, how wide a bandwidth is being considered, and how far away from the carrier the measurement is taken (Figure 8.4).

The measurement of the carrier to noise ratio requires a circuit that can measure the carrier and the desired noise portion independently. Two basic circuits that do this are the direct detection circuit and the superheterodyne circuit.[9] The direct detection circuit can be used to measure AM plus the background noise, and the FM plus background noise separately. The superheterodyne circuit is used to measure the total noise, AM, FM, and background in one or the other sidebands.

The circuit shown in Figure 8.5 is an example of a direct detection circuit used to measure AM noise.[9] The linear detector is used to measure the incom-

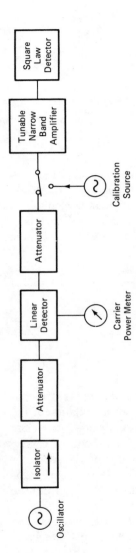

Figure 8.5. Direct detection circuit for AM noise.[9] (Material from the paper "Techniques of Microwave Noise Measurement" by B. G. Bosch and W. A. Gambling[9] is reproduced by permission of the Institution of Electronic and Radio Engineers.)

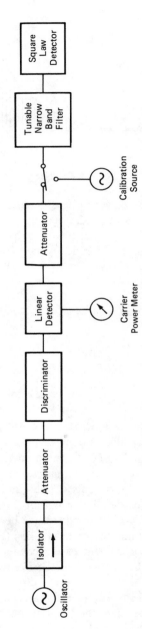

Figure 8.6. Direct detection circuit for FM noise.[9] (Material from the paper "Techniques of Microwave Noise Measurement" by B. G. Bosch and W. A. Gambling[9] is reproduced by permission of the Institution of Electronic and Radio Engineers.)

ing carrier signal, while the remaining noise sidebands are fed into the tunable amplifier. The AM noise in the upper and lower sidebands, that originated by beating the carrier with low frequency signals, are in phase. This produces an output in the square law detector. The FM signals, being in anti-phase on either side of the carrier, cancel one another. Thus this circuit measures only the background plus AM noise. To measure the FM noise in a direct detection circuit, a method must be found to convert the FM signal to an AM signal. One method is to introduce a discriminator in front of the detector used in the AM detector circuit (Figure 8.6). A discriminator is a circuit that has a linear voltage slope with frequency, so that frequency deviations along this slope are converted into amplitude variations. A widely used discriminator is simply a high Q cavity that is tuned at one of its half-power points (Figure 8.7). This technique can be used to measure noise close to the carrier. Because of the high Q of the cavity, the bandwidth of this scheme is limited to a narrow range.

An alternate technique that is applicable to both AM and FM noise measurements is the balanced mixer circuit.[10,11] In the circuit shown in Figure 8.8, the high Q cavity is tuned to suppress only the carrier signal and a small bandwidth around the carrier. The signal entering arm 3 of the magic tee splits evenly into ports 1 and 2 with opposite phase (Figure 8.9). Since the diodes are mounted in opposite polarity as well, the noise signals detected by the diodes are in phase. These signals will be summed in the video amplifier. The carrier signal entering port 4 of the magic tee will split evenly between ports 1 and 2 with the same polarity. Thus, because the detection diodes are in opposite polarity, the noise on this signal will cancel at the video amplifier. This "clean" input signal acts as the local oscillator of the mixer and down-converts the noise signal that entered port 3 to the video frequency range. This may then be handled by a low frequency wave analyzer.

The superheterodyne system with a balanced mixer is essentially the same

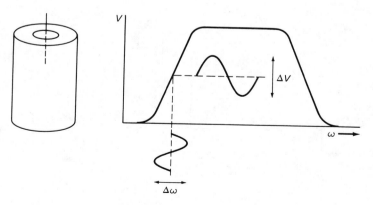

Figure 8.7. Cavity discriminator.

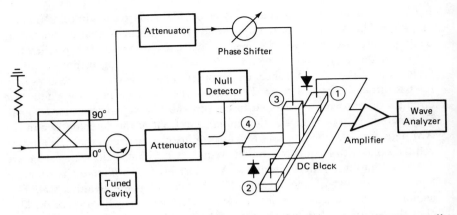

Figure 8.8. Balanced mixer circuit for AM and FM noise measurement. (Ashley[10] and Ondria[11] © 1968 IEEE.)

circuit shown in Figure 8.8. The system differs only in the input to arm 4 of the magic tee which is replaced by a separate external oscillator. Since the relative phases of the oscillator under test and the external local oscillator are uncorrelated, the AM, FM and background noise are all measured in one or the other sidebands. If the AM and FM noise components are correlated, then the noise bands will not be symmetrical about the carrier. Fortunately, this seldom occurs. Noise measurements taken on one sideband usually equal those taken on the opposite sideband.

REFERENCES

1. A. Messiah, *Quantum Mechanics*. Amsterdam: North-Holland Publishing Co., Chapter 12, 1964.

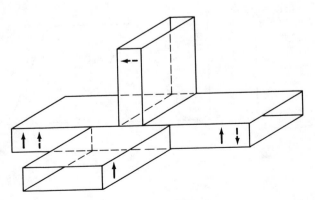

Figure 8.9. Magic tee used in mixer circuit.

2. A. E. Siegman, "Zero-Point Energy as the Source of Amplifier Noise," *Proc. IRE,* Vol. 49, pp. 633–634, March 1961.
3. H. Nyquist, "Thermal Agitation of Electronic Charge in Conductors," *Physical Review,* Vol. 32, p. 110, July 1928.
4. K. Kurokawa, "Actual Noise Measure of Linear Amplifiers," *Proc. IRE,* vol. 49, pp. 1391–1397, March 1961.
5. "IRE Standards on Methods of Measuring Noise in Linear Twoports, 1959," *Proc. IRE,* Vol 48, pp. 60–68, January 1960.
6. "Noise Figure Primer," Hewlett Packard Application Note 57, January 15, 1965.
7. P. J. Khan, private communication, 1970.
8. D. Middleton, "Theory of Phenomenological Models and Measurements of the Fluctuating Output of CW Magnetrons," *IRE Trans. Electron Devices,* Vol. ED-1, pp. 56–89, Feburary 1954.
9. B. G. Bosch and W. A. Gambling, "Techniques of Microwave Noise Measurement," *Journal British IRE,* Vol 21, pp. 503–515, June 1961.
10. J. R. Ashley, C. B. Searles, and F. M. Palka, "The Measurement of Oscillator Noise at Microwave Frequencies," *IEEE Trans. on Microwave Theory and Techniques,* Vol. MTT-16, pp. 753–760, September 1968.
11. J. G. Ondria, "A Microwave System for Measurements of AM and FM Noise Spectra," *IEEE Trans. on Microwave Theory and Techniques,* Vol. MTT-16, pp. 767–781, September 1968.

Chapter 9
Review of Statistical Thermodynamics

9.1 INTRODUCTION

The large number of particles present in a conductor or a semiconductor precludes the possibility of following the path of each or even a representative particle. The many interactions with other particles requires the use of a statistical approach to obtain mathematically tractable results. Once obtained, these results may provide a better understanding of the underlying causes for the macroscopic observations. Of particular interest is the origin of the Fermi-Dirac distribution function. This function governs the rectifying property of *pn* junctions in semiconductors. Also of interest are the Bose-Einstein statistics that are used in the derivation of the black body radiation function. This in turn is used in the Nyquist formula for thermal noise.

Statistical thermodynamics encompasses a large range of topics. Of importance in this chapter is the background to the statistics for electrons and photons. This chapter contains derivations for the Boltzmann factor, the equipartition law for classical statistical mechanics, and the Fermi-Dirac and Bose-Einstein distribution functions. The final section contains a short discussion of the meaning of the statistical thermodynamic point of view.

9.2 STATISTICAL ANALYSIS

Statistical analysis has been employed to find the average energy of a macroscopic system based on a microscopic probability distribution function for the energies of each particle in the system. Once the distribution function $P(\mathcal{E})$ is known, the average energy of the system can be found.

$$\overline{\mathcal{E}} = \frac{\int_0^\infty \mathcal{E} P(\mathcal{E}) \, d\mathcal{E}}{\int_0^\infty P(\mathcal{E}) \, d\mathcal{E}} \tag{9.1}$$

If, as in quantum mechanics, the energies lie in discrete levels, the continuous integrals in (9.1) must be replaced by discrete summations.

$$\overline{\mathcal{E}} = \frac{\sum_{j=0}^{\infty} \mathcal{E}_j P_j}{\sum_{j=0}^{\infty} P_j} \tag{9.2}$$

The basic problem is to find the function P that accurately describes the probability of a particle existing in a given state.

The microscopic state of a gas is defined by the instantaneous states of the individual molecules in the gas (or electrons in a conductor). In classical mechanics the position q and momentum p are assumed to be known exactly. In quantum mechanics the Heisenberg uncertainty principle restricts the accuracy of p and q to $\Delta p \Delta q \geq h/2\pi$, where h is Planck's constant. If two particles were interchanged with one another, the microscopic state would be different, but the macroscopic state would be the same. In fact, a large number of different microscopic states will lead to the same macroscopic state. The probability that a gas is in some macroscopic state is dependent on the number of constituent microscopic states that are at a particular energy level. For example, if in (9.1) $P(p_o, q_o)$ is very large and the rest of $P(p,q)$, $(p \neq p_o, q \neq q_o)$, is very small, the macroscopic energy $\overline{\mathcal{E}}$ will be mainly influenced by the energy $\mathcal{E}(p_o, q_o)$.

9.3 BOLTZMANN FACTOR

A fundamental postulate of statistical thermodynamics is that a macroscopic system in thermodynamic equilibrium is equally likely to be in any state that satisfies the macroscopic conditions of the system. Formally, this is termed the postulate of equal a priori probability. It is used to find the most probable distribution of microscopic states subject to the macroscopic conditions. Consider two macroscopic systems, one very small with energy $\mathcal{E}_1$ and the other very large with energy $\mathcal{E}_2$. When these two systems are brought into thermal contact in such a way that they may exchange energy without interchanging particles between them, as by a rigid heat-conducting wall, the total energy of the two systems can be written as

$$\mathcal{E}_T = \mathcal{E}_1 + \mathcal{E}_2. \qquad \mathcal{E}_1 \ll \mathcal{E}_2 \tag{9.3}$$

This additive property is valid as long as the interaction of the two systems is relatively weak. However, the individual energies are not constant, and in fact each of these systems will have some specified number of available states designated as follows.

$\Omega_1(\mathcal{E}_1)$ = number of states lying between $\mathcal{E}_1$ and $\mathcal{E}_1 + \delta\mathcal{E}_1$
$\Omega_2(\mathcal{E}_2)$ = number of states lying between $\mathcal{E}_2$ and $\mathcal{E}_2 + \delta\mathcal{E}_2$

According to the postulate of equal a priori probability, the probability of system 2 being in energy level $\mathcal{E}_T - \mathcal{E}_{1j}$ is proportional to the number of available energy states it has.

$$P_j = C\Omega_2(\mathcal{E}_T - \mathcal{E}_{1j}) \tag{9.4}$$

The logarithm of this function may be expanded in a Taylor series about the total energy $\mathcal{E}_T$ where it is assumed that $\mathcal{E}_{1j} \ll \mathcal{E}_T$.

$$\ln[\Omega_2(\mathcal{E}_T - \mathcal{E}_{1j})] = \ln[\Omega_2(\mathcal{E}_T)] + \frac{\partial \ln \Omega_2}{\partial \mathcal{E}_2}\bigg|_{\mathcal{E}_T} \frac{\partial \mathcal{E}_2}{\partial \mathcal{E}_T} \mathcal{E}_{1j} + \ldots \tag{9.5}$$

$$\cong \ln[\Omega_2(\mathcal{E}_T)] - \frac{\partial \ln \Omega_2}{\partial \mathcal{E}_2}\bigg|_{\mathcal{E}_T} \mathcal{E}_{1j}$$

The derivative, being evaluated at a fixed energy of the combined system, is a constant that is called β.

$$\beta = \frac{\partial \ln \Omega_2}{\partial \mathcal{E}_2}\bigg|_T \tag{9.6}$$

Since the number of available states increases with energy, the constant β must be a positive number. Consequently, (9.5) becomes

$$\ln[\Omega_2(\mathcal{E}_T - \mathcal{E}_{1j})] = \ln[\Omega_2(\mathcal{E}_T)] - \beta\mathcal{E}_{1j} \tag{9.7}$$

or

$$\Omega_2(\mathcal{E}_T - \mathcal{E}_{1j}) = \Omega_2(\mathcal{E}_T)\exp(-\beta\mathcal{E}_{1j}) \tag{9.8}$$

Since the total probability must be one, the proportionality constant C in (9.4) is found by summing the probabilities of all available states.

$$\sum_j C \exp(-\beta\mathcal{E}_{1j}) = 1$$

or

$$C = \left[\sum_j \exp(-\beta\mathcal{E}_{1j})\right]^{-1} \tag{9.9}$$

Therefore the normalized probability function is determined, and the probability of finding the system at $\mathcal{E}_{1j}$ is

$$P_j = \frac{\exp(-\beta \mathcal{E}_{1j})}{\sum_j \exp(-\beta \mathcal{E}_{1j})} \tag{9.10}$$

The probability function in (9.10) is known as the Boltzmann factor which can be used to find the mean energy $\overline{\mathcal{E}}$ as in (9.2).

The value for β is obtained by finding some thermodynamic relationship between this function and a known macroscopic system. This relationship is found by associating β with the thermodynamic definition of temperature. This may be found by considering again the two weakly interacting systems that exchange energy through a rigid conducting wall. The number of states in the total system is the product of the available states in each system alone. For every state in system 1 there are Ω_2 states in system 2. The total number of states in the combined system is the product of the two.

$$\Omega_T(\mathcal{E}_1) = \Omega_1(\mathcal{E}_1)\Omega_2(\mathcal{E}_T - \mathcal{E}_1) \tag{9.11}$$

This is proportional to the probability of finding the energy near $\mathcal{E}_T$. The maximum probability for a given $\mathcal{E}_T$ is found by taking the derivative of the logarithm of (9.11) and setting it equal to zero.

$$\frac{\partial \ln \Omega_T}{\partial \mathcal{E}_1} = 0 = \frac{\partial \ln \Omega_1}{\partial \mathcal{E}_1} + \frac{\partial \ln \Omega_2}{\partial \mathcal{E}_2} \frac{\partial \mathcal{E}_2}{\partial \mathcal{E}_1} \tag{9.12}$$

or

$$\frac{\partial \ln \Omega_1}{\partial \mathcal{E}_1} = \frac{\partial \ln \Omega_2}{\partial \mathcal{E}_2} = \beta \tag{9.13}$$

For the two systems in thermal equilibrium, (9.13) shows that their thermodynamic temperatures are equal where the absolute temperature scale is defined as

$$\left.\frac{\partial S}{\partial \mathcal{E}}\right|_{\text{const Vol}} = \frac{1}{T} \tag{9.14}$$

and where S is the entropy. In statistical thermodynamics the entropy is defined as

$$S = k \ln \Omega(\mathcal{E}) \tag{9.15}$$

which is a maximum when the probability of the system being in the state $\mathcal{E}$ to $\mathcal{E} + \delta\mathcal{E}$ is maximum. Consequently, from (9.14) and (9.15)

$$\frac{\partial \ln \Omega}{\partial \mathcal{E}} = \frac{1}{kT} \tag{9.16}$$

where the Boltzmann constant k is chosen to correspond with the conventional Celsius temperature scale. Thus, $\beta = 1/kT$. The entropy function for the composite system given by (9.11) and (9.15) is maximum when $S_1 + S_2$ is maximum, which occurs at $T_1 = T_2$.

9.4 EQUIPARTITION LAW

The equipartition law in classical statistical thermodynamics states that the energy of a system splits evenly among the degrees of freedom available to the individual particles. The assumptions for the equipartition law are as follows.

1. The energy of the particles in the system depends on a number of variables such as position q and momentum p. The minimum number of variables describing the particle is the number of degrees of freedom it has. For example, a billiard ball capable of translational motion in three orthogonal directions has three degrees of freedom.
2. The interaction energy between any two particles is zero, so that the total energy can be written as the sum of the separate parts.

$$\mathcal{E}_T = \sum_j \mathcal{E}_j(p_j, q_j) \tag{9.17}$$

3. The particle energy $\mathcal{E}_j$ is proportional to the square of one of its variables, i.e., $\mathcal{E}_j \propto p_j^2$. Consequently, the total energy is written as the sum of the p_j^2 with constant coefficients C_j.

$$\mathcal{E}_T = \sum_j C_j p_j^2 \tag{9.18}$$

4. Each of these variables are continuous from $-\infty$ to $+\infty$.
5. The systems of particles are in thermal equilibrium so that they all have the same average temperature.

It is sufficient to prove this theorem for only one variable, p_j. The average energy from the probability distribution function is

$$\overline{\mathcal{E}}_j = \frac{\int_{-\infty}^{\infty} \mathcal{E}_j \exp(-\beta \mathcal{E}_j) \, dp_j}{\int_{-\infty}^{\infty} \exp(-\beta \mathcal{E}_j) \, dp_j} \tag{9.19}$$

This can be rewritten in terms of the derivative of the numerator as follows.

$$\overline{\mathcal{E}}_j = \frac{-\dfrac{\partial}{\partial \beta} \displaystyle\int_{-\infty}^{\infty} \exp(-\beta \mathcal{E}_j)\, dp_j}{\displaystyle\int_{-\infty}^{\infty} \exp(-\beta \mathcal{E}_j)\, dp_j} \tag{9.20}$$

$$\overline{\mathcal{E}}_j = -\frac{\partial}{\partial \beta} \ln \int_{-\infty}^{\infty} \exp(-\beta \mathcal{E}_j)\, dp_j \tag{9.21}$$

Now from the third postulate, let $\mathcal{E}_j = p_j^2/2m$.

$$\overline{\mathcal{E}}_j = -\frac{\partial}{\partial \beta} \ln \int_{-\infty}^{\infty} \exp(-\beta p_j^2/2m)\, dp_j \tag{9.22}$$

$$= -\frac{\partial}{\partial \beta} \ln \sqrt{\frac{2\pi m}{\beta}} \tag{9.23}$$

The result in (9.23) is obtained from the standard definite integral I, where

$$I^2 = \int_{-\infty}^{\infty} \int_{-\infty}^{\infty} e^{-a(x^2+y^2)}\, dx\, dy$$
$$= \int_0^{2\pi} \int_0^{\infty} e^{-ar^2} r\, dr\, d\theta = \frac{\pi}{a}. \tag{9.24}$$

Performing the indicated differentiation in (9.23) leads to

$$\overline{\mathcal{E}}_j = \tfrac{1}{2} kT \tag{9.25}$$

and for N degrees of freedom, this would be

$$\overline{\mathcal{E}}_j = \frac{N}{2} kT \tag{9.26}$$

The equipartition theorem is valid in the classical case only where the temperature is sufficiently high. Here the difference between quantized energy states is so small that they essentially form an energy continuum.

9.5 QUANTUM MECHANICAL DISTRIBUTION FUNCTIONS

In the following quantum mechanical analysis, it is assumed that each of the particles of the system are independent, so that they do not interact with one another. The problem of finding the energy distribution of the whole therefore reduces to finding the energy levels of the individual particles. Each energy level $\mathcal{E}_j$ in general has n_j particles, where $\{n_j\}$ is the set of occupation numbers. The average energy of the entire system is known once these occupation numbers are known. The energy distribution function is found by taking the average value of the occupation number $\bar{n}_j$ given by

$$\bar{n}_j = \frac{\sum_{\{n_j\}} n_j \exp(-\beta \Sigma n_i \mathcal{E}_i)}{\sum_{\{n_j\}} \exp(-\beta \Sigma n_i \mathcal{E}_i)}. \tag{9.27}$$

To simplify the calculation, $\bar{n}_j$ is considered to be much less than one in order to avoid interactions between particles.

Two basic types of quantum mechanical particles have been discovered: those following Bose-Einstein statistics or Bosons, and those following Fermi-Dirac statistics or Fermions. The first group of particles includes photons and phonons (mechanical vibrations), while the second group, Fermions, includes electrons, protons, and neutrons. These two types of particles are distinguished from one another on the basis of the symmetry of their eigenfunctions for their Hamiltonians. Bosons have a symmetric wave function ψ in the coordinate q whereby if any pair of coordinates are interchanged, the eigenfunction remains unchanged.

$$\psi_B(q_1, q_2, \ldots, q_N) = \psi_B(q_2, q_1, \ldots, q_N) \tag{9.28}$$

On the other hand, Fermions are associated with antisymmetric wave functions whereby if any pair of coordinates are interchanged, the eigenfunction changes sign.

$$\psi_F(q_1, q_2, \ldots, q_N) = -\psi_F(q_2, q_1, \ldots, q_N) \tag{9.29}$$

The implication of these two statements is that, for any j, Bosons have occupation numbers that can take on the values of $n_j = 0, 1, 2, 3, \ldots$ while for Fermions the occupation numbers are limited to just $n_j = 0$ or 1. The occupation numbers for these two types of particles are found subject to the constraint that the total number of particles is N

$$N = \sum_j n_j \tag{9.30}$$

and the total energy is $\mathcal{E}$.

$$\mathcal{E} = \sum_j n_j \mathcal{E}_j \qquad (9.31)$$

There are several approaches for finding the average occupation number $\bar{n}_j$, and the following procedure is based on that given by Reif.[1]* If a particle with energy state $\mathcal{E}_j$ has been removed from the system, then the sum of all the remaining states for the $N - 1$ remaining particles is

$$Z_j(N - 1) = \sum_{\{n_i\}}^{(j)} \exp(-\beta \Sigma n_i \mathcal{E}_i). \qquad (9.32)$$

The superscript on the summation sign indicates the summation extends for all n_i except n_j. This term greatly simplifies the notation in the derivation of the quantum mechanical statistics.

9.5.1 Fermi-Dirac Statistics

Since Fermions obey the Pauli exclusion principle, their occupation number must be either 0 or 1. If the jth state is separated in the expression (9.27) for the average occupation number, then

$$\bar{n}_j = \frac{\sum_{n_j} n_j \exp(-\beta n_j \mathcal{E}_j) \sum_{\{n_i\}}^{(j)} \exp(-\beta \Sigma n_i \mathcal{E}_i)}{\exp(-\beta \mathcal{E}_j) \sum_{\{n_i\}}^{(j)} \exp(-\beta \Sigma n_i \mathcal{E}_i)}. \qquad (9.33)$$

The numerator comprises two terms. If $n_j = 0$ in the first term, then the second term has N particles, and if $n_j = 1$ in the first term then the second has $(N - 1)$ particles. Consequently the numerator of (9.33) can be written in terms of the function defined in (9.32).

$$0 \cdot Z_j(N) + 1 \exp(-\beta \mathcal{E}_j) Z_j(N - 1) \qquad (9.34)$$

A similar argument holds for the denominator of (9.33) so this can be written as

---

*Used with permission of the McGraw-Hill Book Company, 1965.

$$\bar{n}_j = \frac{0 \cdot Z_j(N) + \exp(-\beta \mathcal{E}_j) Z_j(N-1)}{1 \cdot Z_j(N) + \exp(-\beta \mathcal{E}_j) Z_j(N-1)} \quad (9.35)$$

$$= \frac{1}{\exp(\beta \mathcal{E}_j) Z_j(N)/Z_j(N-1) + 1} \quad (9.36)$$

All that remains is finding the ratio $Z_j(N)/Z_j(N-1)$, which can be done by expanding the logarithm of $Z_j(N-1)$ about N. Since $N \gg 1$ the series can be truncated after the second term.

$$\ln[Z_j(N-1)] = \ln[Z_j(N)] - \left.\frac{\partial \ln(Z_j)}{\partial N}\right|_N \quad (9.37)$$

Since the derivative is evaluated at a fixed value N, it is a constant designated as α. Solving (9.37) for α gives

$$\ln[Z_j(N)/Z_j(N-1)] = \alpha$$

or

$$e^\alpha = \frac{Z_j(N)}{Z_j(N-1)} \quad (9.38)$$

Finally, substituting (9.38) into (9.36) gives the occupation number distribution for a Fermion.

$$\bar{n}_j = \frac{1}{\exp(\alpha + \beta \mathcal{E}_j) + 1} \quad \text{Fermi} \quad (9.39)$$

In electronic systems, α is associated with the so-called Fermi energy level, where $\alpha \triangleq -\mathcal{E}_F/kT$. Analysis of

$$\bar{n}_j = \frac{1}{\exp(\mathcal{E}_j - \mathcal{E}_F)\beta + 1} \quad (9.40)$$

shows that the probability of n_j being empty or full is ½ when $\mathcal{E}_j = \mathcal{E}_F$. Furthermore for high temperatures where $(\mathcal{E}_j - \mathcal{E}_F) \gg kT$, the Fermi distribution becomes identical to the Boltzmann distribution corresponding to the classical analysis of a gas.

$$\bar{n}_j = \exp(-\mathcal{E}_j/kT) \quad (9.41)$$

The properties of particles in a Fermi gas are assumed to be the same as free electrons in a crystal. Because of the interaction of electrons in a crystal, the quantum mechanical description given by (9.40) is necessary. The distinction between conductors, semiconductors, and insulators depends on how far the free electrons are from the Fermi energy level. If $(\mathcal{E} - \mathcal{E}_F) \gg kT$, then all the energy levels are empty. If $(\mathcal{E}_F - \mathcal{E}) \gg kT$ then all the energy levels are filled. In either case conduction cannot occur. In semiconductors, the position of the Fermi level relative to the valence and conduction bands characterizes to a large extent the operation of junction devices.

9.5.2 Bose-Einstein Statistics

For Bosons, each of the energy levels can have an arbitrary number of particles, i.e., Bosons are not subject to the Pauli exclusion principle. The average occupation number given by (9.33) must be evaluated for a very large quantity of different occupation numbers. Explicitly, this sum is

$$\bar{n}_j = \frac{0 \cdot Z_j(N) + e^{-\beta \mathcal{E}_j} Z_j(N-1) + e^{-2\beta \mathcal{E}_j} Z_j(N-2) + \cdots}{1 \cdot Z_j(N) + e^{-\beta \mathcal{E}_j} Z_j(N-1) + e^{-2\beta \mathcal{E}_j} Z_j(N-2) + \cdots} \quad (9.42)$$

$$= \frac{Z_j(N)[0 + e^{-(\beta \mathcal{E}_j + \alpha)} + e^{-2(\beta \mathcal{E}_j + \alpha)} + \cdots]}{Z_j(N)[1 + e^{-(\beta \mathcal{E}_j + \alpha)} + e^{-2(\beta \mathcal{E}_j + \alpha)} + \cdots]}$$

where the constant α is given by (9.38). For a specific value of j, the numerator and denominator can be written as a ratio of sums

$$\bar{n}_j = \frac{\sum_{n_j=0}^{N} n_j \exp[-n_j(\beta \mathcal{E}_j + \alpha)]}{\sum_{n_j=0}^{N} \exp[-n_j(\beta \mathcal{E}_j + \alpha)]}. \quad (9.43)$$

This expression is equivalent to

$$\bar{n}_j = \frac{-\dfrac{\partial}{\partial(\beta \mathcal{E}_j + \alpha)} \sum \exp[-n_j(\beta \mathcal{E}_j + \alpha)]}{\sum \exp[-n_j(\beta \mathcal{E}_j + \alpha)]} \quad (9.44)$$

$$= -\frac{\partial}{\partial(\beta \mathcal{E}_j + \alpha)} \ln \left[\sum_{0}^{N} \exp[-n_j(\beta \mathcal{E}_j + \alpha)] \right].$$

The argument of the logarithm is the geometric series in which each term is less than one. Since the total number of particles N is very large, the sum may be approximated by an infinite geometric sum with the value given in (9.45)

$$\bar{n}_j = -\frac{\partial}{\partial(\beta\mathcal{E}_j + \alpha)} \ln\left[\frac{1}{1 - \exp[-(\beta\mathcal{E}_j + \alpha)]}\right] \qquad (9.45)$$

Performing the indicated differentiation, the Bose distribution is obtained.

$$\bar{n}_j = \frac{1}{\exp(\beta\mathcal{E}_j + \alpha) - 1} \quad \text{Bose} \qquad (9.46)$$

Comparison of this with the Fermi distribution shows that both approach the Boltzmann distribution at high energy levels. For the important special case of photons, the parameter $\alpha = 0$. The reason for this is that photons interact with matter so that the total number of photons N is not constant. The function Z must be independent of N, so from (9.38), $\alpha = 0$. Thus the Bose distribution corresponds to the energy distribution of black body radiation of photons first discovered by Planck.

9.6 SOME SECOND THOUGHTS

It seems appropriate here to discuss some of the implications of the second law of thermodynamics. Much has been written on this subject. Indeed Rifkin[2] has recently published a book that deals with the economic and social consequences of entropy as it applies to our physical environment.

Entropy is known intuitively by all from every culture, from the young preschool child to the elderly. Entropy describes the unidirectional fact that ice melts on a hot stove. When a barrier between clear water and ink is removed, the two liquids diffuse into one another. The entropy law states that these liquids can never be separated again without the application of some external energy. The question is, do these and other examples of entropy that we all know and feel relate to the mathematical formulation given by (9.15), based on statistical mechanics? Does it even describe the entropy of physical systems? The results achieved by statistical thermodynamics in the previous sections, seem to correlate well with evidence seen in the semiconductor pn junctions and the noisy photons radiating from a resistor. Nevertheless, when this method is extended to the derivation of the entropy function, the procedure is subject to serious criticism.

In the early days of the development of statistical thermodynamics, the exact formulation went through many conceptual changes. What finally emerged were three basic principles.

1. The thermodynamic function, entropy, corresponds to the probability of the macroscopic state of the system.
2. All microscopic states are equally probable.

3. The increase of entropy of the system means the system has moved to a more probable state.

The first principle says that there ought to be a correspondence between the statistical definition of entropy given by (9.15) and the classical macroscopic definition of entropy given mathematically as

$$\Delta S = \frac{\Delta Q}{T}. \tag{9.47}$$

The above equation states that a change in entropy ΔS results when a quantity of heat ΔQ is transferred from a hotter to a colder body. The transfer of heat occurs at temperature T. Georgescu-Roegen[3] points out that this formula should also be accompanied with the statement that the heat transfer must occur during a positive time interval. Entropy, as given in (9.47), gives the direction in which time must go, although it cannot tell how fast it goes. On the other hand, the statistical definition given in (9.15) is based on a measure of disorder. A rigorous proof of the equivalence of these two definitions has yet to be devised.[4] According to Khinchin[4] the classical formula (9.47) describes a thermodynamic function that is not random. When the temperature is constant and there is no heat transfer, the change in entropy is zero. The statistical formulation allows for a small probability that the entropy would change even if its independent thermodynamic variables remain unchanged.

The second principle states that a macroscopic system in thermal equilibrium is equally likely to have any one of a number of microscopic states subject to the macroscopic constraints on the system. This means that the system can be replicated into many similar systems that are called the microcanonical ensemble. The ergodic hypothesis put forward by Ludwig Boltzmann states that under equilibrium conditions, the time average of a microscopic system is the same as the average of all systems in the ensemble at one time. This hypothesis was used in the derivations in the previous sections. However, d'Abro[5] states that on logical grounds, the ergodic hypothesis is on shaky ground.

In as much as the assumption of determinate values for the probabilities of microscopic states is essential in the construction of a statistical mechanics, we conclude that the kinetic theory can be regarded as valid only insofar as the ergodic hypothesis is justified. Unfortunately, all attempts to demonstrate the correctness of this hypothesis by rigorous mathematical means have failed.

It cannot be said, however, that the ergodic hypothesis is incorrect, but only that it has never been proved. On the other hand, in all the relatively simple situations which have been studied and in which rigorous mathematical means of investigation have been available, the ergodic hypothesis has been proved incorrect.[5]

The third principle highlights the probabilistic foundation of the statistical definition of entropy. This states that while the most probable direction of entropy is toward more disorder, with sufficient time, it is possible that entropy may actually decrease. Entropy does not increase in a unidirectional way. Bronowski finds in this thought a key to the order we find in our world. "Statistics allows order to be built up in some islands of the universe (here on earth, in you, in me, in the stars, in all sorts of places) while disorder takes over in others."[6] But this process has never been observed without the addition of energy from outside the system. The ink and water that have diffused into one another will never, by the autonomous random motions of their molecules, separate on their own into half pure water and half pure ink. It seems that the order in our part of the universe was acquired by an external infusion of low entropy energy, and this entropy is now constantly rising. Whichever hypothesis is chosen for explaining the order we observe, it appears to lie ultimately as a religious choice rather than a scientific question.

An antinomy has arisen in deriving time irreversible thermodynamics from time reversible postulates. This contradiction may be explained by the theory of mutual irreducible modal aspects of meaning that was developed by H. Dooyeweerd.[7] An example of what happens when this irreducibility is violated is the famous paradox given by the Greek philosopher Zeno (c. 500 B.C.). He proposed that an arrow shot from a bow would never reach the target since at any indivisible moment in time, it would be viewed at rest in a definite place and therefore without motion. When he attempted to reduce the "kinematic" modal aspect of meaning to the "spatial" modal aspect of meaning, he encountered an antinomy. Similarly, it may be that the reduction of the irreducible modal meaning of physical energy (entropy) to that of kinematic motion (statistical fluctuations) has resulted in the antinomy found in the derivation of the entropy function from statistical mechanics.

The subject of statistical thermodynamics has engaged some of the best scientists in the world, and it has come up with some remarkable results. While these results have been helpful in understanding many physical phenomena, it seems clear that there is much work remaining.

REFERENCES

1. F. Reif, *Fundamentals of Statistical and Thermal Physics*. New York: McGraw-Hill, pp. 338–343, 1965.
2. J. Rifkin, *Entropy: A New World View*. New York: Viking Press, 1980.
3. N. Georgescu-Roegen, *The Entropy Law and the Economic Process*. Cambridge, Mass.: Harvard University Press, pp. 129–130, 1971.

4. A. I. Khinchin, *Mathematical Foundations of Statistical Mechanics*. Tr. by G. Gamow, New York: Dover, pp. 132–137, 1949.
5. A. d'Abro, *The Rise of the New Physics*. New York: Dover, pp. 392–393, 1957.
6. J. Bronowski, *The Ascent of Man*. Boston: Little, Brown, p. 348, 1973.
7. H. Dooyeweerd, *A New Critique of Theoretical Thought*. Tr. by D. H. Freeman and W. S. Young, Philadelphia: Presbyterian and Reformed Publishing Co., 1969.

Chapter 10
Review of *PN* Junction Theory

10.1 INTRODUCTION

The earliest semiconductor device to be used in microwave applications was the point contact diode. This device was made by attaching a small flexible wire, called a cat whisker, to a silicon semiconductor crystal. Often a metal-semiconductor junction, known as a Schottky barrier, was formed. Somewhat later, the *pn* junction was made by joining two doped semiconductors together. Since then the number of microwave semiconductor devices, using both the Schottky barrier and *pn* junctions, has multiplied. These devices can be broadly classified as two-terminal and three-terminal devices (Table 10.1). Some devices, such as the dual gate FET, have more than three terminals but can be considered as modifications of the elements shown in Table 10.1.

The early point contact diodes were used as detectors and mixers in microwave receivers during World War II. Like the point contact diode, the early *pn* junction could also rectify an AC current and mix two frequencies. Since then the number of applications of semiconductor devices has broadened to include detection, mixing, switching, phase shifting, amplification, etc. A given application is often best performed with only one or two preferred devices. Some of the principal applications are listed in Table 10.2. The design task is

Table 10.1. Sample of Microwave Devices.

2-TERMINAL DEVICES DIODES	3-TERMINAL DEVICES TRANSISTORS
Varactor	Bipolar
Step recovery	Field effect
Tunnel	
Backward	
Schottky barrier	
PIN	
IMPATT	
BARRIT	
Gunn	
TRAPATT	

Table 10.2. Microwave Circuit Functions.

APPLICATIONS
Amplifier/Oscillator using valve action (positive resistance)
Amplifier/Oscillator (negative resistance)
a. By internal semiconductor action
b. By interaction with external circuit
Frequency multiplier
Phase shifter
Electronic switch
Transmit/Receive protector
Variable attenuator
Detector
Mixer
Upconverter
Electronic frequency tuner
Temperature detector

to build a circuit performing a function illustrated in Table 10.2 with a device in Table 10.1.

The purpose of the present chapter is to provide a short review of one of the most pervasive building blocks in modern electronics: the *pn* junction. More detailed analysis can be found in several references[1,2,3,4] but the present material will serve as a background for design of microwave components. The effects of reverse and forward biasing the junction is found later in the chapter. The modern counterpart of the point contact diode, the Schottky barrier diode, will be discussed in chapter 12.

10.2 PN JUNCTION THEORY

The intent of this section is to provide a review of the primary ideas in semiconductor physics relevant to a *pn* junction. The existence of a *pn* junction is based on the existence of valence and conduction bands. These, in turn, are based on quantized energy states for electrons. This quantization of energy states was first realized in the early part of the Twentieth Century with the development of quantum mechanics. The basic ideas of quantum mechanics proceeded from the inability of classical physics to calculate the black body radiation (see chapter 9). The classical theory predicted the energy density in the box approaches infinity as the frequency approaches infinity. This was known as the ultraviolet catastrophe and was resolved by Max Planck in 1901. He postulated that any physical entity that executes sinusoidal motion can possess a total energy, $\mathcal{E}$, that satisfies

$$\mathcal{E} = nhf \qquad (10.1)$$

where f is the frequency of oscillation, h is a constant, and n is a non-negative integer. With this postulate the calculated black body energy density was found to approach zero at infinite frequency as it obviously does in the physical world. Later, Einstein in his work on the photoelectric effect, noted that when a system goes from energy state $\mathcal{E} = nhf$ to $(n-1)hf$, a quantum of energy equal to the difference is emitted.

$$\mathcal{E} = hf \qquad (10.2)$$

DeBroglie in 1924 conjectured that the motion of a particle is governed by the wave propagation of certain pilot waves (or wave functions according to Schrödinger terminology). So particles, including electrons, not only have quantized energy states but, like waves, can display interference phenomena.

Unlike the photons in the black body experiment, no more than two electrons of opposite spin can be in the same energy state since they obey the Pauli exclusion principle. If two or more atoms are brought into close proximity with one another, the electrons in the like energy states must shift in energy to obey the Pauli exclusion principle. When many atoms are brought close together a band of energies is formed (Figure 10.1). Between the energy bands that contain the electrons are forbidden bands that exclude electrons. In an insulator the electrons in the valence band or outermost shell completely fill the allowed energy states, and the energy required to bring an electron to the next allowed energy state is large (e.g., 5 to 9 eV). In a conductor the outermost electron shell is partially filled so that electrons are free to move with small applied fields. The semiconductor is a material that has a small forbidden energy gap ($\cong 1$ eV). Consequently, electrons can be moved up into a conductive energy band with moderate voltages. Furthermore, with the appropriate doping of a small amount of impurities, allowed energy levels may be placed in the once forbidden band. These additional energy levels may occur either near the valence band or near the conduction band (Figure 10.2).

The Pauli exclusion principle together with some statistical mechanical considerations imply that electrons will populate a given set of allowed energy

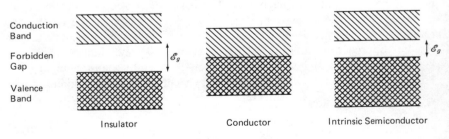

Figure 10.1. Energy band diagram.

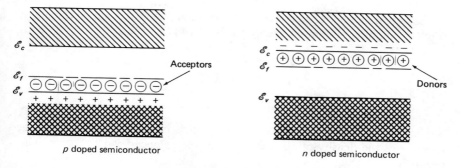

Figure 10.2. Extrinsic semiconductors.

states according to the Fermi-Dirac energy distribution function (see chapter 9),

$$f(\mathcal{E}) = \frac{1}{1 + \exp[(\mathcal{E} - \mathcal{E}_f)/kT]} \tag{10.3}$$

where $\mathcal{E}_f$ is the reference Fermi energy level. At $\mathcal{E} = \mathcal{E}_f$, half of the allowed energy states are filled and half are empty at any temperature. The probability that a low energy electron is in an allowed energy state is close to 1, and the probability that a high energy electron is in an allowed state is close to zero (Figure 10.3c). As the temperature T approaches zero, the transition from a full state to an empty state becomes increasingly more abrupt. Finally, at absolute 0°K, all allowed states below $\mathcal{E}_f$ are empty.

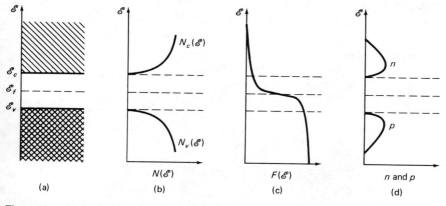

Figure 10.3. (a) Intrinsic semiconductor energy bands, (b) Density of states functions, (c) Fermi distribution, and (d) Hole and electron densities.

The actual number of electrons, n, in the conduction band can be calculated when the number of allowed energy states, $N_c(\mathcal{E})$, is known.

$$n = \int_{\mathcal{E}_c}^{\infty} N_c(\mathcal{E}) f(\mathcal{E}) \, d\mathcal{E} \qquad (10.4)$$

Similarly, the number of empty spaces or holes in the valence band can be found once the density of allowed states in the valence band is known. This results in

$$p = \int_{-\infty}^{\mathcal{E}_v} N_c(\mathcal{E})(1 - f(\mathcal{E})) \, d\mathcal{E}. \qquad (10.5)$$

These integrals have been evaluated in several texts on semiconductors and will not be carried out further here.

The number of electrons in the conduction band or the number of holes in the valence band can be altered by adding a small number of impurity atoms (10^{10} to 10^{18} atoms cm^{-3}) to the semiconductor. For example, application of a very small voltage can produce an excess of electrons in the conduction band when phosphorus is added to the silicon semiconductor. Addition of boron to silicon allows electrons to jump out of the valence band easily, thereby leaving excess holes in the valence band. Phosphorus is a donor impurity and produces an *n*-type semiconductor. Boron is an acceptor impurity and produces a *p*-type semiconductor. The Fermi energy level lies close to the center of an intrinsic semiconductor, close to the conduction band in the *n*-type semiconductor, and close to the valence band in the *p*-type semiconductor (Figure 10.2). When an *n*-type and *p*-type semiconductor are brought close together, their respective

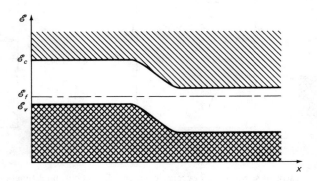

Figure 10.4. Energy levels at thermal equilibrium for a *pn* junction.

Fermi energy levels must line up to maintain the average electron energy on both sides of the junction (Figure 10.4). The difference in electron concentrations between the two sides causes electrons to flow from the *n*-side to the *p*-side, leaving an excess of positively ionized atoms in the *n*-type side. In similar manner an excess of negatively charged ionized atoms are left on the *p*-side. A space charge field thus develops at the junction creating a built-in potential that gives the junction its current rectifying properties. When the *p*-side is forward (or positively) biased relative to the *n*-side, current flows through the semiconductor. When the *p* side is reverse biased, no current flows across the junction.

10.3 REVERSE BIASED *PN* JUNCTION

When a *pn* junction diode is reverse biased, more mobile holes in the *p* region are attracted away from the junction to the negative applied voltage terminal, and more mobile electrons in the *n* region are attracted away from the junction by the applied positive voltage. The overall effect is to widen the space charge region (the region devoid of mobile carriers) and thereby decrease the depletion layer capacitance (Figure 10.5).

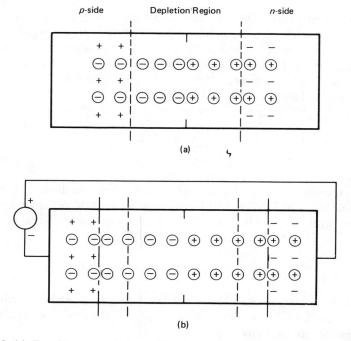

Figure 10.5. (a) Zero bias space charge region where circled charges are ionized atoms and uncircled charges are free carriers. (b) Reverse biased space charge region.

In general, a *pn* junction diode can be described by the following six basic equations. The first two are the current density equations that consist of a drift component and a diffusion component.

$$\mathbf{J}_p = q(\mu_p p \mathbf{E} - D_p \nabla p) \quad (10.6)$$

$$\mathbf{J}_n = q(\mu_n n \mathbf{E} + D_n \nabla n) \quad (10.7)$$

The second set are the continuity equations.

$$-\nabla \cdot \mathbf{J}_p = q\left(\frac{\partial p}{\partial t} + \frac{p - p_n}{\tau_p} - g_p\right) \quad (10.8)$$

$$\nabla \cdot \mathbf{J}_n = q\left(\frac{\partial n}{\partial t} + \frac{n - n_p}{\tau_n} - g_n\right) \quad (10.9)$$

The next expression is Gauss's law where ρ is the net charge density.

$$\nabla \cdot (\epsilon \mathbf{E}) = \rho \quad (10.10)$$

$$\rho = q(p - n + N_d - N_a) \quad (10.11)$$

The hole and electron minority carrier densities are p and n respectively, the equilibrium hole and electron minority carrier densities are p_n and n_p respectively, $\mathbf{J}$ is the current density, N_d and N_a are the donor and acceptor atom densities, and g is the generation term, a term that is often neglected. The calculations that follow are based on a p^+n diode so that the depletion layer extends into the n region from $x = 0$ to $x = W$.

Equation (10.10) can be rewritten as Poisson's equation

$$\frac{\partial^2 \psi}{\partial x^2} = -\frac{qN_d}{\epsilon} \quad (10.12)$$

where the potential ψ is a function of distance within the diode. Solution of this equation gives the fundamental relationship between the depletion layer capacitance and the applied voltage.[5] If the impurity concentration on the *p*-side of the junction is

$$N_d(x) = ax^b, \quad (10.13)$$

then Poisson's equation can be integrated. Using the boundary conditions that the electric field $E(x = W) = 0$ at the edge of the depletion region, the first integration can be performed.

$$\frac{\partial \psi}{\partial x} = -E(x) = \frac{qa}{\epsilon(b+1)} [x^{(b+1)} - W^{(b+1)}] \tag{10.14}$$

Integrating again, subject to the condition that $\psi(0) = 0$, gives

$$\psi(x) = \frac{-qa}{\epsilon(b+1)} \left[\frac{x^{b+2}}{b+2} - xW^{(b+1)} \right]. \tag{10.15}$$

At the edge of the depletion region where $\psi(W) = -(V - \phi)$ (10.15) can be solved for W.

$$W = \left[\frac{\epsilon(b+2)(\phi - V)}{qa} \right]^{1/(b+2)} \tag{10.16}$$

When the applied voltage V is negative, the diode is reverse biased, and the depletion layer widens. The maximum electric field at $x = 0$ is

$$E_{max} = \frac{qaW^{(b+1)}}{\epsilon(b+1)} \tag{10.17}$$

and by eliminating W between (10.16) and (10.17),

$$E_{max} = \frac{b+2}{b+1} \left[\frac{qa(\phi - V)^{b+1}}{\epsilon(b+2)} \right]^{1/(b+2)} \tag{10.18}$$

The stored charge on the varactor with cross sectional area A is $Q = AE_{max}$, as obtained from Gauss's law. The varactor capacitance is thus

$$C = -\frac{\partial Q}{\partial V} = A \left[\frac{qa\epsilon^{b+1}}{(b+2)} \right]^{1/(b+2)} (\phi - V)^{-1/(b+2)}$$

or more concisely

$$C = \frac{C_0}{(1 - V/\phi)^\gamma} \tag{10.19}$$

where $\gamma = 1/(b+2)$ and C_0 is the capacitance at zero bias. The voltage dependence of the capacitance is a function of the doping profile of the varactor (Table 10.3). The graded and abrupt junction diodes have been widely used in frequency multipliers and parametric amplifiers. The hyperabrupt junction

Table 10.3. Doping Profiles for *pn* Junctions.

DESCRIPTION	PROFILE EXPONENT b	CAPACITANCE EXPONENT $\gamma = 1/(b+2)$
Graded junction	1	⅓
Abrupt junction	0	½
Hyperabrupt junction	$\begin{cases} -1 \\ -\tfrac{3}{2} \end{cases}$	1 2

where $\gamma > \tfrac{1}{2}$ is useful in trying to achieve linear frequency tuning. This can be seen for the simple series LC circuit, where the resonant frequency is

$$f_0 = \frac{1}{2\pi\sqrt{LC(V)}} \tag{10.20}$$

$$f_0 = \frac{1}{2\pi\sqrt{LC_0}}(1 - V/\phi)^{\gamma/2}$$

According to this simple analysis, linear tuning could be achieved with $\gamma = 2$. However, in practical circuits, the addition of parasitic capacitance near the varactor chip often dictates that best linearity actually occurs for $1.2 < \gamma < 1.4$.

10.4 FORWARD BIASED *PN* JUNCTION

When a *pn* junction diode is forward biased, the applied voltage reduces the barrier height and allows minority carriers to diffuse more easily across the junction. Thus the majority carrier holes on the *p* side diffuse across the junction to become minority carriers on the *n* side. After a certain time, known as the minority carrier lifetime τ_p, these holes become neutralized and diffuse toward the negative terminal. The external current flow given by the diode equation[6],

$$J = J_0(\exp(qV/kT) - 1) \tag{10.21}$$

shows that the current rises exponentially with applied forward voltage. The derivation of this expression is based on manipulation of the current density equations (10.6) and (10.7) and the continuity equations (10.8) and (10.9). In addition a steady-state assumption ($\partial/\partial t = 0$) is incorporated.

When the steady-state assumption is not used, a complex expression for the current density is obtained. Dividing the resulting current by the applied voltage yields the frequency dependent diffusion conductance and susceptance.

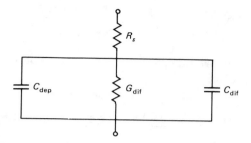

Figure 10.6. Diode model with series resistance, depletion capacitance, diffusion capacitance, and diffusion conductance.

The common approach for this calculation is based on dividing the voltage into a large part and a small part. This approach eliminates all higher order mixing products that occur when a large signal puts the diode in the forward conduction region during at least part of the cycle. During the forward conduction time period, majority carriers are drawn across the junction and become minority carriers. However, if the period of the AC field is shorter than the minority carrier lifetime, most of these carriers will be brought back across the junction before they have a chance to be neutralized. Because the carrier movement lags behind the applied AC field, the charge storage can be modeled as a capacitive susceptance. Thus the complete diode model applicable for both forward and reverse bias is shown in Figure 10.6.

Maneshin and Parygin[7] developed a large signal analysis for the diffusion admittance that is especially appropriate for analysis of frequency multipliers. For the one sided p^+n junction diode, the depletion layer extends in the n region from $x = 0$ to $x = W$. At equilibrium the sum of the drift and diffusion currents at the edge of the depletion region is approximately zero, so $J_n(x = W)$ in (10.7) is set to zero. If the impurity concentration is again $N_d(x) = ax^b$, then (10.7) together with the Einstein relationship, $D_n = \mu_n kT/q$, gives the electric retarding field in the n region.

$$E = -\frac{kT}{q}\frac{1}{N_d}\frac{dN_d}{dx} = -\frac{kT}{q}\frac{b}{x}. \qquad (10.22)$$

The hole current is then given by (10.6):

$$J_p = -qD_p\left(\frac{bp}{x} + \frac{\partial p}{\partial x}\right). \qquad (10.23)$$

This can be combined with the continuity equation (10.8) to yield

$$D_p\left(\frac{\partial^2 p}{\partial x^2} + \frac{b}{x}\frac{\partial p}{\partial x} - \frac{bp}{x^2}\right) - \frac{p - p_n}{\tau_p} = \frac{\partial p}{\partial t}.$$

When a periodic voltage is applied to the diode, the hole concentration can be represented by a Fourier series:

$$p(x,t) = p_0 + \sum_{m=1}^{\infty} (p_{mc} \cos m\omega t + p_{ms} \sin m\omega t).$$

If $p(x,t)$ is substituted into the above equation, and the harmonic balance technique is applied (coefficients of corresponding frequency terms are equated), the following two sets of equations are formed:

$$D_p\left(\frac{d^2 p_{mc}}{dx^2} + \frac{b}{x}\frac{dp_{mc}}{dx} - p_{mc}\frac{b}{x^2}\right) - \frac{p_{mc}}{\tau_p} = m\omega p_{ms}$$

$$D_p\left(\frac{d^2 p_{ms}}{dx^2} + \frac{b}{x}\frac{dp_{ms}}{dx} - p_{ms}\frac{b}{x^2}\right) - \frac{p_{ms}}{\tau_p} = -m\omega p_{mc}.$$

Since

$$\text{Re}[(p_{mc} - jp_{ms})e^{jm\omega t}] = p_{mc} \cos m\omega t + p_{ms} \sin m\omega t$$

the above two equations can be combined into one by defining

$$p_m \stackrel{\Delta}{=} p_{mc} - jp_{ms}$$

where this is not to be confused with $p_n(W)$.

$$x^2\frac{d^2 p_m}{dx^2} + bx\frac{dp_m}{dx} - \left[b + \frac{1 + jm\omega\tau_p}{L_p^2}x^2\right]p_m = 0 \quad (10.24)$$

To transform this into the usual form for a modified Bessel equation the following substitutions are made:

$$\chi = \xi x$$
$$p_m = u\chi^{(1-b)/2}$$

where

$$\xi^2 = \frac{1 + jm\omega\tau_p}{L_p^2}. \quad (10.25)$$

Using the first substitution, (10.24) can be written as

$$\chi^2 \frac{d^2 p_m}{d\chi^2} + b\chi \frac{dp_m}{d\chi} - (b + \chi^2) p_m = 0.$$

The second substitution transforms this expression into

$$\chi^2 \frac{d^2 u}{d\chi^2} + \chi \frac{du}{d\chi} - [(1 + b)^2/4 + \chi^2] u = 0.$$

This is the well-known modified Bessel equation, with the solution

$$p_m(x) = x^{(1-b)/2} [M_1 I_{b'}(\chi) + M_2 K_{b'}(\chi)]$$

where $b' \triangleq (b + 1)/2 = (1 - \gamma)/2\gamma$. Two boundary conditions are now applied to evaluate the constants M_1 and M_2. Since $p_m(x) = 0$ as $x \to \infty$, $M_1 = 0$. The constant M_2 is found by using the hole concentration at the edge of the depletion region, $x = W$. If the resistance of the semiconductor is much less than the resistance of the depletion layer, the entire applied voltage, $V(t)$, appears across the junction. Therefore the hole density at the edge of the depletion region is

$$p(W) = p_n(W) e^{\alpha V}$$
$$= p_n(W) \left[g_0 + \sum_{m=1}^{\infty} (g_{mc} \cos m\omega t + g_{ms} \sin m\omega t) \right]$$

where $\alpha \triangleq q/kT$, and g_0, g_{mc} and g_{ms} are the Fourier coefficients of $e^{\alpha V}$. If W is assumed to be independent of the voltage, and the total minority carrier hole density is $p_m(W) \triangleq p_n(W)(g_{mc} - jg_{ms})$, then

$$M_2 = \frac{p_n(W)(g_{mc} - jg_{ms})}{W^{(1-b)/2} K_{b'}(\xi W)}.$$

The final solution for $p_m(x)$ is thus

$$p_m(x) = p_n(W) \left(\frac{x}{W}\right)^{(1-b)/2} \frac{K_{b'}(\xi x)}{K_{b'}(\xi W)} \frac{1}{\pi} \int_0^{2\pi} e^{\alpha V}(\cos m\omega t - j \sin m\omega t) \, d\omega t. \quad (10.26)$$

The current can be expanded into a Fourier series by using the same technique.

$$J = J_0 + \sum_{m=1}^{\infty} J_{mc} \cos m\omega t + J_{ms} \sin m\omega t$$

If $J_m \stackrel{\Delta}{=} J_{mc} - jJ_{ms}$, the expression for $p_m(x)$ of (10.26) can be substituted into (10.23) to obtain an expression for the current. If $b'' \stackrel{\Delta}{=} (b-1)/2 = (1-3\gamma)/2\gamma$, then

$$J_m(x) = \left(\frac{x}{W}\right)^{-b''} \xi \frac{K_{b''}(\xi x)}{K_{b'}(\xi W)} \frac{qD_p p_n(W)}{\pi} \int_0^{2\pi} e^{\alpha V}(\cos m\omega t - j \sin m\omega t)\, d\omega t \quad (10.27)$$

where the identity $d/dz[z^n K_n(z)] = -z^n K_{n-1}(z)$ has been used. The expression gives the harmonic current components due to the applied voltage $V(t)$. If

$$\alpha_m + j\beta_m \stackrel{\Delta}{=} \xi W \frac{K_{b''}(\xi W)}{K_{b'}(\xi W)}, \quad (10.28)$$

then the coefficients for the cosine and sine terms of the currents are

$$J_{mc} = \frac{qD_p p_n(W)}{\pi W} \int_0^{2\pi} e^{\alpha V}(\alpha_m \cos m\omega t + \beta_m \sin m\omega t)\, d\omega t \quad (10.29)$$

$$J_{ms} = \frac{qD_p p_n(W)}{\pi W} \int_0^{2\pi} e^{\alpha V}(\alpha_m \sin m\omega t - \beta_m \cos m\omega t)\, d\omega t. \quad (10.30)$$

The α_m should not be confused with $\alpha = q/kT$.

If the admittance is defined as

$$\begin{aligned} Y_m &= A \frac{dJ_m}{dV} \\ &= A\alpha J_m \end{aligned} \quad (10.31)$$

where A is the diode area, then there is clearly a resistive and reactive component, the latter being the diffusion susceptance. Although the large nonlinearity of the exponential function is desirable, the loss term limits the usefulness of the diffusion capacitance in low loss applications.

10.5 DIFFUSION ADMITTANCE CALCULATIONS

The analysis of the diffusion admittance provided in the previous section showed that the capacitive susceptance is a complicated function of frequency. At very high frequencies ($\omega \tau_p \gg 1$) the diffusion susceptance, due to the minority carrier holes in the one-sided diode, is independent of τ_p and proportional to $\sqrt{\omega}$. The independence from τ_p in this frequency range occurs because all the minority charge is recovered. Increasing τ_p will not increase the charge storage since all of the charge is already stored. At very low frequencies ($\omega \tau_p \ll 1$), the susceptance depends on $\sqrt{\tau_p}$. Like a parallel plate capacitor it is proportional to ω. The following numerical calculations illustrate more completely the functional trends of the diffusion admittance with variation in W/L_p, τ_p, and ω.

If the diode is excited with an ideal voltage source

$$V(t) = V_b + V_0 \cos \omega_0 t, \tag{10.32}$$

then the integral contained in (10.31) can be readily integrated by using Sonine's expansion

$$e^{z \cos \theta} = I_0(z) + 2 \sum_{n=1}^{\infty} I_n(z) \cos n\theta \tag{10.33}$$

where $I_n(z)$ are modified Bessel functions. Using this expansion, and the orthogonality of the trigonometric functions, the mth harmonic of the current is

$$J_m = \xi \frac{K_{b'}(\xi W)}{K_{b''}(\xi W)} 2q D_p p_n(W) I_m(\alpha V_0) \exp(\alpha V_b). \tag{10.34}$$

Substitution of (10.34) into (10.31) gives the diffusion admittance. It should be recalled from (10.25) that ξ is a complex function of ω and τ_p. Consequently numerical methods based on Lanczos' work[8,9] are employed in the evaluation.

Numerical calculations of the diffusion admittance have been performed for the doping profiles $N(x) = ax^b$ where $b = 0, 1, 3$. These correspond to the abrupt junction $\gamma = \frac{1}{2}$, the graded junction $\gamma = \frac{1}{3}$, and $\gamma = \frac{1}{5}$ respectively, when the diode is reverse biased. Since a p^+n diode is used, the current resulting from the diffusion admittance in the p side can be neglected.

The quality factor for the diode can be defined for the present as B_1/G_1, where the admittance $Y_1 = G_1 + jB_1$ is the ratio of the fundamental current at ω_0 to the applied voltage at ω_0. The calculation of the quality factor is given

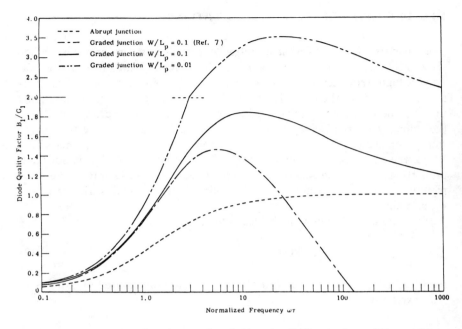

Figure 10.7. Relative loss of an abrupt and graded junction diode under forward bias conditions.

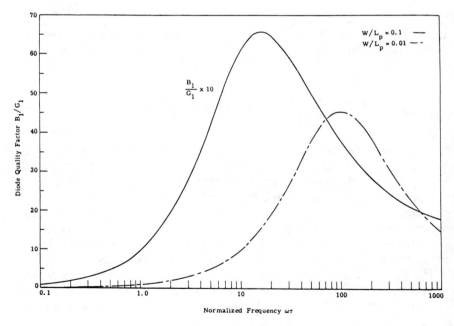

Figure 10.8. Relative loss of diffusion admittance with doping profile $N(x) = ax^3$.

Table 10.4. Minimum Loss vs. Doping Profile.

b	γ	d_n	MAX. B_1/G_1	$\omega\tau_p$ AT MAX. B_1/G_1
0	½	—	1	∞
1	⅓	0.1	1.85	11
1	⅓	0.01	3.53	30
3	⅙	0.1	6.54	17
3	⅙	0.01	45.20	100

in terms of the depletion layer width normalized to the diffusion length $L_p = \sqrt{D_p \tau_p}$.

The plots of the relative loss B_1/G_1 for various doping profiles in Figures 10.7 and 10.8 indicate how the loss in the diffusion admittance can be controlled. The abrupt junction is seen to be inferior to the graded and cubic ($b = 3$) junctions at all frequencies. In general, as b increases, the diode loss decreases. Also, when the base width to diffusion length ratio W/L_p decreases, loss decreases. Furthermore, as W/L_p decreases, the angle $\omega\tau_p$, where the minimum loss occurs, increases. These facts are summarized in Table 10.4.

10.6 DIFFUSION AND DEPLETION ADMITTANCE

The effects of both diffusion admittance and depletion admittance can be combined into a single diode model. At the edge of the depletion layer there are two current components. The first is the displacement current associated with the depletion capacitance, and the second is the diffusion current that passes through the depletion layer. Since these two currents add together, the depletion capacitance and the diffusion admittance are parallel elements (Figure 10.6). The resistor R_s is in series since it represents the bulk resistance of the base region (the n side of the p^+n diode). The diffusion admittance is proportional to $e^{\alpha V}$. It is larger than the depletion susceptance when the diode is forward biased, so that the depletion capacitance does not short circuit the diffusion capacitance.

The impedance of the total diode model of Figure 10.6 was calculated for the representative parameters listed below:

interrogation voltage $V(t) = 0.1 \cos(2\pi f_p t)$
frequency $f_p = 9.5$ GHz
normalized depletion width $W/L_p = 0.1$
minority carriers at W, $p_n(W) = 10^{10}$ cm^{-3}
diode area $A = 4.41 \; 10^{-5}$ cm^2
minority carrier lifetime $\tau_p = 1.84 \; 10^{-10}$ sec
dielectric constant of silicon $\epsilon = 11.8\epsilon_o$

216 MICROWAVE SEMICONDUCTOR CIRCUIT DESIGN

built-in potential $\phi = 0.9$ volt
series bulk resistance $R_s = 1$ ohm
ideality factor $= 2$, so $\alpha = 10\ V^{-1}$
mobility $\mu_p = 401\ cm^2/V$-sec
diffusion constant $D_p = 10$

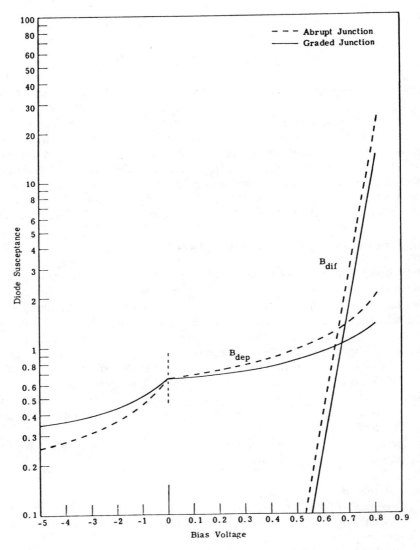

Figure 10.9. The depletion and diffusion susceptance when the minority carrier lifetime is chosen to minimize loss in the graded junction diode, i.e., $\tau = 1.84 \times 10^{-10}$ sec.

Except for the minority carrier lifetime, which was chosen to minimize the diffusion admittance losses, the above diode parameters are typical for silicon parametric varactor diodes.

The relative magnitude of the depletion and diffusion susceptances shown in Figure 10.9 can be compared for various bias voltages for both the abrupt and

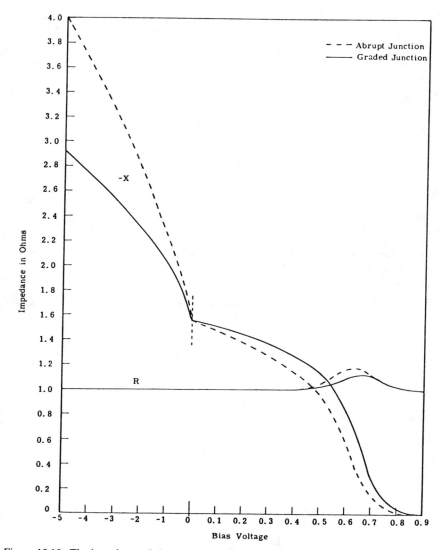

Figure 10.10. The impedance of the total varactor model when the minority carrier lifetime is chosen to minimize loss in the graded junction diode.

graded junction diodes. Clearly the diffusion admittance is important only in the forward bias region. The impedance at the diode terminals is plotted in Figure 10.10. This plot shows that the real part of the diode impedance increases when the bias voltage is between 0.4 and 0.8 volts. Outside of this range the parallel combination of the diode capacitance and diffusion conductance is negligible because of the low conductance at small voltages and the high susceptance at large voltages. If a different value of τ_p is used, instead of the optimum value used above, the resistive part of the diode impedance can become quite large. For example when $\tau_p = 10^{-7}$ sec, this resistance has a maximum value of 6.5 ohms.

REFERENCES

1. R. S. Muller and T. I. Kamins, *Device Electronics for Integrated Circuits*. New York: John Wiley, 1977.
2. S. M. Sze, *Physics of Semiconductor Devices*. New York: John Wiley, 1969.
3. H. A. Watson, *Microwave Semiconductor Devices and Their Circuit Applications*. New York: McGraw-Hill, chapters 1–6. 1969.
4. J. L. Moll, *Physics of Semiconductors*. New York: McGraw-Hill, 1964.
5. M. H. Norwood and E. Shatz, "Voltage Variable Capacitor Tuning: A Review," *Proceedings of the IEEE*, Vol. 56, pp. 788–798, May 1968.
6. W. Shockley, "The Theory of p-n Junctions in Semiconductors and p-n Junction Transistors," *Bell System Technical Journal*, Vol. 28, pp. 435–489, 1949.
7. V. N. Parygin and N. K. Maneshin, "Frequency Multiplication using the Diffusion Capacitance," *Radio Engineering and Electronic Physics*, Vol. 11, pp. 1111–1118, July 1966. By permission of Scripta Publishing Co.
8. C. Lanczos, *Applied Analysis*. Englewood Cliffs, N.J.: Prentice Hall, 1956.
9. C. Lanczos, *Tables of Chebyshev Polynomials $S_n(x)$ and $C_n(x)$*, U.S. Department of Commerce, National Bureau of Standards, Washington, D.C. 1952, pp. V–XXIX.

Chapter 11
Varactor and Step Recovery Diodes

11.1 INTRODUCTION

The varactor diode and step recovery diode (SRD) are of particular interest to microwave circuit designers because of their voltage dependent reactance and low loss. These properties are very useful in the three major applications discussed in this chapter: (1) voltage controlled oscillators, (2) parametric amplifiers and upconverters, and (3) frequency multipliers. In the first application the variable reactance of a varactor diode provides a method of electronically controlling the frequency of an oscillator. In the second application the reactive mixing of the varactor diode is used to develop a negative resistance that results in low noise amplification. In the third application the nonlinearity of the varactor is used to obtain high frequency oscillations by harmonically multiplying the frequency of the excitation signal. More recently, however, the step recovery diode has been found to be more efficient, especially where high orders of multiplication are required. Consequently, the use of the SRD will be emphasized in the last section of this chapter.

11.2 VOLTAGE CONTROLLED OSCILLATOR

A voltage controlled oscillator (VCO) consists of a negative resistance oscillator source, a passive high Q circuit, a load to which the power is delivered, and an externally controlled reactance that tunes the oscillator. Such a circuit can be represented as these four elements in series (Figure 11.1). The tunable impedance can be supplied by magnetically controlled YIG material, a pin

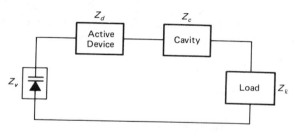

Figure 11.1. Equivalent Circuit of the VCO.

219

diode, or a varactor diode. The YIG tuned oscillator can reach over an octave bandwidth, but is limited in output power by the saturation of the YIG sphere to about 100 mW. The tuning speed of the YIG oscillator is limited to a few hundred Hz, while the varactor tuned oscillator can reach speeds up to approximately 100 MHz.[1] A pin diode can also be used for circuit tuning, but for many applications it is too lossy. The varactor diode has high tuning speed and low loss, and consequently is often used in VCOs. However, its bandwidth is typically limited to about 10% in the 8 to 12 GHz frequency range.

The primary design objectives are to determine the bandwidth, the efficiency, and the output power limitations of the VCO. The latter two requirements are often limited by the loss in the varactor diode itself. Finding the bandwidth limitation is based on the circuit impedance of Figure 11.1 being equal to zero at all frequencies in the band,

$$Z_v + Z_d + Z_c + Z_\ell = 0 \tag{11.1}$$

and on meeting the Kurokawa stability criterion.[2]

Kurokawa's analysis is based on splitting the circuit of Figure 11.1 into the passive circuit $Z(\omega) = Z_v + Z_c + Z_\ell$, and an active device impedance $Z_d(A)$. The passive circuit is a function of frequency, and the device impedance is assumed to be only a function of the RF current amplitude, where the RF current is given by

$$i(t) = A \cos(\omega t + \phi). \tag{11.2}$$

The amplitude A and the phase ϕ are assumed to be slowly varying functions of time. For the desired range of frequencies, the device is assumed to be independent of frequency. Thus the criterion for oscillations shown in Figure 11.2 occurs when the load line of the active device, $Z_d(A)$, intersects the circuit impedance locus $Z(\omega)$, i.e.,

$$Z_d(A) + Z(\omega) = 0. \tag{11.3}$$

The criterion for stable oscillations can be found by calculating the effect of a small change δA on the current amplitude. By performing a truncated Taylor series expansion on (11.3) an expression for the effect of the change in amplitude is obtained.

$$R(\omega_o) + R_d(A) = \delta A \frac{\partial R_d(A_o)}{\partial A} \tag{11.4}$$

$$X(\omega_o) + X_d(A) = \delta A \frac{\partial X_d(A_o)}{\partial A} \tag{11.5}$$

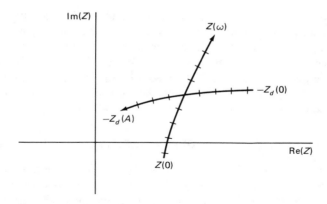

Figure 11.2. Oscillation condition when device load line $Z_d(A)$ intersects circuit impedance $Z(\omega)$. (Kurokawa.[2] Reprinted with permission from The Bell System Technical Journal. Copyright 1969, AT&T.)

A mathematical link must be found between the effect of δA on the device impedance and a change of frequency in the passive circuit. In the usual AC circuit theory, time differentiation corresponds to multiplying by $j\omega$. In this case, time differentiation corresponds to multiplying by $j[\omega + d\phi/dt - j(1/A)(dA/dt)]$. This function is obtained by differentiating (11.2) with respect to time and assuming A and ϕ are time dependent. Finally, expanding the passive circuit impedance in a Taylor series expansion about the resonant frequency ω_o, and substituting (11.4) and (11.5) yields

$$S\delta A + \frac{1}{A_o} \left| \frac{dZ(\omega_o)}{d\omega} \right| \frac{d\delta A}{dt} = 0 \qquad (11.6)$$

where

$$S = \frac{\partial R_d(A_o)}{\partial A} \frac{\partial X(\omega_o)}{\partial \omega} + \frac{\partial R(\omega_o)}{\partial \omega} \frac{\partial X_d(A_o)}{\partial A}. \qquad (11.7)$$

To insure that the solution of this is in the form of a decaying exponential, S must satisfy

$$S > 0. \qquad (11.8)$$

This is Kurokawa's criterion for oscillator stability.

The tuning range of the VCO can be enhanced by increasing the coupling of the varactor to the RF energy. This may be accomplished by lowering the circuit impedance presented to the varactor. There are three penalties for doing this: (1) more power is dissipated in the varactor resistance thereby reducing VCO efficiency, (2) high RF voltage swings on the varactor cause the average capacitance to shift, and (3) self bias caused by rectification reduces the tuning range. Careful design of a VCO thus requires knowing all the circuit impedances illustrated in Fig. 11.1, including the parasitic package elements of the varactor diode.

11.3 PARAMETRIC AMPLIFIER DESIGN THEORY

A second application of varactor diodes is in parametric amplifiers and upconverters in which the varactor becomes the active element that produces a negative resistance. The term parametric is derived by the achievement of amplification, oscillation, upconversion, or harmonic multiplication by choosing a lossless parameter in the system. Mechanical parametric systems were first investigated in the 1830s by Faraday and Lord Rayleigh. In the late 1950s progress in electronic parametric amplifiers was hastened by the development of high quality varactor diodes. A reverse biased varactor diode chip can be accurately modeled as a voltage dependent depletion layer capacitance in series with a small resistance. The addition of diode package elements can be incorporated with the external circuit.

The fundamental relationships between power and frequency for lossless nonlinear reactances were derived by Manley and Rowe[3] in 1956 and have since been named after these two authors. In terms of the pump and signal frequencies, ω_p and ω_s, these are

$$\sum_{m=0}^{\infty} \sum_{n=-\infty}^{\infty} \frac{mP_{mn}}{m\omega_p + n\omega_s} = 0 \qquad (11.9)$$

$$\sum_{m=-\infty}^{\infty} \sum_{n=0}^{\infty} \frac{nP_{mn}}{m\omega_p + n\omega_s} = 0 \qquad (11.10)$$

where P_{mn} is the power flow into the device at frequency $m\omega_p + n\omega_s$. In this formulation the frequencies ω_p and ω_s are either incommensurate or commensurate and incoherent.[4] These relationships are a statement of energy conservation and are useful in understanding the principles of components such as frequency multipliers, reflection parametric amplifiers, frequency upconverters, and frequency downconverters. They also can be used to find the ultimate power gain and conversion efficiency of parametric devices. Although the Manley-Rowe relations illustrate the feasibility of frequency conversion in parametric (nonlinear reactive) devices, they do not contain detailed circuit design information.

11.3.1 Method of Harmonic Balance

Nonlinear circuits can be modeled by a series of nonlinear differential equations that, in general, cannot be solved analytically. However, if a small signal assumption is made, they can be reduced to linear differential equations with time-varying coefficients. The circuit problem may then be solved either by direct solution of the differential equations or by use of the principle of harmonic balance. The latter approach has been the most successful in analysis of parametric circuits. In either case, once the nonlinear differential equations are linearized, analysis of nonlinear effects is no longer possible. These include the ferroresonant effect and higher order instabilities occurring at certain harmonics and fractions of the pump frequency.

For the small signal assumption, the pump voltage (or charge) is much larger than the incoming signal voltage (or charge) or any components at the generated sideband frequencies. Thus the varactor elastance variation is caused solely by the pump source, and the small signal sees a time-varying oscillation at harmonics of the pump frequency. This assumption breaks down when the amplifier begins to saturate, typically when the input signal power > -15 dBm.

The small signal (differential) elastance of a varactor diode can be written in terms of the total charge q_t stored in the diode junction and the total voltage v_t across the junction:

$$S(q_t) = \frac{dv_t}{dq_t}. \tag{11.11}$$

The total applied voltage and charge is the sum of the pump (v_p, q_p) components, plus the signal and sideband components (v'_s, q'_s)

$$v_t = v_p + v'_s$$
$$q_t = q_p + q'_s$$

where $v_p \gg v'_s$ and $q_p \gg q'_s$. Expansion of v_t into a Taylor series yields an incremental voltage-charge relationship

$$v'_s = S(q_p) q'_s \tag{11.12}$$

where

$$S(q_p) = \sum_{m=-\infty}^{\infty} S_m e^{jm\omega_p t}. \tag{11.13}$$

Applying the principle of harmonic balance, two infinite, but equivalent, sets of equations are obtained. One of these sets is written in matrix form.

$$\begin{bmatrix} V_1 \\ V_2^* \\ V_3 \\ V_4^* \\ \vdots \end{bmatrix} = \begin{bmatrix} Z_{11} & \dfrac{jS_1}{\omega_2} & \dfrac{-jS_{-1}}{\omega_3} & \dfrac{jS_2}{\omega_4} & \cdots \\ \dfrac{-jS_{-1}}{\omega_1} & Z_{22}^* & \dfrac{-jS_{-2}}{\omega_3} & \dfrac{jS_1}{\omega_4} & \cdots \\ \dfrac{-jS_1}{\omega_1} & \dfrac{jS_2}{\omega_2} & Z_{33} & \dfrac{jS_3}{\omega_4} & \cdots \\ \dfrac{-jS_{-2}}{\omega_1} & \dfrac{jS_{-1}}{\omega_2} & \dfrac{-jS_{-3}}{\omega_3} & Z_{44}^* & \cdots \\ \vdots & \vdots & \vdots & \vdots & \end{bmatrix} \begin{bmatrix} i_1 \\ i_2^* \\ i_3 \\ i_4^* \\ \vdots \end{bmatrix} \quad (11.14)$$

The notation used in (11.14) is defined as follows:

$$\omega_k = \left[\frac{k}{2}\right]\omega_p + (-1)^{k+1}\omega_s$$

where $k = 1,2,3,\ldots$, $[x]$ is the symbol for the greatest integer less than x, ω_p is the radian pump frequency, ω_s is the radian signal frequency, V_k, i_k and Z_{kk} are the voltages, currents and impedances at ω_k respectively, and $Z_{kk} = -jS_0/\omega_k$ + diode package parasitic elements + circuit impedance. The time origin is ordinarily chosen so that $S_1 = S_{-1}$. This matrix can be written more concisely as

$$\begin{bmatrix} V_1 \\ V_2^* \\ V_3 \\ V_4^* \\ \vdots \end{bmatrix} \begin{bmatrix} Z_{11} & -jX_{12} & jX_{13} & -jX_{14} & \cdots \\ jX_{21} & Z_{22}^* & jX_{23} & -jX_{24} & \cdots \\ jX_{31} & -jX_{32} & Z_{33} & -jX_{34} & \cdots \\ jX_{41} & -jK_{42} & jX_{43} & Z_{44}^* & \cdots \\ \vdots & \vdots & \vdots & \vdots & \end{bmatrix} \begin{bmatrix} i_1 \\ i_2^* \\ i_3 \\ i_4^* \\ \vdots \end{bmatrix} \quad (11.15)$$

where

$$X_{mn} = \frac{-S_k}{\omega_n}$$

and

$$k = (-1)^n \frac{2n-1}{4} - (-1)^m \frac{2m-1}{4}.$$

Two additional assumptions are often invoked to simplify (11.14): (1) S_1 is assumed to be the only significant pumped elastance component, and (2) the open circuit assumptions are employed at the higher sideband frequencies. Most of the nondiagonal terms in (11.14) can be set to zero by means of the first assumption. The open circuit assumption is equivalent to saying that all currents above a given frequency are open-circuited by the external circuit. The simplest analysis of a parametric circuit would assume all currents at frequencies $>\omega_2$ are zero. By this assumption, two currents are left. Since no power is propagaged at the higher frequencies, these voltages may be safely ignored and (11.14) is reduced to a 2 × 2 matrix. In doing numerical calculations these assumptions may be relaxed to incorporate a large number of frequencies which will improve the model accuracy.

The same analysis can be performed to obtain an admittance matrix rather than the impedance matrix of (11.14). The number of equations is reduced by using a short circuit assumption, i.e., voltages above a specified frequency are short-circuited by the external circuit. In general the open circuit assumption is considered superior since the package lead inductance presents a high reactance to the diode chip at high frequencies. If, however, the lead inductance is resonated this argument fails.

11.3.2 Elastance Coefficients

The theory developed thus far can be applied only if the nonlinear characteristic of the varactor diode is precisely known. For a diode heavily doped on one side, and having a doping distribution of

$$N(x) = ax^b \tag{11.16}$$

on the other side, the nonlinear depletion layer elastance is

$$S = S_b(1 - V/\phi)^\gamma \tag{11.17}$$

where $\gamma = 1/(b + 2)$, ϕ is the built-in potential, V is the applied voltage ($V > 0$ for forward bias), and S_b is the zero bias elastance.

One method of finding the harmonic elastance components is to expand (11.17) in a Fourier series by assuming a sinusoidal voltage source and allowing harmonic currents to flow (short circuit assumption). Alternately the expansion can be made by assuming a sinusoidal charge source and allowing harmonic voltages to exist (open circuit assumption). Leeson[6] noticed this expansion could be expressed as a solution of the Laplace integral where the answer was found in terms of gamma functions and the associated Legendre function of nonintegral degree. Oliver[7] has removed the open circuit (or short circuit) assumptions by solving the nonlinear differential equation describing a varactor diode in a particular circuit. He obtained a set of elastance coefficients for (11.13). These procedures give $S_1/S_0 = 0.25$, $S_2/S_0 = 0.15$, $S_3/S_0 = 0.08$, $S_4/S_0 = 0.03$ for $v_p/(\phi - V_b) = 0.9$ where V_b is the varactor breakdown voltage.

11.3.3 Reflection Parametric Amplifier

Parametric amplifiers have found their widest application in very low noise receivers. They can operate at room temperature, but when cooled to cryogenic temperatures, they approach the noise performance of masers. The most common form of electronic parametric circuit is the reflection parametric amplifier circuit (RPA). In this circuit, a high power, high frequency pump signal is reactively mixed with the incoming low frequency signal. The resulting lower sideband is supported in a resonant circuit while the upper sideband signal is terminated in a high impedance. The lower sideband frequency mixes again with the pump signal providing an additional component to the low frequency signal. This double reactive mixing process introduces a 180° phase shift between the incoming and outgoing signal frequency, resulting in the appearance of a negative resistance. The power of course is drawn from the pump source.

Circuit synthesis of a parametric amplifier is greatly simplified by open circuiting all frequencies except ω_p, $\omega_1 = \omega_s$, and $\omega_2 = \omega_{p-s}$. The resulting truncated matrix is:

$$\begin{bmatrix} V_1 \\ V_2^* \end{bmatrix} = \begin{bmatrix} Z_{11} & \dfrac{jS_1}{\omega_2} \\ \dfrac{-jS_{-1}}{\omega_1} & Z_{22}^* \end{bmatrix} \begin{bmatrix} i_1 \\ i_2^* \end{bmatrix} \qquad (11.18)$$

where Z_{11} is the circuit impedance at the signal frequency. This includes the diode bulk series resistance R_s, the average varactor elastance S_0, the diode package and mount parasitic reactances, and the external circuit. But it excludes the signal source impedance. Solving for the input impedance of the RPA yields

$$Z_{in} = V_1/i_1 \qquad (11.19)$$
$$= Z_{11} - \frac{|S_1|^2}{\omega_1 \omega_2 Z_{22}^*} + \frac{jS_1 V_2^*}{\omega_2 Z_{22}^*}.$$

This expression illustrates the origin and magnitude of the negative resistance. The equivalent circuit shown in Figure 11.3 results directly from the above 2

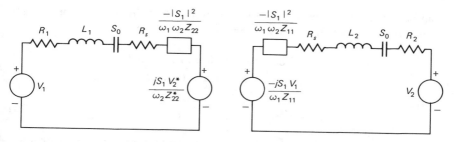

Figure 11.3. Equivalent circuit for the reflection parametric amplifier. *(Kahn[10] © 1971 IEEE.)*

× 2 matrix in (11.18). The power gain is obtained from the reflection coefficient

$$G_{11} = \left| \frac{R_1 - Z_{in}}{R_1 + Z_{in}} \right|^2 \tag{11.20}$$

which at resonance reduces to

$$G_{11} = \left| \frac{R_1 - R_s + \frac{|S_1|^2}{\omega_1 \omega_2 (R_2 + R_s)}}{R_1 + R_s - \frac{|S_1|^2}{\omega_1 \omega_2 (R_2 + R_s)}} \right|^2 \tag{11.21}$$

$$G_{11} = \left(\frac{(x-1)(y+1) + K}{(x+1)(y+1) - K} \right)^2 \tag{11.22}$$

where

$x = R_1/R_s$ normalized signal resistance
$y = R_2/R_s$ normalized idler resistance
$r = \omega_2/\omega_1$ frequency ratio
$Q_{11} = S_{-1}/(\omega_1 R_s)$ dynamic Q at ω_1
$Q_{21} = S_1/(\omega_2 R_s)$ dynamic Q at ω_2
$K \stackrel{\Delta}{=} Q_{11} Q_{21} = Q_{21}^2 r = Q_{11}^2/r.$

The amount of gain is chosen by the pump power (as it affects S_1) and the input matching impedance R_1.

11.3.4 Noise Figure for the RPA

The noise figure of an amplifier is a measure of the degradation of the signal to noise ratio caused by the amplifier. Under certain conditions the noise figure can be expressed as the ratio of the Signal/Noise at the input to the Signal/Noise at the output. A better definition is

$$F = \frac{\text{Total noise power at the output}}{\text{Output noise power due to the source only}}$$

$$F_m = \frac{N_m}{kT_0 \Delta f G_{m1}} \qquad (11.23)$$

The noise power delivered to the load can be analyzed as two components:

1. Noise from sources at $f_r (r \neq m)$ where $m = 1$ for the RPA, and
2. Noise from internal sources of the amplifier at f_m.

If T_a is the ambient temperature and T_d is the temperature of the varactor diode, then the first noise component is

$$N_{mr} = |Y_{mr}|^2 R_{Lm} \{4kB[R_s T_d + R_r T_a]\} \qquad (11.24)$$

where Y_{mr} is the (m,r) component of the inverse of (11.14). The second component is generated at f_m from the resistance $R_{mm} - R_{Lm}$:

$$N_{mm} = |Y_{mm}|^2 R_{Lm} \, 4kB[R_s T_d + R_m T_a - R_{Lm} T_a]. \qquad (11.25)$$

Summing these two contributions gives the total output noise power

$$N_m = N_{mr} + N_{mm}$$
$$= kB \left\{ 4R_{Lm} \left[\sum_{r=1}^{n} |Y_{mr}|^2 (R_s T_d + R_r T_a) - |Y_{mm}|^2 R_{Lm} T_a \right] \right\} \qquad (11.26)$$

where $n - 1$ is the number of sidebands considered. For the lower sideband upconverter considered later, a third noise component must be added to that in (11.26). For the RPA the noise figure is

$$F_1 = \frac{N_1}{kT_a \Delta f G_{11}} \qquad (11.27)$$

that can be reduced to

$$F_1 = 1 + \frac{4x(1+y)(1+y+Q_{21}^2)}{[(x-1)(y+1)+K]^2} \qquad (11.28)$$

$$= 1 + (1 - 1/G_{11}) \left[\frac{Q_{21}^2 + y + 1}{K - (y+1)} \right]. \qquad (11.29)$$

The expression for the gain (11.22) and the noise figure (11.29) provide the basis for design of the RPA. Ordinarily, the design variables are $r = f_2/f_1$, $x = R_1/R_s$ and $y = R_2/R_s$ while the desired performance specifications are given by the noise figure F_1, the gain G_{11}, and the bandwidth Δf. Optimizing the circuit for maximum bandwidth has been treated by Khan[8] and will not be considered further here. Since any desired gain can be achieved by an appropriate choice of source resistance, the optimization criteria will be based on minimizing the noise figure. The gain is chosen at a high value by letting the denominator in (11.22) approach zero.

$$K = (x + 1)(y + 1) \tag{11.30}$$

Putting K in (11.29), the high gain noise figure relationship is

$$F_1 = 1 + \frac{1}{x} + \frac{x+1}{rx}. \tag{11.31}$$

Inspection of this relation shows F is minimum when the normalized source resistance x is maximum. The noise figure must also be subject to the constraint of (11.30). Hence $y = 0$ and $x = K - 1$, and the noise figure reduces to

$$F_1 = 1 + \frac{r^2 + Q_{11}^2}{r(Q_{11}^2 - 1)}. \tag{11.32}$$

The remaining parameter available to the designer is the frequency ratio r. If the noise figure is further optimized with respect to this variable, the optimum r is found to be

$$r = -1 + \sqrt{1 + Q_{11}^2} \tag{11.33}$$

and the resulting minimum noise figure is

$$F_{min} = 1 + \frac{2[1 + \sqrt{1 + Q_{11}^2}]}{Q_{11}^2}. \tag{11.34}$$

If for example the diode quality factor is given as $Q_{11} = 10$, then $F_{min} = 1.22$, $r = 9.05$, $R_1 = 10.05 R_s$, and $R_2 = 0$.

11.3.5 Lower and Upper Sideband Upconverters

The lower sideband upconverter (LSUC) suppresses the upper sideband signal and provides an output at the lower sideband frequency. The LSUC, like the

RPA, is a negative resistance device; however, the gain of this amplifier includes a component due to the ratio of the output frequency to the input frequency.

Within the family of parametric components, a comparison between the widely used RPA and LSUC shows the following differences:

1. Although the noise figures for both devices are identical at infinite gain, they differ for finite gain: the noise figure increases with gain in the RPA, but decreases with gain in the LSUC.
2. For gain below a threshold value, the gain sensitivity to pump amplitude variations is lower in the LSUC since only part of its gain is derived from the negative resistance. Above this gain value, the RPA exhibits greater stability.[9]
3. The gain bandwidth product is larger for the LSUC than for the RPA, by a factor approximately equal to the square root of the output to input frequency ratio.
4. For optimum performance, the RPA requires a circulator at the input port while the LSUC requires an isolator; this provides an advantage to the LSUC since a circulator usually has higher insertion loss than an isolator.
5. The input and output of the RPA are at the same frequency, whereas the output of the LSUC is at a higher frequency. However, in both cases the signal must be downconverted to an IF frequency using a mixer, and the difference in the amplifier output frequency does not appreciably affect the total system performance.

The transducer power gain expression was found in chapter 7 to be the most convenient for use in parametric upconverters since the expression depends on both the load and source resistances. Because both these resistances physically affect the upconverter operation, it is important to have them included in the gain definition. When the transducer power gain is applied to frequency conversion devices it is given by

$$G_{mn} = \frac{\text{average power delivered to the load at } \omega_m}{\text{average available power from the source at } \omega_n}.$$

If the source voltage V_n is conducted through a source resistance R_{gn} and the resulting output current i_m is absorbed by a load resistance R_{Lm}, the transducer power gain is

$$\begin{aligned} G_{mn} &= \frac{|i_m|^2 R_{Lm}}{|V_n|^2/4R_{gn}} \\ G_{mn} &= |Y_{mn}|^2 4 R_{Lm} R_{gn} \end{aligned} \quad (11.35)$$

where Y_{mn} is the (m,n) element of the inverse of the impedance matrix (11.14).

Synthesis of a LSUC to obtain a specified gain presently is limited to using only the signal and lower sideband frequencies. The open circuit assumption is used for all frequencies $>\omega_2$, so the infinite impedance matrix (11.14) is reduced to a 2 × 2 matrix. Since there is only one voltage source V_1, the matrix may be reduced to a lower triangular form even when more than two frequencies are considered. Thus a complete matrix inversion is unnecessary to find Y_{mn}. After some straightforward algebra the gain for the LSUC is

$$G_{21} = \frac{4Q_{11}^2 R_{g1} R_{L2} R_s^2}{|Z_{11}Z_{22}^* - Q_{11}Q_{21}R_s^2|^2} \tag{11.36}$$

and for the upper sideband upconverter (USUC) the gain is

$$G_{31} = \frac{4Q_{11}^2 R_{g1} R_{L3} R_s^2}{|Z_{11}Z_{33} + Q_{11}Q_{31}R_s^2|^2} \tag{11.37}$$

where $Q_{ij} = S_j/(\omega_i R_s)$, R_s is the diode series bulk resistance, and Z_{mn} is a matrix element in (11.14). At midband, these two expressions may be rewritten in terms of the normalized impedances x and y just as was done for the RPA.

$$G_{21} = \frac{4xyrK}{[(1+x)(1+y) - K]^2} \tag{11.38}$$

$$G_{31} = \frac{4xyrK}{[(1+x)(1+y) + K]^2} \tag{11.39}$$

The symbols used in (11.38) and (11.39) for the LSUC and USUC respectively are given below at the midband radian frequencies ω_{10}, ω_{20}, and ω_{30}.

	LSUC	USUC
$x =$	R_{g1}/R_s	R_{g1}/R_s
$y =$	R_{L2}/R_s	R_{L3}/R_s
$K =$	$Q_{11}Q_{21}$	$Q_{11}Q_{31}$
$r =$	ω_{20}/ω_{10}	ω_{30}/ω_{10}

The applications and design criteria of the LSUC and USUC are different. The LSUC is a negative resistance amplifier capable of oscillating, while the USUC is a positive resistance amplifier whose gain is limited to a value $<r$. Since the LSUC is capable of infinite gain, the optimum design is concerned with obtaining a minimum noise figure for a specified gain. The results obtained by Khan[10] for the LSUC are

$$x = \sqrt{1 + rK} \tag{11.40}$$
$$y = 1 - K/(1 + x) \tag{11.41}$$
$$F_{\min} = 1 + 2[1 + \sqrt{1 + Q_{11}^2}]/Q_{11}^2 \tag{11.42}$$

$$G_{21} = \frac{Q_{11}^2(1 + Q_{11}^2)^{1/2}}{[1 + (1 + Q_{11}^2)^{1/2}][1 + (1 + Q_{11}^2)^{1/2} - Q_{11}^2/r]}. \quad (11.43)$$

These values for the load and source resistances (x and y) will give a minimum noise figure and locally maximum gain for a specified frequency ratio r. Khan goes on to find the value of r that would give infinite gain with the corresponding noise figure, as well as giving x and y for maximum gain bandwidth product. These results show that $x > y$ provides broader bandwidth than when $y > x$.

In the positive resistance USUC the noise power developed in the output load is different from that in the negative resistance LSUC. Consequently the noise figure is derived directly from (11.23) through (11.26), and at resonance is given by

$$F_3 = 1 + (1/x)[1 + (x + 1)/Q_{11}^2]. \quad (11.44)$$

The gain is maximum for a given value of normalized source resistance when

$$y = 1 + K/(1 + x) \quad (11.45)$$

and maximum for a given value of normalized load resistance when

$$x = 1 + K/(1 + y). \quad (11.46)$$

Maximum gain occurs when $x = y = \sqrt{1 + K}$ and is given by

$$G_{31} = r\frac{\sqrt{K + 1} - 1}{\sqrt{K + 1} + 1}. \quad (11.47)$$

As expected from the Manley-Rowe relations, the gain is proportional to the frequency ratio r and is degraded by the finite Q of the varactor.

11.4 FREQUENCY MULTIPLIERS

Two important applications for frequency multipliers are highly stable microwave signal sources and alternatives to fundamental frequency millimeter sources. In the first application, the low frequency of a crystal controlled oscillator or surface acoustic wave oscillator provides the fundamental frequency input to the multiplier while the output is taken at the desired harmonic frequency. An example of the second case is an X-band Gunn fundamental oscillator used as the input source. The desired millimeter wave frequency is developed at the output. In all cases, the frequency multiplier will introduce loss, restrict bandwidth, and introduce unwanted frequency harmonics that may in turn produce parametric oscillations. The basic design problem is to provide a single output frequency at the nth harmonic with the greatest possible efficiency and adequate power-handling capability.

11.4.1 The Nonlinear Device

The two devices most commonly used as the nonlinear element in multipliers are the varactor diode and the step recovery diode (SRD). As discussed in chapter 10, the nonlinear depletion layer capacitance of the varactor was found to have a voltage dependence given by

$$C(V) = C_0(1 - V/\phi)^{-\gamma} \qquad (11.48)$$

so that capacitance varies smoothly with the instantaneous applied RF voltage. In contrast, the nonlinearity of the SRD is provided by an abrupt change in diode reactance when all the stored minority carriers are depleted. The efficiency of the varactor multiplier at the nth harmonic has been found to be proportional to $1/n^2$ while that for the SRD multiplier has been found to be proportional to $1/n$. Since the SRD waveform is rich in harmonics and has good efficiency performance, it has largely replaced the older varactor multiplier. The analysis of the varactor diode multiplier has been extensively reviewed by Scanlan[11] and will not be considered further here. In the next section a contrast will be drawn between the varactor and SRD multiplier, and some guidelines for SRD multiplier design will be presented.

11.4.2 Comparison of Varactor and SRD Multipliers

The SRD offers improved performance over the varactor in two respects: (1) higher power-handing capability at a given frequency, and (2) greater linearity of power output with changes in power input.[12] The power capacity is increased by increasing the maximum back voltage which demands a higher resistive material. In the conventional varactor, increasing the resistivity means increasing the series bulk resistance which reduces efficiency. For the SRD, the resistivity near the junction can be made comparatively high. When a forward voltage is applied, the depletion layer spreads into the high resistive intrinsic region near the junction, leaving only the low resistivity regions to contribute to the bulk resistance. Since considerable forward voltage may be applied, more power can be handled without undue dissipation.

The second advantage is a result of the step recovery character of the diode. The harmonic current generation is not dependent on signal level (ΔC is small), but is a function of the incoming waveform and the transition time of the diode. If self-biasing is used, the output current wave remains constant for a large input power range. This is an important feature when a SRD is used in an AM circuit.

C. B. Burckhardt[13] has numerically computed the output power and efficiency for multipliers having $\gamma = 0, 0.33, 0.4,$ and 0.5 with and without idlers. He showed that power-handling capability as well as efficiency increases when the diode is overdriven. This effect is most pronounced with low values of γ as in the SRD.

The primary problem in building higher order varactor multipliers is their need for complicated idler circuits. These idler circuits are necessary because the capacitance voltage curve of a varactor is not sufficiently nonlinear to generate higher order harmonics (>3) with sufficient amplitude. The idler circuits provide a method of upconverting the applied signal to the desired harmonic. The SRD has been sold on its ability to generate the desired harmonic without the need of complicated idler circuits. Burckhardt found for a quadrupler, sextupler, and octupler when $\gamma = 0$ that the efficiency was about as high as a nominally driven abrupt junction 1-2-3-4 quadrupler, 1-2-4-6 sextupler, and a 1-2-4-8 octupler respectively. However, the SRD multipliers had smaller power-handling capacity than their counterparts with idlers. For the quadrupler case, an idler at 2ω improved both the efficiency and power-handling capacity of the SRD multiplier.

Gerald Schaffner[14] has found experimentally that the predictions given by Burckhardt are too high for efficiency and too low for power output. The latter is due to a too conservative assumed overdrive. It is safe to multiply Burckhardt's calculated power by 4.

11.4.3 SRD Physics

When a diode is driven in the forward direction during part of the RF cycle, the charge carriers that cross the junction become minority carriers. These will recombine after a minority carrier lifetime, τ, that depends on the type of doping and host material. If these minority charges are swept back across the junction before they have time to combine, charge in effect has been stored using the diffusion capacitance mechanism. Obviously, diffusion capacitance arises only when the carrier lifetime is greater than the RF period of oscillation. The doping profile of the SRD is fabricated to enhance bunching of the stored charge near the junction. The ideal SRD is characterized by the waveforms in Figure 11.4.[15]

An ideal SRD would have the following properties: (1) long minority lifetime, (2) infinite capacitance in the forward direction, (3) infinitely fast switching time, (4) zero capacitance in the reverse direction, and (5) small package parasitic elements. In practice there must be some compromise between properties 2 and 3, since as stored charge increases, the retarding field decreases and the distribution of minority carriers becomes more diffuse. The more

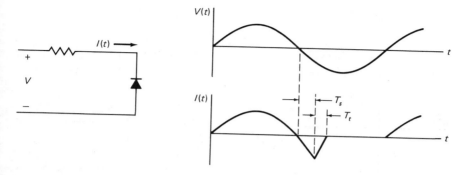

Figure 11.4. Step recovery diode current waveform with storage time T_s and transition time T_t.

charge that is stored, the slower the transition time becomes, resulting in lower multiplier efficiency.

The doping profile for a linearly graded junction varactor and a SRD are shown in Figure 11.5 and the resulting variation in capacitance is shown in Figure 11.6. The capacitance of the SRD is relatively insensitive to voltage, because a low initial voltage spreads the depletion layer considerably. Further increase in voltage has little effect on capacitance.[16]

The charge storage time, T_s, is the portion of the recovery transient during which the conductance remains high. This can be calculated by assuming that all of the charge is pulled out or recombined during this time. The charge and currents must obey the continuity equation

$$\frac{dQ}{dt} = I - \frac{Q}{\tau} \qquad (11.49)$$

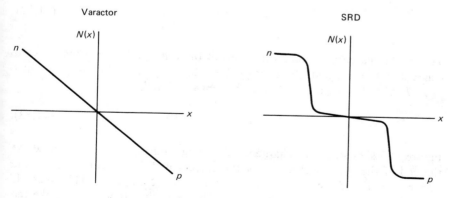

Figure 11.5. Doping profile of a varactor and SRD.

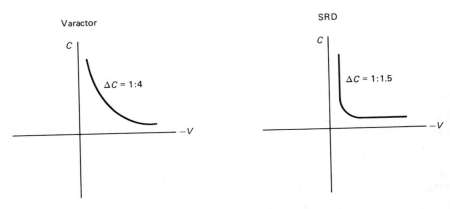

Figure 11.6. Capacitance variation of varactor and SRD.

where I is the conduction current across the junction and Q/τ is the recombination current. Moll et al.[15] estimate the storage time by assuming a rectangular current wave where $I = I_f$ is the forward current that changes to $I = -I_r$ at time $t = 0$. The solution to (11.49) is

$$Q = Ke^{-t/\tau} + I\tau. \qquad (11.50)$$

If at $t = 0^+$, $Q = Q_0$ and $I = -I_r$, then

$$Q = (Q_0 + I_r\tau)e^{-t/\tau} - \tau I, \quad t > 0. \qquad (11.51)$$

Since at $t = T_s$, the stored charge has been depleted, then $Q = 0$. Equation (11.51) may then be solved for T_s.

$$T_s = \tau \ln\left(1 + \frac{Q_0}{\tau I_r}\right) \qquad (11.52)$$

The initial charge Q_0 is found in terms of the time $|T_f| > \tau$ when forward conduction begins again. By solving (11.50) for the two conditions $Q = 0$ at $t = -T_f$ and $Q = Q_0$ at $t = 0$, the charge Q_0 is found.

$$Q_0 = I_f\tau(1 - e^{-T_f/\tau}) \qquad (11.53)$$

Figure 11.7 shows how the charge profile changes during the storage phase. At the end of the storage phase all of the current is diffusion current since the minority carrier density at the junction is approaching zero. This phase is marked by an abrupt end of the current flow. The time required to make the

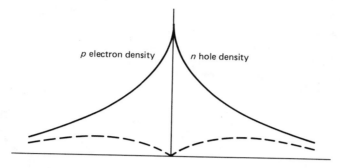

Figure 11.7. Initial charge density (———) and charge density at T_s (---) for the minority carriers.

jump from $i(T_s)$ to $i = 0$ is the transition time T_t that ranges from 100 ps to several nanoseconds. The actual drop of reverse current from $-I_r$ to 0 is complicated but may be approximated by $I_r \exp(-t/T_t)$. Moll et al.[15] derived an approximation for T_t.

$$T_t = \frac{D_p + D_n}{2D_p D_n} \left(\frac{Q_0/A}{4qb(N_0 b)} \right)^{2/3} \tag{11.54}$$

In this expression D_p and D_n are the diffusion constants, A is the diode area, q is the electronic charge, and $N_0 b$ is the impurity gradient at the junction. Large charge storage requires a long transition time and thus less multiplication efficiency. Also, as $N_0 b$ increases, T_t and the reverse breakdown voltage decreases. This analysis has treated the lifetime, τ, as a known constant. However, care must be taken in choosing τ, for it depends on temperature, diode size, charge concentration, junction gradient, resistivity of the semiconductor, and method of preparation.

11.4.4 Analysis of a SRD Multiplier Circuit

As in most nonlinear circuit problems, some compromise must be made between a tractable solution and a completely rigorous solution. In an attempt to bridge this gap, Johnston and Boothroyd[17] have chosen a set of simplifying assumptions that provide closed form expressions for the input and output resistances. From these expressions, the efficiency and power-handling capability are derived. Their analysis of the circuit in Figure 11.8 is subject to the following three limitations: (1) only a fundamental and one harmonic current are applied to the SRD, (2) the charge on the diode changes sign only at $\omega t = \pm m\pi/n$ for a times n multiplier, and (3) the total diode charge changes only

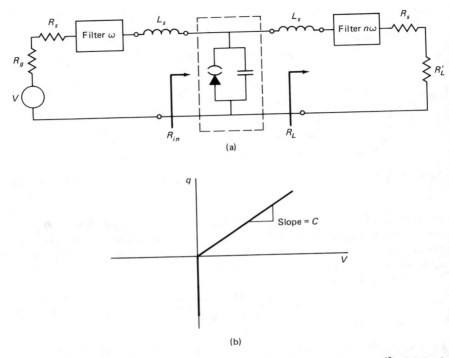

Figure 11.8. Step recovery diode multiplier model of Johnston and Boothroyd[17]: (a) circuit model, and (b) q-V characteristic of nonlinear element. *(Johnston and Boothroyd, "Charge Storage Frequency Multipliers," Proc. IEEE, Vol. 56, pp. 167–176, Feb. 1968.)*

twice per fundamental cycle. From the first limitation, the total charge stored in the diode is

$$q = Q_0 + Q_1 \cos \omega t - Q_n \sin n\omega t. \tag{11.55}$$

The conduction period is defined by m, and in accordance with the second and third limitations, m is a positive integer $<n$. The direct charge Q_0 is adjusted by the bias so that $q(\omega t = m\pi/n) = 0$ or

$$q = Q_1 \left(\cos \omega t - \cos \frac{m\pi}{n} \right) - Q_n \sin n\omega t \tag{11.56}$$

and

$$i = \frac{dq}{dt} = -\omega Q_1 \sin \omega t - n\omega Q_n \cos n\omega t. \tag{11.57}$$

VARACTOR AND STEP RECOVERY DIODES   239

The fundamental voltage across the diode in phase with the fundamental current is found by a Fourier analysis

$$V_1 = \int_{-m\pi/n}^{m\pi/n} -q \sin \omega t \, d\omega t \qquad (11.58)$$

$$= \frac{nQ_n}{C} \left[\frac{\sin(m\pi - m\pi/n)}{n(n-1)\pi} - \frac{\sin(m\pi + m\pi/n)}{n(n+1)\pi} \right] \qquad (11.59)$$

$$V_1 \triangleq \frac{I_n}{\omega C} K_1 \qquad (11.60)$$

where $I_1 = \omega Q_1$ and $I_n = n\omega Q_n$.

Similarly for the nth harmonic voltage

$$V_n = \frac{1}{\pi} \int_{-m\pi/n}^{m\pi/n} -q \cos n\omega t \, d\omega t \qquad (11.61)$$

$$V_n = -\frac{I_1}{\omega C} \left[\frac{\sin(m\pi + m\pi/n)}{\pi(n+1)} + \frac{\sin(m\pi - m\pi/n)}{\pi(n-1)} \right] \qquad (11.62)$$

$$V_n \triangleq -\frac{I_1}{\omega C} K_2. \qquad (11.63)$$

As a result of the conduction and phase angle relationship, V_1 is a function only of I_n, and V_n is a function only of I_1. Moreover, expansion of K_1 and K_2 shows that they are both equal.

$$K = K_1 = K_2 = \frac{-2}{\pi(n^2-1)} \sin \frac{m\pi}{n} \cos m\pi. \qquad (11.64)$$

The harmonic voltage is developed across the resistive load

$$V_n = -R_L I_n \qquad (11.65)$$

while the input resistance is

$$R_{in} = \frac{V_1}{I_1} = \left(\frac{K}{\omega C}\right)^2 \frac{1}{R_L}. \qquad (11.66)$$

Numerical values of K range from 0.2 for a doubler to 0.00025 for an $n = 20$ multiplier. To assure that the diode changes state only once per cycle

240 MICROWAVE SEMICONDUCTOR CIRCUIT DESIGN

$$\left|\frac{dq_n}{dt}\right| \leq \left|\frac{dq_1}{dt}\right| \text{ at } \omega t = \pm m\omega/n. \quad (11.67)$$

This results in

$$|nQ_n \cos m\pi| \leq |Q_1 \sin (m\pi/n)| \quad (11.68)$$

or

$$\frac{I_n}{I_1} \leq \sin (m\pi/n). \quad (11.69)$$

Since R_s is not part of the model but part of the load and generator resistors, the SRD model here is lossless. Therefore, the input power must equal the output power.

$$R_{in}I_1^2 = R_L I_n^2 \quad (11.70)$$

Consequently, the minimum load to input resistance ratio is

$$\frac{R_L}{R_{in}} \geq \frac{1}{\sin^2 (m\pi/n)}. \quad (11.71)$$

The efficiency of the multiplier is degraded by the presence of the series bulk resistance R_s, and it can be calculated in terms of two new resistances.

$$R'_{in} \stackrel{\Delta}{=} R_{in} + R_s \quad (11.72)$$

$$R'_L \stackrel{\Delta}{=} R_L - R_s \quad (11.73)$$

The efficiency is simply the ratio of the power dissipated in the load to the

$$\eta = \frac{R'_L I_n^2}{R_{in} I_1^2} \quad (11.74)$$

$$= \frac{1 - R_s/R_L}{1 + R_s/R_{in}} \quad (11.75)$$

$$= \frac{1 - R_s/R_L}{1 + R_s R_L (\omega C/K)^2}$$

The load resistance for maximum efficiency with respect to the load resistance is found by differentiating η. The optimum load is therefore

$$R_L = \frac{K}{\omega C}\left[\sqrt{\left(\frac{\omega}{\omega_{cn}}\right)^2 + 1} - \frac{\omega}{\omega_{cn}}\right]^{-1} \qquad (11.76)$$

where $\omega_{cn} = K/(R_s C)$ and where R_L must also satisfy the minimum resistance ratio of (11.71). Substituting (11.76) into (11.75) gives

$$\eta = 1 + 2\frac{\omega}{\omega_{cn}}\left[\frac{\omega}{\omega_{cn}} - \sqrt{1 + \left(\frac{\omega}{\omega_{cn}}\right)^2}\right]. \qquad (11.77)$$

The efficiency reaches close to 98% for $\omega/\omega_{cn} = 0.01$ and drops to near zero for $\omega/\omega_{cn} = 5.0$. Thus a large $R_s C$ product degrades the efficiency.

The maximum power conversion is limited by power dissipation, peak current flow, or reverse breakdown voltage. The first limitation is readily found from the efficiency. The second limitation may be increased by placing an external capacitor in parallel with the diode. At lower frequencies, the power limitation is most commonly caused by the diode breakdown voltage which is readily found in terms of the time when the diode charge is maximum. Johnston and Boothroyd found that maximum efficiency occurred when $m = n/2$ and maximum output power when $m \to 0$.

REFERENCES

1. W. Foster and F. A. Myers, "Gunn Effect Technology," *Microwave Systems News*, Vol. 5, pp. 56–64, Oct/Nov. 1975.
2. K. Kurokawa, "Some Basic Characteristics of Broad-Band Negative Resistance Oscillator Circuits," *Bell System Technical Journal*, Vol. 48, pp. 1937–1955, 1969.
3. J. M. Manley and H. E. Rowe, "Some General Properties of Nonlinear Elements: Part I—General Energy Relations," *Proc. IRE*, Vol. 44, pp. 904–913, July 1956.
4. B. D. O. Anderson, "When do the Manley-Rowe Relations Really Hold?", *Proc. IEE*, Vol. 113, pp. 585–587, April 1966.
5. A. P. Boole, "Application of Complex Symbolism to Linear Variable Networks," *IRE Transactions on Circuit Theory*, Vol. CT-2, pp. 32–35, March 1955.
6. D. B. Leeson, "Capacitance and Charge Coefficients for Varactor Diodes," *Proc. IRE*, Vol. 50, p. 1854, August 1962.
7. D. E. Oliver, *Impedance Characteristics of Pumped Varactors*, Report No. 7695-189, prepared for USAEL Contract No. DA28-043-AMC-01870(E), Cooley Electronics Laboratory, The University of Michigan, November 1967.
8. P. J. Khan, "Optimum Design of Varactor Diode Parametric Amplifiers," *Conference on High Energy Frequency Generation and Amplification*, Ithaca, N.Y., pp. 421–437, Aug. 1967.
9. P. J. Khan, *Design of Parametric Lower-Sideband Upconverters using Semiconductor Diodes*, Ph.D. Thesis, University of Sydney, pp. 214–215, October 1963.
10. P. J. Khan, "Optimum Design of Low-Noise Lower-Sideband Parametric Up-Converters," *IEEE Trans. on Electron Devices*, Vol. ED-18, pp. 924–931, October 1971.
11. J. O. Scanlan, "Analysis of Varactor Harmonic Generators," *Advances in Microwaves*, Vol. 2, ed. L. Young. New York: Academic Press, 1967.

12. G. Schaffner, "High-Power Varactor Diodes Theory and Application," *Motorola Application Note AN 147*, Phoenix, Arizona.
13. C. B. Burckhardt, "Analysis of Varactor Frequency Multipliers for Arbitrary Capacitance Variation and Drive Level," *Bell System Technical Journal*, Vol. 44, pp. 75–92, April 1965.
14. G. Schaffner, "Selecting Varactor Diodes," *Motorola Application Note AN 260*, Phoenix, Arizona.
15. J. L. Moll, S. Krakauer, R. Shen, "P-N Junction Charge Storage Diodes," *Proceeding of the IRE*, Vol. 50, pp. 43–53, Jan. 1962.
16. G. Schaffner, "Charge Storage Varactors Boost Harmonic Power," *Electronics*, Vol. 37, pp. 42–47, July 13, 1964.
17. R. H. Johnston and A. R. Boothroyd, "Charge Storage Frequency Multipliers," *Proc. of the IEEE*, Vol. 56, pp. 167–176, Feb. 1968.

Chapter 12
Schottky-Barrier Diode Applications

12.1 SCHOTTKY-BARRIER DEVICE PHYSICS

When a *p*-type semiconductor is joined to an *n*-type semiconductor, the junction that is formed can rectify an AC current and provide a voltage variable capacitance. Similarly, when a semiconductor and metal are joined, a junction is formed that has the same general properties as the *pn* junction. The metal-semiconductor junction is called the Schottky-barrier diode. However, there are some important distinctions between these types of junctions. In comparison to the *pn* junction, the Schottky-barrier diode usually has a steeper forward *IV* slope, lower series resistance, lower forward turn-on voltage, and lower breakdown voltage.[1] However, the Schottky-barrier diode is a majority carrier device and does not have the minority carrier lifetime characteristic found in the *pn* junction. This property, together with its ruggedness and reproducibility, has made it a first choice in most detector and mixer designs.

12.2 ENERGY-BAND DIAGRAMS

The difference between a metal, semiconductor, and insulator was shown in Figure 10.1 to be related to the relative band gap between the valence band and the conduction band. In particular, the Fermi level for the metal is immersed in a continuum of allowed energy states, while in the semiconductor, the Fermi level lies between two bands of allowed states. Before further discussion of the Schottky-barrier, one additional feature needs to be added to the energy diagrams in Figure 10.1. This is the energy level, $\mathcal{E}_0$, needed by an electron to just escape from the surface of the material. The energy difference between the free electron energy and the average electron energy (or Fermi) level is called the work function Φ (Figure 12.1).

$$q\Phi = \mathcal{E}_0 - \mathcal{E}_f \qquad (12.1)$$

Since the Fermi energy $\mathcal{E}_f$ varies with the type and amount of doping in the semiconductor, so does the work function. Another energy quantity, known as the electron affinity $q\chi$, is independent of semiconductor doping. It is defined as the energy difference between the bottom of the conduction band and the free electron energy.

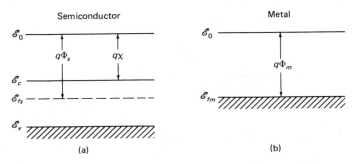

Figure 12.1. Energy band diagram for (a) a semiconductor, and (b) a metal.

$$q\chi = \mathcal{E}_0 - \mathcal{E}_c \qquad (12.2)$$

Since $\mathcal{E}_c = \mathcal{E}_f$ in a metal, the electron affinity in a metal is the same as its work function.

As shown in chapter 10, the average energy of the electrons on both sides of the junction must be the same when two differently doped semiconductors are joined together. Just as the Fermi levels on both sides of the junction line up, so too the Fermi levels on either side of the metal-semiconductor junction line up. This is the first of the three features needed to construct an energy level diagram for the Schottky barrier. The second feature is that the free energy level $\mathcal{E}_0$ must change continuously in going from one side of the junction to the other. If this were not so, energy conservation would be violated. More energy could be extracted from the system than was put into it by appropriate emission and reabsorption of the electrons a short distance away. The third feature is that the electron affinity χ remains constant throughout the semiconductor. In summary, the constraints on the energy level diagram are as follows:

1. $\mathcal{E}_f$ is constant,
2. $\mathcal{E}_0$ changes continuously, and
3. χ is constant.

The resulting energy diagram is shown in Figure 12.2.

The potential barrier that impedes the flow of electrons between the metal and the semiconductor conduction band is the electron affinities of the metal ($= \Phi_m$) and the semiconductor.

$$\Phi_B = \Phi_m - \chi \qquad (12.3)$$

When a p-type semiconductor is used, such as in the low barrier Schottky-barrier detector diodes, then

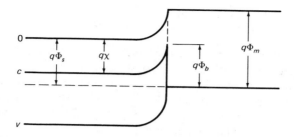

Figure 12.2. Ideal band structure for a Schottky-barrier diode.

$$\Phi_{Bp} = \mathcal{E}_g - q(\Phi_m - \chi) \tag{12.4}$$

where $\mathcal{E}_g$ is the semiconductor energy gap. This result provides a basic understanding of junction effects, but Cowley and Sze[2] have found that better correlation with experimental results are obtained when the presence of surface states is accommodated in the theory for the barrier potential. The details for this modification are found in references 2 and 3.

12.2.1 Reverse Biased Junction

When an n-type semiconductor-metal junction is formed, electrons flow into the metal from the semiconductor. An equilibrium is established between the electrons in the metal and in the semiconductor. The electrons in the metal seek to overcome the barrier height $\Phi_m - \Phi_s$ and recombine with ionized donors in the semiconductor. The electrons in the semiconductor seek to overcome the average thermal energy and spill over into the metal. Hence, there is a cluster of electrons in the metal and a cluster of ionized donors in the semiconductor. The charge separation provides a capacitance just as that found in the pn junction depletion layer capacitance. The Schottky-barrier junction capacitance has the same expression as the one-sided pn junction (10.19).

$$C = \frac{C_0}{(1 - V/\phi)^\gamma} \tag{12.5}$$

Goodman[4] has shown that this capacitance is more correctly modeled as

$$C = \frac{C_0}{\left[1 - \dfrac{V - kT/q}{\phi}\right]^\gamma} \tag{12.6}$$

where kT/q arises from contributions of the mobile carriers to the electric field.

12.2.2 Forward Biased Junction

The theory developed from the energy band structure shown in Figure 12.2 did not include the effects caused by image force barrier lowering. This occurs when an electron, placed a short distance above the metal surface, induces a positive image charge in the metal. Also not included were the effects of surface states and interfacial energy states between the metal and the semiconductor. A more realistic energy level diagram appears in Figure 12.3. It is on the basis of this band picture that the theory for the forward conduction of a Schottky-barrier diode has been developed. The two most widely used theories that describe the forward bias characteristics are the thermionic theory by Bethe[5] and the diffusion theory by Schottky[6]. These two theories differ primarily in the effect the depletion layer, lying between x_m and L, has on the collisions between electrons. According to the thermionic theory, this layer is assumed to be sufficiently narrow so that these collisions can be neglected. The diffusion theory assumes that the electron collisions are so numerous that the electron flow can be described entirely by a diffusion process. Crowell and Sze[7] have developed a theory that incorporates both of these mechanisms. In all three of these theories, the result is the same as the diode equation (10.21) described in Section 10.4.

$$J = J_s(e^{qV/kT} - 1) \tag{12.7}$$

The difference between this equation and (10.21) is the value of J_s. Furthermore, an empirical ideality factor $1 \leq n \leq 2$ is used to better the match between theory and experiment.

$$J = J_s(e^{qV/nkT} - 1) \tag{12.8}$$

In the synthesized theory, Crowell and Sze recognize that the energy bands change so rapidly in the region $0 \leq x \leq x_m$ that the usual diffusion equations do not hold. Instead a thermionic recombination velocity v_R is defined with the assumption that this region acts as a sink for electrons. In the depletin region $x_m \leq x \leq L$ an effective diffusion velocity v_D is defined. These two velocities are

$$v_R = \frac{A^* T^2}{qN_c} \tag{12.9}$$

$$v_D = \left[\int_{x_m}^{L} \frac{q}{\mu kT} e^{-q\phi(x)/kT} \, dx \right]^{-1} \tag{12.10}$$

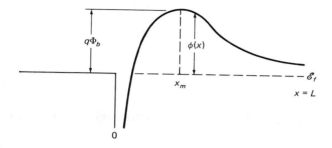

Figure 12.3. Rounding of the electron potential energy because of the ionized donors and the image force.

where A^* is the effective Richardson constant, N_c is the effective density of states in the conduction band, and μ is the electron mobility. The saturation current is then[3,7]

$$J_s = \frac{qN_c v_R}{1 + v_R/v_D} \exp(-q\Phi_B/kT). \quad (12.11)$$

Refinements can be made in the expression for v_R to account for the backscattering of electrons and quantum mechanical tunneling as the electrons attempt to cross the barrier. Refinements in $\phi(x)$ can also be made in the expression for v_D to account for the barrier lowering shown in Figure 12.3. In most room temperature applications $v_R \ll v_D$, and (12.11) reduces to the saturation current obtained by the thermionic theory. Conversely, when $v_D \ll v_R$, the standard Schottky diffusion result is obtained.

12.2.3 Minority Carrier Storage Time

A major advantage Schottky-barrier devices have over other semiconductor devices in detector and mixer designs is their small minority carrier storage time. This is defined as the ratio of the minority carrier charge between the edge of the depletion region L and the end of the epitaxial layer L_e.

$$\tau_s = \frac{1}{J} \int_L^{L_e} qp(x) \, dx \quad (12.12)$$

The minority carrier density at the edge of the depletion region can be given in terms of the forward bias current, the intrinsic semiconductor carrier density n_i, and the donor impurity density N_d.

$$p_n(L) = p_{no}(e^{qV/kT} - 1) \qquad (12.13)$$

$$= \frac{n_i^2}{N_d}(e^{qV/nkT} - 1) \qquad (12.14)$$

$$= \frac{n_i^2}{N_d}\frac{J}{J_s} \qquad (12.15)$$

The minority carrier hole density in the neutral region of the semiconductor as a function of distance is

$$p_n = p_{no} + p_{no}(e^{qV/nkT} - 1)\exp[-(x - L)/L_p] \qquad (12.16)$$

where the equilibrium hole density is p_{no} and the diffusion length is $L_p = \sqrt{D_p \tau_p}$. Under high current conditions the first term in (12.16) is negligible. Furthermore, it is assumed that at the edge of the epitaxial layer, $e^{-(L_e - L)/L_p} \cong 0$. Consequently, when (12.16) and (12.15) is substituted into (12.12), the storage time is obtained.

$$\tau_s = \frac{qn_i^2 \sqrt{D_p \tau_p}}{N_d J_s} \qquad (12.17)$$

The requirements for small storage time are a low resistivity material that is governed by N_d, a low barrier height in order to raise J_s, and a large band gap semiconductor to lower n_i. Sze[3] shows that for a Au-Si diode, the storage time typically falls in the range between 1 and 0.01 ns.

12.3 DETECTOR CIRCUITS

A video detector circuit is used in the front end of a microwave receiver to directly rectify the incoming high frequency signal. If this is an amplitude modulated signal, the video modulating signal or carrier envelope is produced at the output of the detection circuit. Cowley and Sorensen[8] have compared several types of detector circuits using different devices. These include the Schottky-barrier diode, point contact diode, thermionic detector, tunnel diode, backward tunnel diode, and space-charge-limited dielectric diode. In addition to their repeatable characteristics during the manufacturing process, the Schottky-barrier diodes have lower $1/f$ noise and better burnout ratings than the point contact diode. However, there are some important applications where the point contact diode is best suited.

12.3.1 Criteria for Schottky-Barrier Diode Selection

The Schottky-barrier diode may be either a high barrier kind using a metal/ n-type semiconductor junction or a low barrier kind using a metal/p-type semiconductor. The low barrier devices have the higher power sensitivity and can be used to measure power levels as low as -70 dBm. High barrier diodes must be DC biased to attain their maximum sensitivity.

The primary design parameters for a detector circuit are the RF bandwidth which translates to a short rise time in pulse applications, maximum RF sensitivity, good RF input match, low noise, and wide dynamic range. Many of these requirements are conflicting so the final detector design will be a compromise among these parameters.

For low level detection the diode is a square law device. This means the increase in output voltage across an open circuited diode is proportional to the input power,

$$\Delta v = \gamma P \tag{12.18}$$

or the increase in output current through a short circuited diode is proportional to the input power.

$$\Delta i = \beta P \tag{12.19}$$

The actual value of γ or β is dependent on the bias, the load resistance, the signal level and the RF frequency. The quality of a detector depends on the circuit, as well as the diode itself. A diode will exhibit a square law characteristic for input powers of up to approximately -20 dBm. This dynamic range may be increased by raising the bias current, but this will reduce the sensitivity. When the input power exceeds the square law value (i.e., -20 dBm $< P <$ $+20$ dBm), the diode is operating in the linear range. The detector will provide an output in this range, but the measurement system will need to be recalibrated to account for the detected voltage or current rather than the power.

An approximate expression for the current sensitivity is obtained by finding the ratio of the added current to the RF power. When an RF signal is applied to the diode terminals, a DC current plus harmonics of the signal frequency flow through the diode. If the applied RF voltage is

$$v = V_0 + V_1 \cos \omega t \tag{12.20}$$

then substituting (12.20) in to the diode equation (12.8)

$$J = J_s \left[\exp\left(\frac{q}{nkT}(V_0 + V_1 \cos \omega t)\right) - 1 \right] \tag{12.21}$$

will give the harmonic currents. Using the generating function for the modified Bessel function $I_k(z)$, these currents are given by

$$J = -J_s + J_s \exp\left(\frac{qV_0}{nkT}\right) \left[I_0\left(\frac{qV_1}{nkT}\right) + 2\sum_{\nu=1}^{\infty} I_\nu\left(\frac{qV_1}{nkT}\right) \cos \nu \omega t \right] \tag{12.22}$$

The DC current at the output of the video circuit is given by

$$J = J_s[\exp(qV_0/nkT)I_0(qV_1/nkT) - 1] \tag{12.23}$$
$$\cong J_s\{\exp(qV_0/nkT)[1 + \tfrac{1}{4}(qV_1/nkT)^2 + \ldots] - 1\} \tag{12.24}$$

and when the DC bias voltage is zero, this current is proportional to the square of the RF voltage.

$$J \cong \frac{J_s}{4}\left[\frac{qV_1}{nkT}\right]^2 \tag{12.25}$$

The average RF power per unit diode area is obtained by multiplying (12.22) by $V_1 \cos \omega t$ and integrating over one RF period.

$$\bar{P} = \int_0^{2\pi} 2J_s \exp\left(\frac{qV_0}{nkT}\right) I_1\left(\frac{qV_1}{nkT}\right) V_1 \cos^2 \omega t \, d\omega t \tag{12.26}$$

$$= V_1 J_s \exp\left(\frac{qV_0}{nkT}\right) I_1\left(\frac{qV_1}{nkT}\right) \tag{12.27}$$

From (12.23) the incremental current is $J(V_1) - J(0)$.

$$J = J_s \exp\left(\frac{qV_0}{nkT}\right)\left[I_0\left(\frac{qV_1}{nkT}\right) - 1\right] \tag{12.28}$$

Consequently the current sensitivity is

$$\beta = \frac{I_0\left(\dfrac{qV_1}{nkT}\right) - 1}{V_1 I_1\left(\dfrac{qV_1}{nkT}\right)} \tag{12.29}$$

which for small voltages will be of the form

$$\beta = \frac{1 + C_1 V_1^2}{1 + C_2 V_1^2}. \tag{12.30}$$

The RF voltage V_1 across the diode is dependent on the junction resistance R_j. This term is obtained by taking the derivative of the diode current with respect to the DC bias voltage.

$$R_j = \left(\frac{\partial I}{\partial V_0}\right)^{-1} = \frac{nkT}{q(I_0 + I_s)} \quad (12.31)$$

Since (12.31) is inversely proportional to the DC current, the voltage V_1 and consequently β will be dependent on the bias. The voltage sensitivity γ is related to β through the diode video resistance $R_v = R_j + R_s$.

$$\gamma = \beta R_v \quad (12.32)$$

Another often used figure of merit for a detector is the tangential sensitivity (TSS). This is measured by determining the pulse power required to raise the lowest noise peaks in the pulse to the top of the noise peaks on either side of the pulse (Figure 12.4). This measurement is usually done by displaying the detected pulse on an oscilloscope and is somewhat dependent on the judgment of the operator. It also is a function of the detector circuit, bias, and bandwidth of the video amplifier, so comparison of different diodes should be done with the same circuit under similar operating conditions.

The noise equivalent power (NEP) is still another detector characterization that removes the video bandwidth dependence. It is defined as the RF input power required to produce, in a one Hertz bandwidth, an output signal-to-noise ratio equal to one.[8] In terms of the TSS, it is

$$\text{NEP} = \frac{\text{TSS}}{2.5\sqrt{\Delta f}} \quad (12.33)$$

where Δf is the video bandwidth.

12.3.2 Detector Circuit

The video detector circuit must provide a means of matching the RF source to the detector diode, providing a path for the video power to the output, and biasing the diode for optimum operation. The circuit shown in Figure 12.5 is a circuit that provides all three of these functions. At RF frequencies the input choke coil L_c is an open circuit and the output bypass capacitor C_b is a short

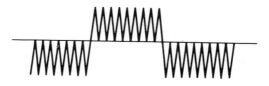

Figure 12.4. Display of a tangential sensitivity measurement.

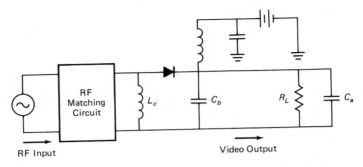

Figure 12.5. Video detector circuit.

circuit. The capacitance C_a represents the input capacitance of the video amplifier that follows the detector. Thus the input circuit is that shown in Figure 12.6, where the equivalent circuit for the diode package has been included. At video frequencies the choke L_c acts as a short circuit thereby isolating the video frequencies from the RF circuit. The bypass capacitor then appears as an open circuit. Thus the equivalent circuit at the video frequency is shown in Figure 12.7.

The RF matching circuit provides an impedance match between the RF source which is usually 50 ohms and the diode junction resistance which typically falls in the range of 1000 to 2000 ohms. Because of the presence of parasitic diode, package and mount reactances, the actual impedance transformation may be somewhat less. However, this improvement in matching requirement comes at the cost of reduced bandwidth. The junction resistance R_j can be lowered by increasing the DC bias current (12.31), but this comes with a sacrifice of some sensitivity.

For broadband, moderate sensitivity applications, a shunt resistance may be all that is necessary to help match the junction resistance. For higher sensitivity, a shunt inductance L_c has been suggested in a Hewlett-Packard Applica-

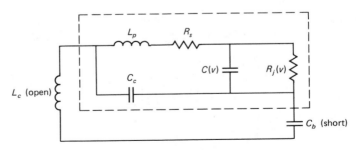

Figure 12.6. RF equivalent circuit of the detector.

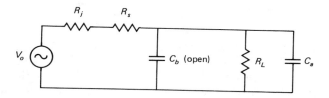

Figure 12.7. Video frequency equivalent circuit of the detector.

tion Note.[9] The shunt inductance can be realized in microstrip or stripline as a short circuited transmission line at right angles to the main transmission line.

12.4 MIXERS

The detector circuit discussed in the previous section is used to convert an RF carrier signal to a slowly varying video signal. The microwave mixer circuit converts an RF signal to a frequency intermediate between the RF and the video signal. This intermediate frequency (IF) is usually chosen to fall in the range of 20 to 100 MHz. The frequency is low enough to build high quality IF amplifiers relatively inexpensively and yet high enough to avoid the flicker noise that is inversely proportional to frequency ($1/f$ noise). The IF frequency is obtained by injecting, in addition to the signal, a local oscillator (LO) or pump frequency ω_p. The difference or IF frequency $\omega_0 = \omega_s - \omega_p$ is a result of the diode nonlinearity (Figure 12.8). Mathematically, the sum and difference frequencies of two sinusoids is obtained when the two sinusoids are multiplied together. The volt-ampere relation of a nonlinear device can be expanded into a power series in the following form.

$$i = a_0 + a_1 v + a_2 v^2 + \cdots \qquad (12.34)$$

Consequently if the input to the nonlinear device consists of a sum of two sinusoids,

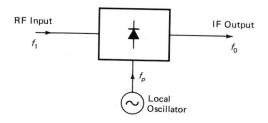

Figure 12.8. Block diagram of a mixer circuit.

$$v = V_p \cos \omega_p t + V_1 \cos \omega_1 t \qquad (12.35)$$

then the resulting current will have components of the form $\cos(n\omega_p + m\omega_1)t$.

The most common choice today for the nonlinear device is the Schottky-barrier diode. The frequency at which the $1/f$ noise becomes excessive is lower for the Schottky-barrier diode than either the *pn* or point contact diodes. Varactor diodes are not used in frequency downconverting because of the high conversion loss predicted by the Manley-Rowe relations.

In a practical mixer, the LO power is much larger than either the RF or IF power, so the infinite number of sidebands predicted by (12.34) is limited to those around the LO (Figure 12.9). A convenient numbering system for these frequencies was given by Saleh.[10]

$$\omega_n = n\omega_p + \omega_0 \qquad (12.36)$$

The frequencies of primary importance in most mixer designs are defined as follows:

ω_p	local oscillator frequency
$\omega_0 = \omega_1 - \omega_p$	intermediate frequency
ω_1	RF signal frequency
$\omega_{-1} = -\omega_p + \omega_0$	image frequency
$\omega_2 = 2\omega_p + \omega_0$	sum frequency

Terminating the circuit with the correct impedance at each of these frequencies can significantly alter its performance.

The primary design concerns for a mixer are to minimize the noise figure and the conversion loss L. The conversion loss is defined as

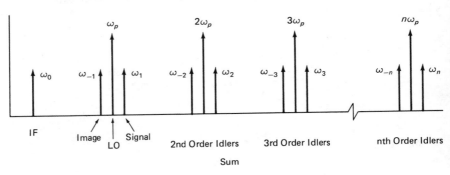

Figure 12.9. Frequency spectrum generated by the mixer.

$$L = \frac{\text{available input signal power, } P_1}{\text{available output IF power, } P_0} \quad (12.37)$$

This definition of conversion loss is dependent on the load at the signal port. Consequently, L is a function of both the circuit and the intrinsic properties of the nonlinear device.

Mixer designs generally fall into three categories: single-ended, balanced, and double-balanced. An equivalent circuit for the single-ended mixer is shown in Figure 12.10. This type usually has only one diode, and it has been extensively analyzed by Torrey and Whitmer.[11] The output is isolated from the input by the shunt tank circuit, but the output contains not only the IF frequency ω_0 but also the local oscillator frequency ω_p, the signal ω_1, the image ω_{-1}, and the sum ω_2. It is usually not difficult to filter out these higher frequencies since $\omega_0 \ll \omega_p, \omega_1, \omega_{-1}, \omega_2$.

The "single-balanced" or simply "balanced" mixers, shown in Figure 12.11, use at least two diodes.[12] Because of the circuit symmetry of the local oscillator excitation, the local oscillator voltage in all these forms does not appear at the output. However, the signal frequency source alternatively sees the output load and a short circuit at the local oscillator rate. Since the average output impedance is not zero at the signal frequency, this frequency is present at the output load.

The "double-balanced" mixers shown in Figure 12.12 are capable of isolating both the signal and local oscillator voltages from the output. Hence, the term double-balanced is used. In the ring circuit, all the diodes are pointed in the same direction, while in the star circuit[13] two diodes point toward the central node and two away from it. During one-half of the local oscillator cycle, half of the diodes are in their high resistance state and half in their low resistance state. During the other half of the local oscillator cycle, the diodes are in the opposite state. Thus the mixer can be considered as a symmetrical switch turning on and off at the local oscillator frequency. On the average, the signal voltage at the output is zero.

This can be seen explicitly by evaluating the insertion loss from the ring equivalent circuit in Figure 12.13. If the diode forward bias resistance is R_f and the reverse bias resistance R_b, the output voltages vary between V_0^+ and V_0^- at a radian frequency of ω_p.

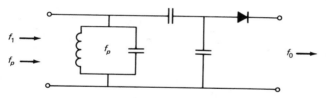

Figure 12.10. Single-ended mixer circuit.

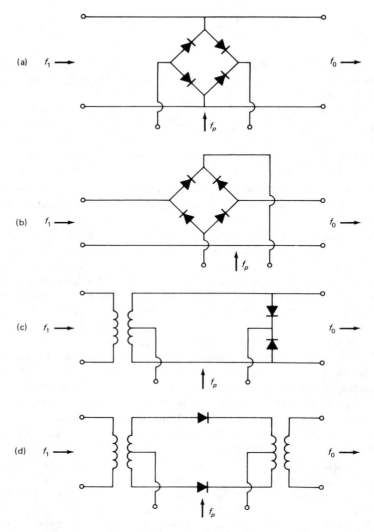

Figure 12.11. Single-balanced mixer circuits: (a) 4-diode shunt circuit, (b) 4-diode shunt circuit, (c) 2-diode shunt circuit, and (d) 2-diode series circuit.

$$V_0^+ = V_1 \cos(\omega_1 t) \frac{(R_b - R_f)R_0}{2R_0R_1 + 2R_bR_f + (R_f + R_b)(R_1 + R_0)} \quad (12.38)$$

$$V_0^- = -V_0^+ \quad (12.39)$$

or,

$$V_0 \stackrel{\Delta}{=} V_1 S(t) \cos(\omega_1 t) \quad (12.40)$$

SCHOTTKY-BARRIER DIODE APPLICATIONS 257

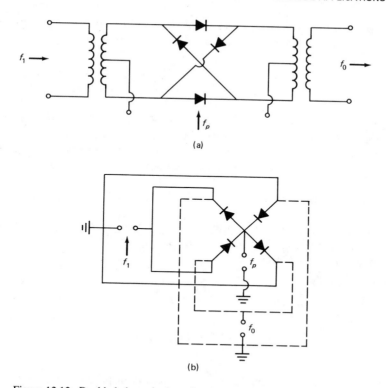

Figure 12.12. Double-balanced mixer circuits: (a) ring mixer, and (b) star mixer.

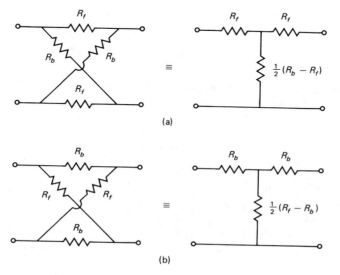

Figure 12.13. Equivalent circuit of a ring mixer in the two states.

The switching function $S(t)$ under ideal conditions would be a squarewave which can be expanded in a Fourier series.

$$S(t) = \frac{4}{\pi} K_1 \left[\cos(\omega_p t) - \frac{1}{3} \cos(3\omega_p t) + \cdots \right] \quad (12.41)$$

Since $S(t)$ contains no DC term, the output voltage as expressed by (12.40) will contain no signal frequency term. By means of circuit balance and symmetrical switching, the star mixer can likewise be shown to provide high isolation between the IF and the two high frequency ports.

12.4.1 Single-Ended Mixers

The analysis of the single-ended mixer given by Torry and Whitmer[11] considered only the signal, IF, and image frequencies. If the voltages and currents developed in the mixer are

$$V(t) = \sum_{n=-\infty}^{\infty} V_n \exp[j(n\omega_p + \omega_0)t] \quad (12.42)$$

$$i(t) = \sum_{n=-\infty}^{\infty} I_n \exp[j(n\omega_p + \omega_0)t] \quad (12.43)$$

then the relationship between the three frequencies of primary interest is given by the following matrix relationship.

$$\begin{bmatrix} V_1 \\ V_0 \\ V^*_{-1} \end{bmatrix} = \begin{bmatrix} z_{11} & z_{12} & z_{13} \\ z_{21} & z_{22} & z_{23} \\ z_{31} & z_{32} & z_{33} \end{bmatrix} \begin{bmatrix} I_1 \\ I_0 \\ I^*_{-1} \end{bmatrix} \quad (12.44)$$

The truncation of the infinite series has been accomplished by assuming all other higher frequencies are open circuited. It is convenient to reduce (12.44)

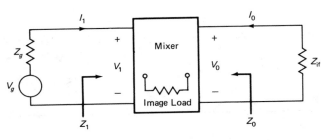

Figure 12.14. Two-port equivalent mixer circuit.

to a 2 × 2 matrix that can be represented by the circuit in Figure 12.14 where the image impedances have been expressed in terms of the RF and IF impedances.

$$\begin{bmatrix} V_1 \\ V_0 \end{bmatrix} = \begin{bmatrix} Z_{11} & Z_{12} \\ Z_{21} & Z_{22} \end{bmatrix} \begin{bmatrix} I_1 \\ I_0 \end{bmatrix} \qquad (12.45)$$

$$Z_{11} = z_{11} - \frac{z_{13}z^*_{13}}{z^*_{11} + z^*_{-1}}$$

$$Z_{12} = z_{12} - \frac{z_{13}z^*_{12}}{z^*_{11} + z^*_{-1}}$$

$$Z_{21} = z_{21} - \frac{z^*_{13}z^*_{21}}{z^*_{11} + z^*_{-1}}$$

$$Z_{22} = r_{22} - \frac{z^*_{21}z^*_{12}}{z^*_{11} + z^*_{-1}}$$

The image impedance is $z^*_{-1} = -V^*_{-1}/I^*_{-1}$, and the output resistance is $r_{22} = Re(z_{22})$. Torrey and Whitmer obtained the conversion loss and optimized it with respect to the signal source impedance Z_g. The optimum conversion loss is

$$L = \left| \frac{Z_{12}}{Z_{21}} \right| \frac{1 + \sqrt{1-\epsilon}}{1 - \sqrt{1-\epsilon}} \qquad (12.46)$$

where

$$\epsilon = \frac{2|Z_{12}Z_{21}|}{2R_{11}R_{22} + |Z_{12}Z_{21}| - Re(Z_{12}Z_{21})} \qquad (12.47)$$

The resulting signal source impedance is

$$Z_g = R_{11}\left\{ 1 - \frac{Re(Z_{12}Z_{21})}{R_{11}R_{22}} - \left[\frac{Im(Z_{12}Z_{21})}{2R_{11}R_{22}}\right]^2 \right\}^{1/2} \\ - j\left\{ X_{11} - R_{11}\left[\frac{Im(Z_{12}Z_{21})}{2R_{11}R_{22}}\right] \right\}. \qquad (12.48)$$

The local oscillator wave shape and the optimum image impedance are the two remaining parameters needed to minimize conversion loss. Power at the image and harmonic sidebands should be reactively terminated to reduce loss. Furthermore, substantially improved conversion loss is obtained if the image impedance is properly adjusted. This occurs when the phase at the IF frequency ω_0 resulting from the mixing of ω_{-1} with ω_p and the mixing of $\omega_p + \omega_0$ with ω_p add constructively.

Saleh extended Torrey and Whitmer's analysis to include not only the Z-mixer described above, but also those he called the Y-, G-, and H-mixers. In the Y-mixer, all the out-of-band frequencies are short circuited. In the H-mixer, all the odd order out-of-band idlers are open circuited and all the even order out-of-band idlers are short circuited. The G-mixer is the dual of the H-mixer. Saleh found the optimum waveform for the Y- and Z-mixers was a squarewave whose duty cycle depended on the ratio of the minimum to maximum resistance of the mixer diodes. For 0 dB conversion loss, this ratio is zero and the waveform is a series of impulses. However, for the G- and H-mixers, the waveform for minimum conversion loss is a squarewave with a duty cycle of ½ for all resistance ratios. Furthermore, for a given diode R_{min}/R_{max}, the G- and H-mixers provide lower conversion loss than the Y- or Z-mixers. The so called image enhanced mixers are ones where the image frequency is either short circuited or open circuited. According to Saleh's analysis, all the other odd idler impedances should be the same. The even idler frequencies should be terminated with the dual termination.

12.4.2 Practical Microwave Mixer Examples

The practical embodiment of a microwave mixer takes on forms quite different from those shown diagrammatically in Figures 12.10–12.12. For example, a single-ended mixer could be implemented as shown in Figure 12.15. The coupler allows both the signal and local oscillator voltages to enter the diode along a common transmission line. The DC blocking capacitance could be implemented in microstrip or stripline with either a chip capacitor soldered to the circuit or with the three-finger coupled line shown in Figure 12.16. In coaxial lines the capacitance in the center conductor can be realized as a quarter wave series stub shown in Figure 12.17. Impedances at the image and sum frequencies can be controlled by removal of the isolator and addition of the appropriate lowpass and bandpass filters.[14]

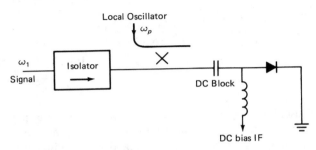

Figure 12.15. Single-ended microwave mixer circuit.

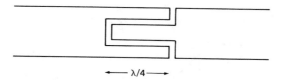

Figure 12.16. DC block in stripline or microstrip.

As shown in Figure 12.18, the single-balanced mixer can be implemented with the aid of 180° hybrid couplers (or magic tee). The signal and local oscillator voltages are added and subtracted in the Σ and Δ arms respectively.

$$v_s = V_p \cos \omega_p t + V_1 \cos \omega_1 t \qquad (12.49)$$
$$v_d = V_p \cos \omega_p t - V_1 \cos \omega_1 t \qquad (12.50)$$

The output of the mixer in the Σ arm is

$$\begin{aligned}v'_s &= k_1 v_s + k_2 v_s + \cdots \\ &= k_1 V_p \cos \omega_p t + k_1 V_1 \cos \omega_1 t + k_2 V_p V_1 [\cos \omega_2 t + \cos \omega_0 t] \quad (12.51)\\ &\quad + k_2 V_p^2 \cos^2 \omega_p t + k_2 V_1^2 \cos^2 \omega_1 t + \cdots\end{aligned}$$

Similarly, the output from the mixer in the Δ arm is

$$\begin{aligned}v'_d &= k_1 V_p \cos \omega_p t - k_1 V_1 \cos \omega_1 t - k_2 V_p V_1 [\cos \omega_2 t + \cos \omega_0 t] \\ &\quad + k_2 V_p^2 \cos^2 \omega_p t + k_2 V_1^2 \cos^2 \omega_1 t \quad (12.52)\end{aligned}$$

These two signals are added and subtracted in a second 180° hybrid coupler to provide outputs to the Σ' and Δ' arms.

$$v''_s = v'_s + v'_d \qquad (12.53)$$
$$v''_d = v'_s - v'_d \qquad (12.54)$$

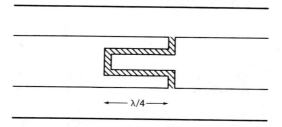

Figure 12.17. DC block in coaxial line.

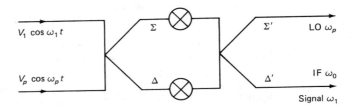

Figure 12.18. Single-balanced microwave mixer.

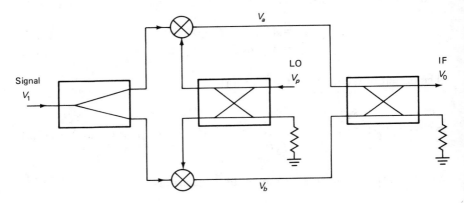

Figure 12.19. Balanced mixer with image rejection.

Clearly, the Σ' arm contains the frequencies ω_p, $2\omega_p$, and $2\omega_1$ while the Δ' arm contains the frequencies ω_0, ω_1, and ω_2. The local oscillator power and the IF power are directed toward different ports on the basis of circuit symmetry.

Another form of the balance mixer is shown in Figure 12.19. The outputs of the mixers V_a and V_b differ because of the 90° phase difference of the local oscillator power going into each mixer. An additional 90° phase shift between V_a and V_b is provided by the second 90° hybrid coupler. Again the output contains the IF and signal voltages while the local oscillator signal is dissipated in the load.

REFERENCES

1. J. C. Irvin and N. C. Vandewal, "Schottky-barrier Devices," in H. A. Watson, ed. *Microwave Semiconductor Devices and their Circuit Applications*. New York: McGraw-Hill, pp. 340–360, 1969.
2. A. M. Cowley and S. M. Sze, "Surface States and Barrier Height of Metal-Semiconductor Systems," *Journal of Applied Physics*, Vol 36, pp. 3212–3220, 1965.
3. S. M. Sze, *Physics of Semiconductor Devices*. New York: John Wiley, 1969.

4. A. M. Goodman, "Metal Semiconductor Barrier Height Measurement by the Differential Capacitance Method-One Carrier System," *Journal of Applied Physics*, Vol. 34, pp. 329–338, 1963.
5. H. A. Bethe, "Theory of the Boundary Layer of Crystal Rectifiers," *MIT Radiation Laboratory, Report 43-12*, 1942.
6. W. Schottky, *Naturwiss*, Vol. 26, p. 843, 1938.
7. C. R. Crowell and S. M. Sze, "Current Transport in Metal-Semiconductor Barriers," *Solid State Electronics*, Vol. 9, pp. 1035–1048, 1966.
8. A. M. Cowley and H. O. Sorensen, "Quantitative Comparison of Solid State Microwave Detectors," *IEEE Transactions on Microwave Theory and Techniques*, Vol. MTT-14, pp. 588–602, Dec., 1966.
9. "Impedance Matching Techniques for Mixers and Detectors," *Hewlett-Packard Application Note 963*, Sep. 1974.
10. A. A. M. Saleh, *Theory of Microwave Mixers*. Cambridge, Massachusetts: MIT Press, 1971.
11. H. C. Torrey and C. A. Whitmer, *Crystal Rectifiers*. New York: McGraw-Hill, 1948.
12. D. G. Tucker, *Modulators and Frequency-Changers*. London: MacDonald and Co., 1953.
13. R. B. Mouw, "A Broad-band Hybrid Junction and Application to the Star Modulator," *IEEE Transactions on Microwave Theory and Techniques*, Vol. MTT-16, pp. 911–918, Nov. 1968.
14. H. E. Elder and V. J. Glinski, "Detector and Mixer Diodes and Circuits," in H. A. Watson, ed., *Microwave Semiconductor Devices and their Circuit Applications*. New York: McGraw-Hill, p. 384, 1969.

Chapter 13
Circuits Using PIN Diodes

13.1 INTRODUCTION

PIN diodes can control the magnitude and phase of large microwave signals with high speed and efficiency. PIN diodes have five characteristics that make them attractive in a number of applications. They are capable of handling large power levels with little change in diode temperature, require little power to activate them, have a response time measured in hundreds of nanoseconds, allow compact circuit design and, like other solid state devices, are highly reliable. The development of PIN diodes has been stimulated by the need for fast and economical phase shifters for phased array antennas. Because of the PIN diode phase shifter, antennas can produce single as well as multiple tracking beams without being hindered by mechanical inertia.

The following section provides a review of the physical concepts pertinent to PIN diodes. Operational limitations caused by the basic device physics are also discussed. Design information for attenuators, switches, limiters is found in sections 13.3 and 13.4. In the last section of the chapter are found designs for four types of phase shifters.

13.2 PIN DIODE OPERATION

The PIN diode is similar to the *pn* diode described in chapter 10 except that an undoped intrinsic region has been inserted between the *p* and *n* regions. This central region is usually a lightly doped *p*-layer (π-layer) or a lightly doped *n*-layer (ν-layer). A zero biased or reverse biased PIN diode can be modeled as a capacitance created by the depletion capacitance of the intrinsic region. When forward biased, the intrinsic region fills with mobile charges so the diode in this state can be modeled as a low resistance. The switching of the diode between a low loss capacitance and a small resistance enables it to control the direction of flow of large microwave power levels.

13.2.1 Reverse Bias Characteristics

The actual PIN diode operation is much more complicated than that described above. If the central region were truly intrinsic, the depletion layer would extend across the full width W of the intrinsic layer at zero and reverse bias

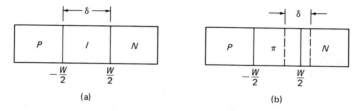

Figure 13.1. The PIN and PπN diode structure.

(Figure 13.1). In the more practical diode structure, the central region is a lightly doped π-layer. The few mobile charge carriers present in the π-layer at zero bias begin to be swept out as the magnitude of the reverse bias increases. This action enlarges the region depleted of charge. When the depletion region encompasses the entire π-layer, the diode has reached its punch-through condition. Any further increase in the reverse bias removes charge carriers from the highly doped p and n regions, and the depleted layer width remains largely unaffected. If the π-layer is made too long, avalanche breakdown will occur before the punch-through condition occurs. However, most PIN diodes manufactured today are made with a punch-through voltage $\cong$ one-tenth the breakdown voltage.[1]

The depletion layer width can be approximated as a one-sided abrupt junction πn^+ diode. From (10.17) the depletion layer width is

$$\delta = \left[\frac{2\epsilon(\phi - V)}{qN_a} \right]^{1/2} \tag{13.1}$$

and the depletion layer capacitance is

$$C_j = \frac{A\sqrt{qN_a\epsilon/2}}{\phi - V}. \tag{13.2}$$

The values for the undepleted intrinsic layer capacitance and resistance are

$$C_i = \frac{\epsilon A}{W - \delta} \tag{13.3}$$

$$R_i = \frac{\rho(W - \delta)}{A} \tag{13.4}$$

where ρ is the material resistivity. Therefore, the reverse biased PIN diode can be represented by the model in Figure 13.2. Ordinarily, R_i and C_i are neglected as they have a minor effect under reverse bias.

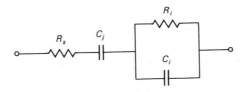

Figure 13.2. PIN diode model.

These calculations, based on the step junction model, fairly accurately predict the value for the capacitance, but give only a qualitative idea of the reverse bias resistance. Senhouse's[2] numerical calculations of a diffused PIN diode model compared much better with experimental results. Because of the inability to know with high accuracy such diode parameters as carrier diffusion length, effective cross sectional area, widths of the various regions, etc., close correlation between his experimental and numerical results was not possible. However, the correlation was much better than that obtained from the step junction model.

Unfortunately, numerical calculations often give little insight into designing PIN diodes. Olson[3] published a noniterative, approximate analytical technique that obtains the diode capacitance and resistance as a function of reverse bias. These calculations are based on a given set of layer thicknesses and carrier concentrations. His approximate technique does provide important insight into PIN diode design and is accurate enough to be within the technological capability of reproducing the design specifications in a physical device.

13.2.2 Forward Bias Characteristics

To understand how the PIN diode is used as a limiter, Leenov[4] and Chaffin[5] calculated the forward bias diode characteristics. The resistance of a forward biased PIN diode was shown to be governed by the number density of mobile charges in the intrinsic region. This is simply the integral of the resistivity over the length of the intrinsic region.

$$R_i = \frac{1}{A} \int_{-W/2}^{W/2} \frac{dx}{\sigma(x)} \tag{13.5}$$

where the conductivity is

$$\sigma(x) = q(\mu_n n(x) + \mu_p p(x)). \tag{13.6}$$

CIRCUITS USING PIN DIODES

The problem reduces to finding the functional relationship of the carrier density with distance. The fundamental equations (10.6)–(10.9), repeated below, provide a basis for finding the mobile charge densities.

$$\mathbf{J}_p = q(\mu_p p \mathbf{E} - D_p \nabla p) \tag{10.6}$$

$$\mathbf{J}_n = q(\mu_n n \mathbf{E} + D_n \nabla n) \tag{10.7}$$

$$-\nabla \cdot \mathbf{J}_p = q\left(\frac{\partial p}{\partial t} + \frac{p - p_n}{\tau_p}\right) \tag{10.8}$$

$$\nabla \cdot \mathbf{J}_n = q\left(\frac{\partial n}{\partial t} + \frac{n - n_p}{\tau_n}\right) \tag{10.9}$$

Simplification of these expressions results from the nearly zero electric field in the heavily injected intrinsic region. Furthermore, the time derivatives are zero because of the assumed DC steady-state bias. For the electron carriers, (10.7) may be substituted into (10.9) to yield

$$\frac{\partial^2 n}{\partial x^2} = \frac{n - n_p}{\tau_n D_n} \tag{13.7}$$

or

$$\frac{\partial^2 n}{\partial x^2} = \frac{n}{L^2} \tag{13.8}$$

where in the heavily injected region $n \gg n_p$, and the diffusion length L is $\sqrt{\tau_n D_n}$. The solution is

$$n(x) = K_1 e^{x/L} + K_2 e^{-x/L}. \tag{13.9}$$

If the PIN diode is symmetrical, then it may be assumed that $p(x) = n(x)$. Consequently, at the center of the intrinsic region the total carrier density will reach a minimum.

$$\frac{\partial}{\partial x}[p(x) + n(x)]\bigg|_{x=0} = 2\frac{dn(x)}{dx}\bigg|_{x=0} = 0 \tag{13.10}$$

Furthermore, $K_1 = K_2$ and the number density becomes

$$n(x) = 2K_1 \cosh(x/L). \tag{13.11}$$

Making use of the boundary condition at the interface of the intrinsic and n regions where $x = W/2$, the constant K_1 is eliminated.

$$n(x) = \frac{n(W/2) \cosh(x/L)}{\cosh(W/2L)} \qquad (13.12)$$

Solution for $n(x)$ in terms of the DC bias current is found by recognizing that the total current anywhere in the intrinsic region is the same. At $x = W/2$ the hole current is assumed to be zero so that

$$D_p \frac{\partial p}{\partial x}\bigg|_{x=W/2} = \mu_p E p(W/2) \qquad (13.13)$$

$$= \mu_n E n(W/2) \qquad (13.14)$$

where use is made of the assumption that $p(x)$ and $n(x)$ have the same functional relationship in a symmetrical device. The total DC current is found from (10.7).

$$I_o = J_n A = Aq\left(\mu_n n E + D_n \frac{\partial n}{\partial x}\right)\bigg|_{x=W/2} \qquad (13.15)$$

$$= \frac{2AqD_n n(W/2) \sinh(W/2L)}{L \cosh(W/2L)}$$

The electron density may be obtained in terms of the DC bias current.

$$n(x) = \frac{I_o L \cosh(x/L)}{2AqD_n \sinh(W/2L)} \qquad (13.16)$$

The indicated integration in (13.5) may be performed using (13.16) and the conductivity $\sigma = 2q\mu_n n(x)$.

$$R_i = \frac{2D_n \sinh(W/2L)}{\mu_n I_o} \text{Arctan}[\sinh(W/2L)] \qquad (13.17)$$

An often quoted approximate relation for R_i is obtained by assuming $W/2L \ll 1$.

$$R_i = \frac{W^2}{2\mu_n I_o \tau_n} \qquad (13.18)$$

Thus, the forward bias resistance of the diode is independent of the diode area, proportional to the square of the intrinsic region width, and inversely proportional to the diode current. The capacitance C_i can be estimated from (13.3) by letting $\delta = 0$. Ordinarily under forward bias, C_j can be neglected.

In switching applications the charge injected by the RF power must not exceed the charge injected by the DC bias. An approximation for the RF charge can be calculated if the RF current is assumed to be sinusoidal, and that positive charge injection occurs during the positive half cycle.

$$q_{RF} = \int_0^{T/2} I_{RF} \sin \omega t \, dt$$
$$= I_{RF} T/\pi \quad (13.19)$$

The DC charge is approximately

$$q_{DC} = I_o \tau \quad (13.20)$$

where τ is the carrier lifetime. The DC bias current must exceed the effects of the RF modulation.

$$I_o > \frac{I_{RF} T}{\pi \tau} \quad (13.21)$$

For a 100 W signal in a 50 ohm transmission line, where the PIN diode carrier lifetime is 2 μs, the DC current must exceed 0.225 mA. A practical switch design, however, should use a bias current that exceeds this minimum amount by a substantial margin.

13.2.3 PIN Diode Limitations

Since the PIN diode is often used as a switch or attenuator, two diode parameters of primary interest are its power-handling capability and switching speed. When the diode is biased in either one of its two states, it reflects most of the power or allows most of it to pass by unaffected. Thus, the power-handling capability of the diode switch is much larger than the power dissipated in the diode. Ultimately the power-handling capability is determined by the temperature rise of the diode itself as long as the voltage is lower than the breakdown voltage. Mortenson[6] has evaluated diode temperature rise when microwave energy is dissipated in the diode under both forward and reverse bias conditions. For the forward biased diode, he found that thermal dissipation occurs

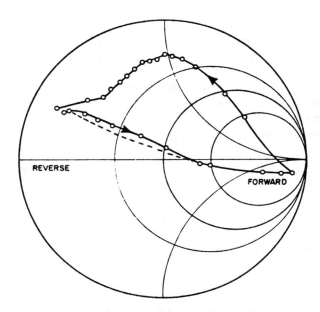

Figure 13.3. Reflection coefficient for PIN diode. Solid line = transient measurement (points spaced 200 ns). Broken line = static measurement. *(Galvin and Uhlir[7] © 1964 IEEE.)*

throughout the intrinsic region. For the reverse biased diode, most of the heat is generated near the PI or IN junction planes. For RF pulses shorter than approximately 2 μs, the thermal response for the reverse biased diode is faster than the forward biased case. However, for longer pulses the added temperature drop in the intrinsic region makes the forward biased diode hotter. Consequently, in controlling high power levels, the pulse width becomes a factor in determining whether a shunt or series switch is superior.

Switching speed is also an important parameter in the design of many circuits using PIN diodes. Defining switching speed is complicated by its dependence on both the diode and the driver circuit. The switching speed is related to how quickly the charge in the intrinsic region is removed or injected. Galvin and Uhlir[7] have shown graphically the hysteresis effect in going from forward to reverse or from reverse to forward bias (Figure 13.3). When the diode is switched from the reverse to forward biased state, the charge carriers begin to flow in the intrinsic region almost immediately. However, in going in the other direction (forward to reverse), charge removal occurs first at the PI and IN junctions. This leaves the central region filled with charge and surrounded on either side by depletion regions. During this time, the PIN diode is modeled as two depletion admittances in series (Figure 13.4).

Switching speed is increased when the forward current is as small as practical and a large reverse current spike is applied to the diode to speed discharge

CIRCUITS USING PIN DIODES 271

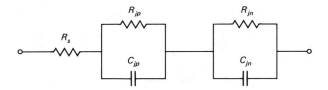

Figure 13.4. Equivalent circuit of PIN diode during the forward to reverse bias transient.

of the intrinsic layer. Fast switching is enhanced by the choice of a diode with a small cross sectional area, thin doping layers, and a short carrier lifetime. Therefore, high switching speed is achieved at the expense of the power-handling capability, so that a compromise between these characteristics must be made.

13.3 PIN DIODE ATTENUATORS

When a fast-acting electrically variable attenuator is required, PIN diodes are often used. The resistance of a PIN diode may be varied over the range from approximately 0.1 to 10^9 ohms. Since the diode resistance varies inversely with the forward bias current (13.18) and directly with the reverse applied voltage, the resistance value can be controlled with a suitably designed bias source. Exploiting this variable resistance requires the isolation of the DC supply from the RF circuit and accommodating the parasitic diode package and mount reactances in the design.

The basis of an attenuator design is the lumped symmetrical tee circuit described in chapter 2 section 2.3 (Figure 13.5). For the matched, symmetrical circuit it was shown that

$$\alpha = \frac{V_2}{V_1} \tag{13.22}$$

$$R_1 = \frac{2R_0(1 - \alpha)}{1 + \alpha} \tag{13.23}$$

$$R_2 = \frac{R_0^2 - R_1^2/4}{R_1}. \tag{13.24}$$

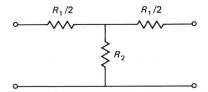

Figure 13.5. Symmetrical T equivalent circuit for an attenuator.

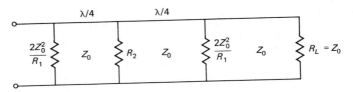

Figure 13.6. Realization of the T attenuator in a high frequency microwave circuit.

At lower microwave frequencies an attenuator can be made by putting PIN diodes in place of resistors in the symmetrical T circuit, but at higher frequencies the distance between diodes becomes an appreciable fraction of a wavelength. Furthermore, a series mounted diode often presents heat-sinking problems at elevated power levels. The diodes can be separated and all mounted in shunt by introducing a quarter wavelength transmission line or other impedance inverter between each of the diodes. Such a circuit is shown in Figure 13.6. It may be easily verified that input impedances of the circuit in Figure 13.6 are Z_0 when R_1 and R_2 are given by (13.23) and (13.24). Required resistance values for several levels of attenuation in a 50 ohm line are given in Table 2.1.

Several of the more detailed practical design aspects are covered in the literature. Hunton and Ryals[8] consider the diode-mounting structure and its consequences on bandwidth, while Parris[9] describes a 3 GHz attenuator that exhibits very low phase shift with changes in frequency.

13.4 SWITCHES AND LIMITERS

The high power-handling capability of the PIN diode has made this device the best choice for most switching circuits and power limiters where the ultimate switching speed is not required. The most desirable switch would be one that was a short circuit under forward bias and an open circuit under reverse bias. Because of the diode junction capacitance, parasitic package reactance, and mount reactance, it is better to think of the PIN diode switch as a device that provides two different reactances for the two-bias states. The circuit design must incorporate these reactances to provide the desired high isolation in the off state and low insertion loss in the on state.

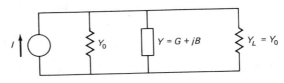

Figure 13.7. Shunt-mounted switch with admittance Y.

The PIN diode may be mounted in shunt across the transmission line or in series with the transmission line. Most practical switches use the shunt configuration. This is because the typical package parasitic reactances of the shunt-mounted switch is less deleterious to bandwidth than the series mounted switch[10] and because of the superior heat-sinking in the shunt-mounted position.

The insertion loss through a line with a shunt-mounted switch is defined as the ratio of the available power from the source to the power delivered to the load (Figure 13.7). The available power is the maximum power that could be delivered to the load if the shunt admittance $Y = 0$. This power is

$$P_a = |I|^2/8Y_0 \qquad (13.25)$$

where the I is the peak current from the AC source. The actual power delivered to the load is

$$P_L = \frac{|I|^2 Y_0/2}{(2Y_0 + G)^2 + B^2} \qquad (13.26)$$

and the transducer power loss is

$$\frac{P_a}{P_L} = (1 + G/2Y_0)^2 + (B/2Y_0)^2. \qquad (13.27)$$

The minimum insertion loss is 0 dB if both the shunt conductance and susceptance are zero. Since there are always parasitic susceptances, these must be tuned out to achieve minimum insertion loss. The PIN diode with its package can be modeled as shown in Figure 13.8 for the reverse and forward bias states. At low frequencies the diode is capacitive when reverse biased and inductive when forward biased, while at high frequencies the opposite occurs. The reverse bias susceptance can be tuned out with an added shunt transmission line that

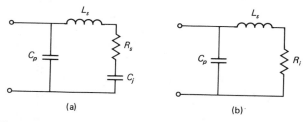

Figure 13.8. The equivalent circuit for a packaged *PIN* diode under (a) reverse bias and (b) forward bias.

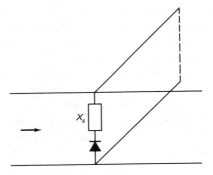

Figure 13.9. Circuit to tune out reactance of *PIN* diode switch.

is either open or short circuited. In the forward bias case a capacitance or inductance may be added in series with the diode to tune it (Figure 13.9).

A PIN diode limiter is often used to protect a sensitive receiver from a nearby high power transmitter. A sudden application of high RF energy will inject electrons and holes into the intrinsic region of the diode causing it to have a low resistance. This causes the high power to be reflected back. Higher isolation can be achieved with an external bias source.[4] In this case the imminent arrival of a high power pulse must either be predicted from a known timing cycle or sensed by a lightly coupled detector so as to turn the PIN diode bias source on at the appropriate time. Typically two diodes can provide up to 40 dB isolation with only 10 mA of bias current. The incident pulse energy required to burn out a typical PIN diode switch is approximately

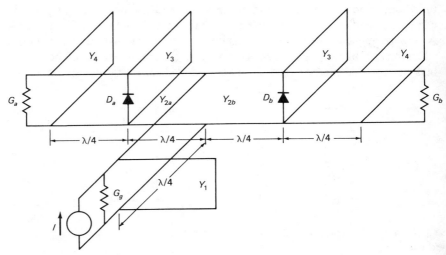

Figure 13.10 Circuit for SPDT switch. (*Fisher, R. E.*,[11] © 1965 IEEE.)

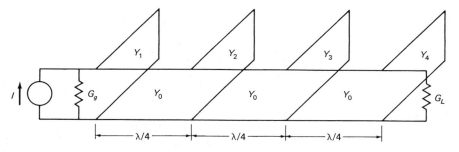

Figure 13.11. Quarter wave coupled bandpass filter.

38 MW-μs. It can handle up to 500 W of continuous power or up to 100 kW of peak power.[10]

A PIN diode switch is used in a wide number of applications to control the destination of a microwave signal. One example is the single-pole, double-throw switch (SPDT), which directs the signal into one of two output loads. These loads are depicted in Figure 13.10 as G_a and G_b. Fisher[11] showed that a SPDT switch could be designed using the maximally flat quarter-wave coupled bandpass filter (Figure 13.11). The design for the filter circuit was worked out by Mumford.[12] His technique is especially useful in this application, because all the through lines have the same characteristic admittance Y_0. When, for example, diode D_a is forward biased, and diode D_b is reverse biased, then the Y_{2a} line is a short circuit stub. This becomes the second stub in the bandpass filter with the termination G_b. For the symmetrical switch $Y_{2a} = Y_{2b}$ which in turn must equal Y_0. These lines must function both as a shorted stub and a through line, depending on which way the diodes are biased. Consequently, Mumford's tables are searched for the case where $Y_2 \cong Y_0$.

For example, a symmetric filter design in a 50 ohm transmission line system, using four stubs from Mumford's tables, has $Z_1 = 1/Y_1 = 125$ ohms and $Z_2 = 1/Y_2 = 45.09$ ohms (rather than 50 ohms). The theoretical input impedance of this filter is shown in Figures 13.12 and 13.13. Replacing the 45 ohm value for Z_2 with 50 ohms would closely approximate the exact design. An alternative approach is to force $Z_2 = 50$ ohms and use a computer optimization program to vary the other three stubs for best match across a given bandwidth. For a 60% bandwidth it is found that $Z_1 = 70.72$, $Z_2 = 50.00$, $Z_3 = 49.11$, and $Z_4 = 78.03$ with the resulting filter response shown in Figure 13.14 and 13.15.

The second aspect of the switch design is to find the appropriate value for the Y_3 in Figure 13.10, so that the stub plus the diode closely resemble a shorted stub required by the filter design. Fisher[11] equated the susceptance slopes at f_o of the stub, loaded with the capacitance of a reverse biased diode, with the prototype stub Y_3 (Figure 3.16). If v is the velocity of the electromagnetic wave in the transmission medium, then

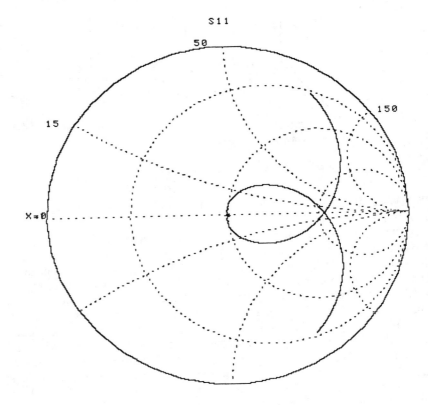

Figure 13.12. Input impedance of filter designed by Mumford's tables.

$$\left.\frac{\partial B}{\partial \omega}\right|_{\omega_o} = C + \frac{Y_a d_a}{v} \csc^2\left(\frac{\omega_o d_a}{v}\right) = \frac{Y_3 d_3}{v} \csc^2\left(\frac{\omega_o d_3}{v}\right) \quad (13.28)$$

Since at resonance $\omega_o d_3/v = \pi/2$, the cosecant term on the right can be set to unity. Also, at center frequency, stubs in Figure 13.16 must have zero input susceptance.

$$\frac{\omega_o C}{Y_a} = \cot\left(\frac{\omega_o d_a}{v}\right) \quad (13.29)$$

Substituting (13.29) into (13.28) yields the desired value of Y_3.

$$Y_3 = \frac{2}{\pi}\left\{\omega_o C + Y_a \operatorname{Arctan}\left(\frac{\omega_o C}{Y_a}\right)\left[1 + \left(\frac{\omega_o C}{Y_a}\right)^2\right]\right\} \quad (13.30)$$

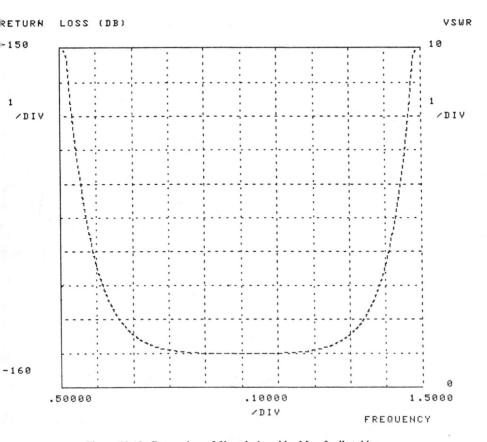

Figure 13.13. Return loss of filter designed by Mumford's tables.

Solution of this expression for Y_a in terms of the known capacitance, frequency, and Y_3 requires the use of a numerical procedure such as Newton's method. For example, with a center frequency of 10 GHz, a diode capacitance of 0.2 pF, and $Y_3 = 1/49.11 = 0.0204$ mhos, the value of Y_a is $1/78.00 = 0.01282$ mhos. For combinations of C and Y_3 that give no solution to (13.29), White and Mortenson[13] have suggested using either series or shunt stubs at the input junction to improve the range of realizable admittances in the the two output arms.

13.5 DIODE PHASE SHIFTERS

The introduction of phased array antennas with the capability of beam-steering and multiple beams have provided a major application of phase shifter and time delay networks. The ideal phase shifter network shifts the phase of the transmitted signal the same amount at all frequencies. The ideal time delay

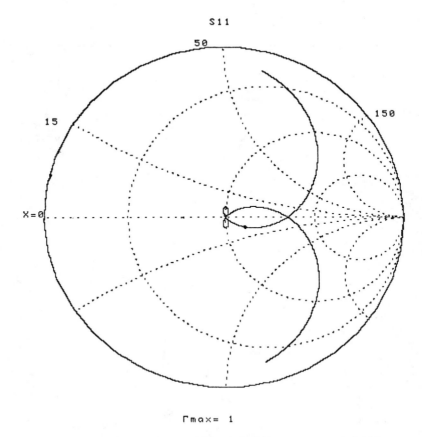

Figure 13.14. Input impedance of filter designed by computer optimization.

network delays the arrival of the signal at the output a specified amount of time. Both types of circuits can be used for beam-steering. For pulsed systems with short pulse lengths, the transit time effects may demand the use of the more complex time delay circuit. Since no diode will alternate between a perfect short and perfect open circuit, the time delay circuit can be approximated only over a limited frequency range.

The phase shifter or time delay circuit can be implemented with either a transmission or reflection type circuit. The transmission circuits can be realized as a (1) switched line circuit, (2) a lowpass/highpass circuit, or (3) a loaded line circuit. The reflection circuit can be realized with either a circulator or a 3 dB coupler. The relative merits of these four basic circuits illustrated in Figure 13.17 are discussed below.

The switched line circuit is the one usually specified when a time delay func-

CIRCUITS USING PIN DIODES 279

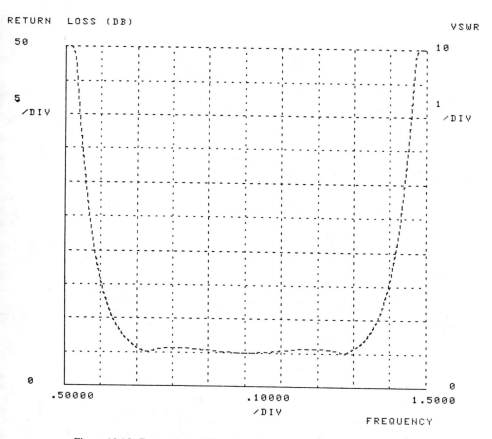

Figure 13.15. Return loss of filter designed by computer optimization.

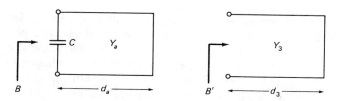

Figure 13.16. Equivalent circuit of stub in SPDT switch.

280 MICROWAVE SEMICONDUCTOR CIRCUIT DESIGN

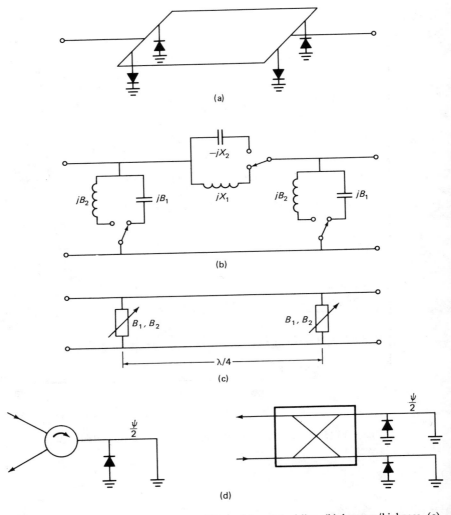

Figure 13.17. Four types of diode phase shifters: (a) switched line, (b) lowpass/highpass, (c) loaded line, and (d) circulator or hybrid coupler reflection circuit.

tion is needed. However, when a true phase shift rather than a time delay is needed, the path lengths of both lines are made equal. The phase shift is accomplished with the addition of a passive coupled line phase shifter known as a C-section. In the switched line circuit, the diode contribution to the insertion loss is the same for both bias states, and phase shifts up to 180° per bit is readily achieved. However, its disadvantages include the need for four diodes for each bit, and complementary bias voltages for each bit. Also, the insertion loss for all bits less than 180° are the same as the 180° bit. For the other circuits in Figure 13.17, the diode loss decreases with reduced phase shift.

The lowpass/highpass circuit can provide up to 180° phase shift with very broad bandwidth. It is similar to the loaded line phase shifter where the quarter wavelength transmission line is replaced by a lumped reactance. The lumped elements in this design make it useful primarily up to UHF frequencies where the compact design inherent in lumped elements could not be achieved with transmission lines. A disadvantage of this approach is the high frequency limitation caused by the lumped elements. In addition this circuit requires three diodes per bit, and at least one of these must be series mounted creating a heat-sinking problem for high power applications.

The loaded line phase shifter is capable of handling more power than any of the other designs in Figure 13.17. This circuit requires only two diodes per bit and both of these can be shunt-mounted to provide maximum heat-sinking. They are, however, restricted to phase shifts of up to approximately 22½° per bit by the bandwidth. Thus 30 diodes would be required to achieve 180° phase shift in 22½° increments.

The reflection circuit using either a circulator or coupler provides less insertion loss per degree of phase shift than any of the other circuits. It uses the minimum number of diodes per bit, and any phase shift up to 180° can be achieved. Also the input match depends on the coupler or circulator rather than the diode switch itself. However, the power-handling capability of the coupler type phase shifter is less than that obtained for the loaded line phase shifter.

13.5.1 Switched Line Phase Shifter

The switched line phase shifter requires the use of two SPDT switches such as those discussed in section 13.4. The primary design problem occurs in making the switches and calculating the desired line lengths in each of the two arms. Conversion of the time delay circuit in Figure 13.17(a) to a true phase shifting circuit is achieved by adding a C-section to one of the two arms (Figure 13.18). The analysis of a C-section given by Cristal[14] shows that the phase shift is the ratio of the odd to even part of the polynomial F.

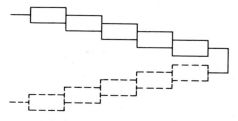

Figure 13.18. Passive C-section phase shifter constructed from a cascade of coupled lines.

$$\beta = \frac{F_o(S)}{jF_e(S)}\bigg|_{S=j\tan\theta} = \frac{-B_1 \tan\theta(1 - B_3^2 \tan^2\theta) \cdots}{(1 - B_2^2 \tan^2\theta) \cdots (1 - B_n^2 \tan^2\theta)} \quad (13.31)$$

The positive real coefficients satisfy the following conditions.

$$1 > B_1 \quad (13.32)$$
$$B_2 > B_3 > \cdots > B_n.$$

The input impedance of an analogous stepped impedance transformer terminated with an open circuit is

$$Z = -\frac{F_e(S)}{F_o(S)}. \quad (13.33)$$

From this function the even mode characteristic impedances may be extracted using Richard's theorem. By setting $Z_{0e}/Z_0 = Z(1)$, a reduced impedance function Z' is left.

$$Z' = Z(1) \frac{Z - SZ(1)}{Z(1) - SZ} \quad (13.34)$$

This process is repeated until all the even mode impedances are extracted. The odd mode characteristic impedances are found from $Z_{0o}/Z_0 = Z_0/Z_{0e}$ and once these are known the C-section can be implemented in any one of several kinds of transmission line media. The design lies basically in choosing the appropriate set of B_i elements. For a single-section 90° phase shifter Cristal uses $B_1 = \frac{2}{3}$ which gives $Z_{0e}/Z_0 = \frac{5}{2}$. This results in a phase shifter with a bandwidth of approximately 2:1.

13.5.2 Lowpass/Highpass Circuit

Garver[16] described a phase shifter that alternately switches between a lowpass and highpass filter characteristic (Figure 13.17(b)). This circuit is analyzed using the ABCD matrix defined in Figure 13.19.

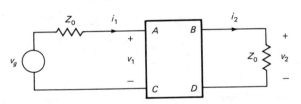

Figure 13.19. The ABCD matrix representation of a 2-port circuit.

CIRCUITS USING PIN DIODES 283

$$\begin{bmatrix} v_1 \\ i_1 \end{bmatrix} = \begin{bmatrix} A & B \\ C & D \end{bmatrix} \begin{bmatrix} v_2 \\ i_2 \end{bmatrix} \quad (13.35)$$

The voltage transmission coefficient is obtained from the ratio of the power delivered to the load to the available power from the source.

$$S_{21} = \sqrt{\frac{v_2^2/2Z_0}{v_g^2/8Z_0}} \quad (13.36)$$

In terms of the ABCD parameters defined in (13.35),

$$S_{21} = \frac{2}{A + B/Z_0 + CZ_0 + D}. \quad (13.37)$$

The circuit switches from a lowpass state ($i = 1$) to a highpass state ($i = 2$) in such a way that the passbands of both states include the desired phase shifter bandwidth. For the cascade of a shunt susceptance B_i, series reactance X_i, and a second shunt susceptance B_i, the total ABCD matrix is

$$\begin{bmatrix} A & B \\ C & D \end{bmatrix} = \begin{bmatrix} 1 & 0 \\ jB_i & 1 \end{bmatrix} \begin{bmatrix} 1 & jX_i \\ 0 & 1 \end{bmatrix} \begin{bmatrix} 1 & 0 \\ jB_i & 1 \end{bmatrix} \quad i = 1,2$$

$$= \begin{bmatrix} 1 - B_iX_i & jX_i \\ jB_i(2 - B_iX_i) & 1 - B_iX_i \end{bmatrix} \quad (13.38)$$

In terms of these susceptances and reactances the transmission coefficient is

$$S_{21} = \frac{2}{2(1 - B_iX_i) + j[X_i/Z_0 + B_iZ_0(2 - B_iX_i)]} \quad (13.39)$$

and the transmission phase is

$$\theta_i = -\text{Arctan}\left[\frac{X_i/Z_0 + 2B_iZ_0 - B_i^2Z_0X_i}{2(1 - B_iX_i)}\right] \quad i = 1,2. \quad (13.40)$$

The phase shift is easily found since switching from state 1 to state 2 changes the sign of the B_i and X_i.

$$\psi = \theta_1 - \theta_2 = -2\,\text{Arctan}\left[\frac{X_i/Z_0 + 2B_iZ_0 - B_i^2X_iZ_0}{2(1 - B_iX_i)}\right] \quad (13.41)$$

At the center frequency of the ideal phase shifter, $|S_{21}| = 1$. Using this in (13.39) the value for the series reactance is obtained.

$$X_i = \frac{2B_i Z_0^2}{1 + B_i^2 Z_0^2} \qquad (13.42)$$

Substitution of this into (13.41) gives the expression for phase shift with zero insertion loss.

$$\psi = 2 \text{ Arctan} \left[\frac{2B_i Z_0}{Z_0^2 B_i^2 - 1} \right] \qquad (13.43)$$

The synthesis problem of obtaining the B_i and X_i from the required ψ is easily solved from (13.42) and (13.43).

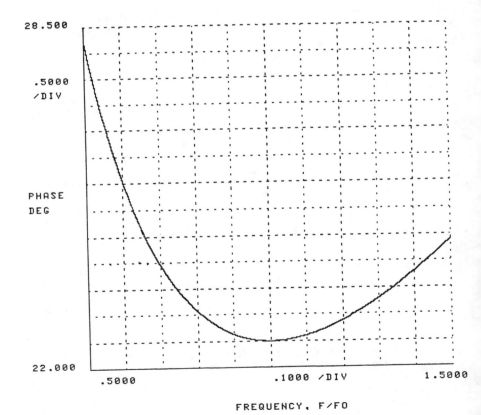

Figure 13.20. Theoretical response of a 22.5° lowpass/highpass phase shifter.

$$B_i = \frac{1}{Z_0} \tan (\psi/4) \qquad (13.44)$$

$$X_i = Z_0 \tan (\psi/2) \qquad (13.45)$$

Analysis of the lowpass/highpass phase shifter using ideal switches and lumped capacitors and inductors is shown in Figures 13.20 and 13.21 for a 22½° and 45° design. If the bandwidth is defined where the phase error is 2.5°, then the bandwidth of the 22½° and 45° phase shifters are 106% and 64% respectively. Parasitic reactances of practical circuit elements and diodes will reduce the achievable bandwidth below these values.

A dual of the π circuit of Figure 13.17(b) is a T circuit consisting of one switchable shunt susceptance B_i between two series switchable reactances X_i. The solution of the T circuit is the same as (13.39)–(13.45) except all X_i/Z_0 and B_i/Y_0 are interchanged.

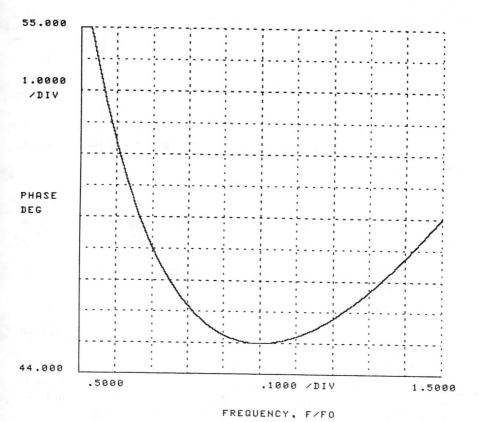

Figure 13.21. Theoretical response of a 45° lowpass/highpass phase shifter.

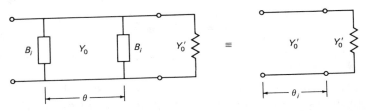

Figure 13.22. Equivalence of loaded line phase shifter with a matched transmission line.

13.5.3 Loaded Transmission Line Circuit

The loaded line phase shifter can be constructed by replacing the series reactance element of the lowpass/highpass circuit with a transmission line that is usually a quarter wavelength long. The analysis is based on equating the ABCD matrices of the loaded line phase shifter with an equivalent transmission line of characteristic admittance Y_0' (Figure 13.22). Only the A and B elements of these two circuits are needed, as the C and D elements are redundant.

$$A = \cos\theta - B_i/Y_0 \sin\theta = \cos\theta_i' \tag{13.46}$$

$$B = \frac{\sin\theta}{Y_0} = \frac{\sin\theta_i'}{Y_0'} \tag{13.47}$$

By eliminating the line length of the equivalent circuit θ_i' between (13.46) and (13.47), a solution for the desired characteristic admittance of the loaded line circuit is found.

$$Y_0' = Y_0[1 - (B_i/Y_0)^2 + 2(B_i/Y_0)\cot\theta]^{1/2} \qquad i = 1,2 \tag{13.48}$$

The line length of the equivalent transmission line for the two diode states is then

$$\theta_i' = \text{Arccos}\,[\cos\theta - (B_i/Y_0)\sin\theta] \qquad i = 1,2 \tag{13.49}$$

and the phase shift is

$$\psi = \theta_1' - \theta_2'. \tag{13.50}$$

The expressions (13.48) and (13.50) represent a set of three equations and four unknowns (θ, Y_0, B_1, B_2) that can be solved in terms of Y_0' (chosen to insure that SWR = 1) and the phase shift ψ. Obviously, the solution for B_1, B_2, and Y_0 will have to be a function of θ. First (13.48) is solved for B_i.

$$B_1 = Y_0 \cot \theta + [Y_0'^2 \csc^2 \theta - Y_0^2]^{1/2} \tag{13.51}$$
$$B_2 = Y_0 \cot \theta - [Y_0'^2 \csc^2 \theta - Y_0^2]^{1/2} \tag{13.52}$$

The trigonometric identity

$$\tan(\theta_i'/2) = \sqrt{\frac{1 - \cos \theta_i'}{1 + \cos \theta_i'}} \tag{13.53}$$

is substituted into

$$\tan(\psi/2) = \frac{\tan(\theta_1'/2) - \tan(\theta_2')}{1 + \tan(\theta_1'/2)\tan(\theta_2'/2)}. \tag{13.54}$$

The resulting expression can be simplified by making use of the product, sum, and difference of (13.51) and (13.52).

$$B_1 B_2 = Y_0'^2 - Y_0^2 \tag{13.55}$$
$$B_1 + B_2 = 2Y_0' \cot \theta \tag{13.56}$$
$$B_1 - B_2 = 2[Y_0'^2 \csc^2 \theta - Y_0^2]^{1/2} \tag{13.57}$$

Consequently, (13.54) reduces to

$$\tan(\psi/2) = \frac{[1 - (Y_0'/Y_0)^2 \sin^2 \theta]^{1/2}}{(Y_0'/Y_0) \sin \theta} \tag{13.58}$$

or

$$\sec(\psi/2) = (Y_0'/Y_0) \csc \theta. \tag{13.59}$$

Substituting this back into (13.51) and (13.52) gives the solution for the required shunt susceptances in terms of the desired phase shift and θ.

$$B_1 = Y_0' [\sec(\psi/2) \cos \theta + \tan(\psi/2)] \tag{13.60}$$
$$B_2 = Y_0' [\sec(\psi/2) \cos \theta - \tan(\psi/2)] \tag{13.61}$$

These equations provide theoretically perfect match and zero phase error for a lossless diode with any shunt susceptance spacing. If the shunt susceptance is a single, lumped reactance, then the maximum bandwidth as a function of θ may be calculated (Figures 13.23 and 13.24). For this ideal case, the bandwidth as a percent of the center frequency drops from 45% to 12% when θ

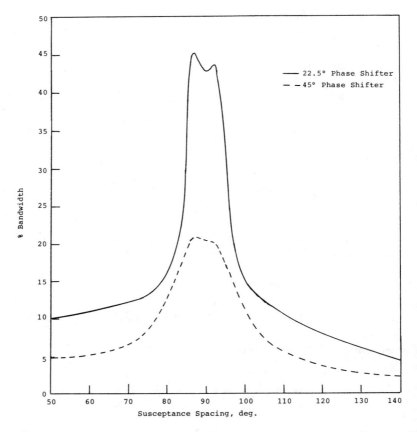

Figure 13.23. Bandwidth of phase shifter as a function of susceptance spacing θ (*Davis, W. A.,*[17] © 1974 IEEE.)

decreases from 90° to 75°. The bandwidth is defined as the range of frequencies where SWR < 1.2 and the phase error < 2.5°. When minimum phase error and SWR are required at midband, clearly $\theta = 90°$ is the best choice.[17]

An actual PIN diode switches between two susceptances B_{d1} and B_{d2} that are in general different from the required susceptances B_1 and B_2 in (13.60) and (13.61). However, a transmission line of characteristic admittance Y_0'' and of length ϕ can be made to provide the required B_i from the given diode B_{di} (Figure 13.25). From the transmission line equation, the desired susceptance for the phase shifter is

$$jB_i = Y_0'' \frac{j(B_{di} + Y_0'' \tan \phi)}{Y_0'' - B_{di} \tan \theta} \qquad (13.62)$$

CIRCUITS USING PIN DIODES   289

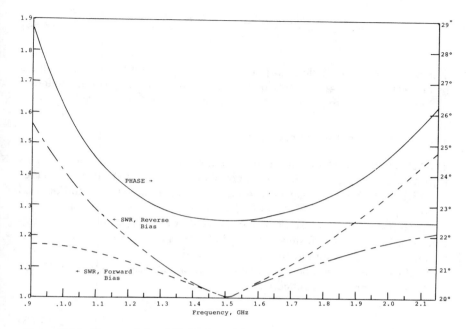

Figure 13.24. Characteristics of 22.5° phase shifter with $\theta = 90°$ *(Davis[17] © 1974 IEEE.)*

The line length, which is the same for both diode states, is found in terms of the diode and required susceptances.

$$\tan \phi = \frac{Y_0''(B_i - B_{di})}{Y_0''^2 - B_i B_{di}} \quad i = 1,2 \tag{13.63}$$

Equating this function for $i = 1$ to the same function for $i = 2$ and solving for the characteristic admittance gives

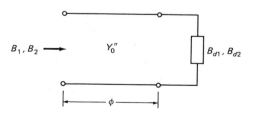

Figure 13.25. Circuit used to obtain desired values of B_i for loaded line phase shifter with diode susceptance B_{di}, $i = 1,2$.

$$Y_0''^2 = \frac{B_1 B_{d1}(B_2 - B_{d2}) - B_2 B_{d2}(B_1 - B_{d1})}{B_2 - B_1 + B_{d1} - B_{d2}}. \quad (13.64)$$

When the measured diode susceptances are B_{di}, the transmission line defined by (13.63) and (13.64) will provide the desired B_i. If these values are found to be impractical, a cascade of two transmission lines may be required.

13.5.4 Reflection Type Phase Shifter

The reflection type phase shifter must isolate the input from the output. This is most commonly done with either a circulator or a 3 dB hybrid coupler (Figure 13.17(d)). A switch that provides a perfect open and short could be used in the manner shown in Figure 13.26(a) to give a time delay circuit with a phase shift ψ. Burns and Stark[18] have shown that any phase shift can be realized by using the two transmission lines indicated in Figure 13.26(c). This latter circuit will be treated at the end of this section.

A *PIN* diode switches between two low loss impedances, as shown in Figure 13.26(b), and provides a reflection coefficient of

$$\Gamma = \frac{jX_i - Z_0}{jX_i + Z_0}. \quad (13.65)$$

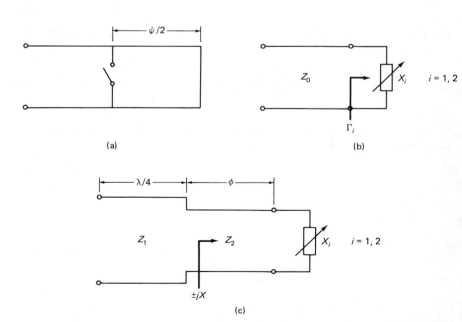

Figure 13.26. Terminating switches for the reflection phase shifter: (a) ideal switch, (b) switch between lossless reactances, X_1, X_2, and (c) switch two transmission lines.

CIRCUITS USING PIN DIODES 291

The reflection phase angle is

$$\theta_i = -2 \text{ Arctan } (X_i/Z_0) \qquad i = 1,2 \qquad (13.66)$$

The phase shift between the two diode states is

$$\psi = 2 [\text{Arctan } (X_1/Z_0) - \text{Arctan } (X_2/Z_0)]. \qquad (13.67)$$

If both reactance states are assumed to be either two values of capacitance or two values of inductance, then the condition for frequency independent phase shift is

$$\frac{\partial \psi}{\partial \omega} = \frac{X_1/Z_0}{1 + (X_1/Z_0)^2} - \frac{X_2/Z_0}{1 + (X_2/Z_0)^2} = 0 \qquad (13.68)$$

or

$$\frac{X_1}{Z_0} = \frac{Z_0}{X_2}. \qquad (13.69)$$

Substituting (13.69) into (13.67) provides a solution of X_1 for the desired phase shift.

$$X_1 = Z_0 [\tan (\psi/2) \pm \sec (\psi/2)] \qquad (13.70)$$

The numerical values of X_1 and X_2 given in Table 13.1 show clearly the difficulty of realizing a 180° phase shifter using this method. However, if one reactance state is an inductance and the other a capacitance, then the sign of X_1 will be opposite to that of X_2. In this case the frequency independent condition is

$$\frac{X_1}{Z_0} = -\frac{Z_0}{X_2}. \qquad (13.71)$$

Table 13.1. Reactance Values for Reflection Phase Shifter.

ψ	+ SIGN		− SIGN	
	X_1/Z_0	X_2/Z_0	Z_1/Z_0	X_2/Z_0
135°	5.027	0.199	−0.199	−5.027
90°	2.414	0.414	−0.414	−2.414
45°	1.497	0.668	−0.668	−1.497
22.5°	1.219	0.821	−0.821	−1.219

Substitution of (13.71) into (13.67) shows that $\psi = 180°$. The $180°$ phase shifter is thus best achieved with a capacitance-inductance switching circuit. The two-step transmission line transformer shown in Figure 13.26(c) is an alternative approach to achieving arbitrary phase shift. The design of the first transformer section is based on the dual of (13.63) and (13.64) where the two desired reactances are of equal magnitude but of opposite sign. The input impedance of the quarter wavelength section is then

$$jX_i = Z_1^2/(\pm jX) \qquad i = 1,2 \qquad (13.72)$$

and the required value of Z_1 is obtained by substituting this into (13.67).

$$Z_1 = \sqrt{XZ_0 \tan(\psi/4)} \qquad (13.73)$$

The actual value of X is arbitrary, and it can be chosen for ease of realizability. If the diode is considered an ideal switch as was done by Burns and Stark,[18] then from (13.63) and (13.64) $Z_2 = |X|$ and $\phi = \pi/4$. Furthermore, they chose $Z_2 = Z_0$ so that (13.73) reduces to*

$$\frac{Z_1}{Z_0} = \sqrt{\tan(\psi/4)}. \qquad (13.74)$$

However, as was seen above, the assumption of an ideal switch is not necessary to make use of their basic idea.

REFERENCES

1. K. E. Mortenson and J. M. Borrego, *Design Performance and Applications of Microwave Semiconductor Control*. Dedham, Mass: Artech House, 1972, pp. 29–31.
2. L. S. Senhouse, "Reverse Biased *PIN* Diode Equivalent Circuit Parameters at Microwave Frequencies," *IEEE Trans. on Electron Devices*, Vol. ED-13, pp. 305–314, March 1966.
3. H. M. Olson, "Design Calculations of Reverse Bias Characteristics for Microwave *p-i-n* Diodes," *IEEE Trans. on Electron Devices*, Vol. ED-14, pp. 418–428, August 1967.
4. D. Leenov, "The Silicon *PIN* Diode as a Microwave Radar Protector at Megawatt Levels," *IEEE Trans. on Electron Devices*, Vol. ED-11, pp. 53–61, February 1964.
5. R. J. Chaffin, *Microwave Semiconductor Devices: Fundamentals and Radiation Effects*. New York: John Wiley and Sons, 1973, pp. 182–209.
6. K. E. Mortenson, "Analysis of the Temperature Rise in *PIN* Diodes Caused by Microwave Pulse Dissipation," *IEEE Trans. on Electron Devices*, Vol. ED-13, pp. 305–314, March 1966.
7. R. Galvin and A. Uhlir, "Transient Microwave Impedance of *PIN* Switching Diodes," *IEEE Trans. on Electron Devices*, Vol. ED-11, p. 441, September 1964.

---

*The expression in the reference is apparently in error by a power of ½.

8. J. K. Hunton and A. G. Ryals, "Microwave Variable Attenuators and Modulators Using PIN Diodes," *IRE Trans. On Microwave Theory and Techniques,* Vol. MTT-10, pp. 262–273, July 1962.
9. W. J. Parris, "*P-I-N* Variable Attenuator with Low Phase Shift," *IEEE Trans. on Microwave Theory and Techniques,* Vol. MTT-20, pp. 618–619, September 1972.
10. M. R. Barber, K. F. Sodomsky, and A. Zacharias, "Microwave Switches, Limiters and Phase Shifters," *Microwave Semiconductor Devices and Their Circuit Applications.* ed. H. A. Watson. New York: McGraw-Hill, chapter 10, 1969.
11. R. E. Fisher, "Broadbanding Microwave Diode Switches," *IEEE Trans. on Microwave Theory and Techniques,* Vol. MTT-13, p. 706, September 1965.
12. W. W. Mumford, "Tables of Stub Admittances for Maximally Flat Filters Using Shorted Quarter-Wave Stubs," *IEEE Trans. on Microwave Theory and Techniques,* Vol. MTT-13, pp. 695–696, September 1965.
13. J. F. White and K. E. Mortenson, "Diode SPDT Switching at High Power with Octave Microwave Bandwidth," *IEEE Trans. on Microwave Theory and Techniques,* Vol. MTT-16, pp. 30–36, January 1968.
14. E. G. Cristal, "Analysis and Exact Synthesis of Cascaded Commensurate Transmission-Line *C*-Section All-Pass Networks," *IEEE Trans. on Microwave Theory and Techniques,* Vol. MTT-14, pp. 285–291, June 1966.
15. E. G. Cristal, "Correction," *IEEE Trans. on Microwave Theory and Techniques,* Vol. MTT-14, pp. 495–499, October 1966.
16. R. V. Garver, "Broad-Band Phase Shifters," *IEEE Trans. on Microwave Theory and Techniques,* Vol. MTT-20, pp. 314–323, May 1972.
17. W. A. Davis, "Design Equations and Bandwidth of Loaded Line Phase Shifters," *IEEE Trans. on Microwave Theory and Techniques,* Vol. MTT-22, pp. 561–563, May 1974.
18. R. W. Burns and L. Stark, "PIN Diodes Advance High-Power Phase Shifting," *Microwaves,* Vol. 11, pp. 38–48, November 1965.

Chapter 14
Avalanche Devices and Circuits

14.1 INTRODUCTION

When a *pn* junction diode is reverse biased beyond its breakdown voltage, a reverse current will flow that is limited only by the external circuit. The sudden increase in reverse current is a result of the generation of a large number of charge carriers in the semiconductor from the avalanche mechanism. In 1958 W. T. Read[1] published a classic paper describing how a microwave oscillator or amplifier could be made using the avalanche process in a specially doped diode. This profile became known as the Read profile. In 1964 after a long research effort, Johnston, DeLoach, and Cohen[2] fortuitously found a *pn* junction diode fabricated 10 years previously that produced the long sought microwave oscillations. Only a short time later, the first Read profile device was operated.[3] These devices have come to be known as IMPATT diodes (IMPact ionization Avalanche Transit Time), and the story of their early development has been retold by DeLoach.[4] Their technological importance comes from their ability to produce more microwave power than any other semiconductor device at this time at frequencies as high as 300 GHz.

This chapter begins with a qualitative description of the operation of an IMPATT diode. Section 14.3 provides an extended treatment of the quasistatic approximation of the large signal analysis of the Read type IMPATT diode. This is followed by the small signal analysis of a Read diode. The choice of this order of presentation avoids the subtle error in Read's original work that results in a factor of 2 instead of 3 in the equation for the avalanche current. Section 14.5 through 14.8 briefly discusses some practical factors that must be considered in using an IMPATT diode. These are temperature effects, spurious oscillations, noise, and efficiency. A brief discussion of TRAPATT diodes is found in Section 14.9. The final sections provide design information for two commonly used power-combining circuits suitable for IMPATT diodes.

14.2 IMPATT DIODE OPERATION

A microwave oscillator must somehow transfer energy from an external source such as a DC power supply to a microwave field. When charge carriers are moving in the same direction that the external field is pushing them, these carriers acquire energy at the expense of the field. If the carriers are moving in the direction opposed by the field, the carriers give up energy to the field. This condition is established for the DC field in an oscillating IMPATT diode for half of the RF cycle. In this way energy is transferred from the DC field to the AC field. During the other half of the cycle when the AC and DC fields are in the same direction, the supply of charge carriers is cut off. This avoids transferring energy from the AC to the DC field in the second half of the RF cycle. Providing the mechanism for turning on and off the charge carriers at the right time was the problem addressed by Read[1] in his proposed diode-doping profile (Figure 14.1).

Each end of the Read profile is a heavily doped ohmic contact that minimizes the series resistance between the semiconductor and the external wire. When the junction between the n^+ and p regions is reverse biased beyond breakdown potential, an avalanche of electrons and holes is created in the region between 0 and x_a. The remainder of the p region from $x = x_a$ to $x = b$ and all of the intrinsic region is called the drift region.

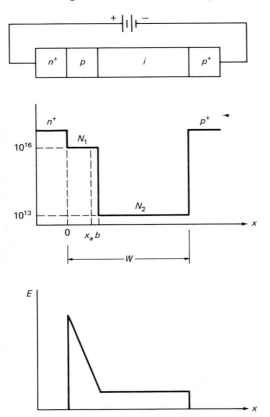

Figure 14.1. Read diode-doping structure and electric field profile.

DeLoach[5] has given a graphic picture of the effect of a small AC voltage originating from some random noise fluctuation on the IMPATT diode biased at breakdown voltage. On the right side of Figure 14.2 is the graph of the AC voltage and resulting current; the left side shows the electric field and charge density at the beginning of each of the four time quadrants. As the AC voltage raises the total electric field above the critical breakdown field, E_c, an ava-

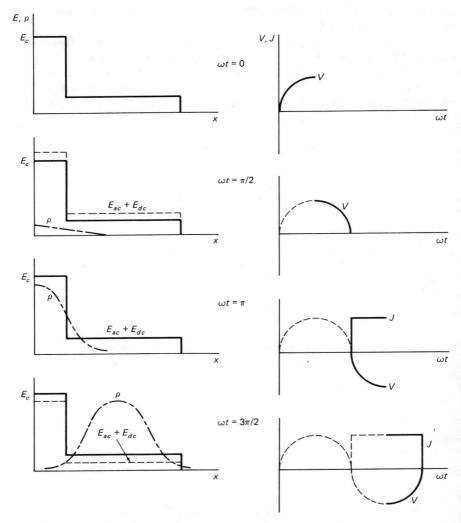

Figure 14.2. The left column shows the internal electric field and charge density ρ; and the right column, the external voltage and current in each quadrant. (DeLoach[5] from *Microwave Semiconductor Devices and Their Circuit Applications*, ed. by H. A. Watson. Copyright © 1969 McGraw-Hill. Used with permission of McGraw-Hill Book Company.)

lanche of hole-electron pairs is generated. The holes move into the i region while the electrons are quickly swept out by the DC bias source. The DC and AC fields at this time are in the same direction so that energy is being lost by the AC field. Since the avalanche has only begun, a small number of carriers has actually been created by the time $\omega t = \pi/2$. The applied AC voltage starts to decrease in the range $\pi < \omega t < 3\pi/2$. However, the avalanche process continues to build up because the total field remains above E_c. The clump of holes obtained from the avalanche region now begins traveling through the diode drift region under the force of the DC field. These holes however are now traveling against the direction of the AC field, so they give up energy to the AC field. Their traveling through the drift region produces an external current J that has the opposite polarity of the applied voltage (Figure 14.2). This implies a negative resistance has been developed. The drift region is made just long enough for the charge clumps to reach the end of the drift region as the AC voltage starts going positive again. The time required for the holes to travel across the drift region is the transit time which is made to be half of the oscillation period. Because of the internal scattering and the high fields employed, the holes traverse the drift region at a velocity approximately independent of the electric field. This is the saturated drift velocity. According to this model, there has been a phase delay of the current by $\pi/2$ caused by the avalanche process and another $\pi/2$ phase delay caused by the drift region.

14.3 READ DIODE LARGE SIGNAL ANALYSIS

As a one-port high power device, it is more natural to think of the IMPATT diode as a negative resistance oscillator than as an amplifier. Today its primary application is in high power solid state sources and locked oscillators in both the pulsed and CW mode. In these applications, a large signal analysis is required to obtain such characteristics as efficiency, power saturation, and harmonic content. Therefore, the large signal analysis will be discussed first with the small signal approximation given in the following section.

The Read diode profile (p^+nin^+ or its complement n^+pip^+) provides higher efficiency than the flat profile diode. The latter device was shown by Misawa[6] to have a distributed avalanche region while the Read diode separates the avalanche and drift zones into two distinct regions in the diode. Although it is more difficult to fabricate, the Read structure has received the most research attention in the last few years.

The rate of generation of holes and electrons caused by impact ionization is

$$g = \alpha_n n \mu_n + \alpha_p p \mu_p \qquad (14.1)$$

where α is the ionization coefficient and μ is the carrier mobility. The ionization rates are dependent on both the electric field and the junction temperature. An empirical relationship for the ionization coefficient is

$$\alpha(E) = A \exp\left[-(b/E)^m\right]. \tag{14.2}$$

For GaAs, Holway et al.[7] have found that $m = 2$, and for T in °C

$$A = 1.61\ 10^5 + 112.7(T - 25)\ cm^{-1} \tag{14.3}$$

and

$$b = 5.41\ 10^5 + 525(T - 25)\ V/cm. \tag{14.4}$$

The microscopic ionization coefficient can be related to the observed macroscopic multiplication factor. For an n^+pip^+ Read diode, the avalanche occurs across the n^+p junction (Figure 14.3). The multiplication factors for holes and electrons respectively are

$$M_p \triangleq \frac{J_p(x_a)}{J_p(0)} = \frac{J}{J_{ps}} \tag{14.5}$$

$$M_n \triangleq \frac{J_n(0)}{J_n(x_a)} = \frac{J}{J_{ns}} \tag{14.6}$$

where J is the total current while J_{ps} and J_{ns} are the reverse saturation currents. The edge of the avalanche region has been chosen by convention to be the point where 95% of the carrier multiplication has occurred. As the holes proceed from left to right, the hole current increases because of the ionization of more holes by the amount $\alpha_p J_p\ dx$. Also more electrons traveling to the left leave holes to travel to the right, adding the additional term $\alpha_n J_n\ dx$.

$$dJ_p = \alpha_p J_p\ dx + \alpha_n J_n\ dx. \tag{14.7}$$

Similarly, for the electron-initiated avalanche from the right

$$dJ_n = -\alpha_n J_n\ dx - \alpha_p J_p\ dx. \tag{14.8}$$

The total current, which is independent of position, is the sum of the hole and electron currents.

$$J(t) = J_n(x,t) + J_p(x,t) \tag{14.9}$$

Substituting this into (14.7) and (14.8) two differential equations for the hole and electron currents are obtained.

$$\frac{dJ_p}{dx} - (\alpha_p - \alpha_n)J_p = \alpha_n J \tag{14.10}$$

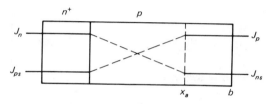

Figure 14.3. Electron and hole currents in the avalanche region of an n^+pip^+ IMPATT diode.

$$\frac{dJ_n}{dx} - (\alpha_p - \alpha_n)J_n = -\alpha_p J \tag{14.11}$$

Before solving these equations with the use of an integrating factor, it will be convenient to define the function $R(x_1,x_2)$.

$$R(x_1,x_2) \triangleq \int_{x_1}^{x_2} (\alpha_p - \alpha_n) \, dx \tag{14.12}$$

In terms of $R(x_1,x_2)$, (14.10) becomes

$$\frac{d}{dx}(e^{-R(0,x)}J_p) = \alpha_n e^{-R(0,x)}, \tag{14.13}$$

which is solved subject to the boundary condition $J_p(0) = J_{ps}$.

$$J_p(x) = e^{R(0,x)} \int_0^x J\alpha_n e^{-R(0,x')} \, dx' + J_{ps} e^{R(0,x)} \tag{14.14}$$

Evaluation of this expression at $x = x_a$ where $J_p(x_a) = J$ provides the sought after ratio M_p.

$$\frac{1}{M_p} = e^{-R(0,x_a)} - \int_0^{x_a} \alpha_n e^{-R(0,x)} \, dx \tag{14.15}$$

The value α_p may be added and subtracted inside the integral to give

$$\frac{1}{M_p} = e^{-R(0,x_a)} + \int_0^{x_a} (\alpha_p - \alpha_n) e^{-R(0,x)} dx - \int_0^{x_a} \alpha_p e^{-R(0,x)} \, dx. \tag{14.16}$$

Since

$$d[e^{-R(0,x)}] = -(\alpha_p - \alpha_n)e^{-R(0,x)}\,dx,$$

the first integral is readily evaluated.

$$\frac{1}{M_p} = 1 - \int_0^{x_a} \alpha_p e^{-R(0,x)}\,dx \qquad (14.17)$$

In similar fashion, the electron-initiated multiplication factor may be solved from (14.11). However, the integration must proceed from x_a to x, and the boundary condition that is used is $J_n(x_a) = J_{ps}$.

$$J_n(x)e^{R(x,x_a)} = -\int_{x_a}^x \alpha_p J e^{R(x',x_a)}\,dx' + J_{ns} \qquad (14.18)$$

Evaluation of $J_n(0) = J$ gives the sought ratio M_n.

$$\frac{1}{M_n} = e^{R(0,x_a)} - \int_{x_a}^x \alpha_p e^{R(x,x_a)}\,dx \qquad (14.19)$$

$$= e^{R(0,x_a)}\left[1 - \int_0^{x_a} \alpha_p e^{-R(0,x)}\,dx\right] \qquad (14.20)$$

Proceeding as before, the electron-initiated multiplication factor is

$$\frac{1}{M_n} = 1 - \int_0^{x_a} \alpha_n e^{R(x,x_a)}\,dx. \qquad (14.21)$$

Furthermore, from (14.17) and (14.20), the ratio of the two factors is the simple function

$$\frac{M_p}{M_n} = e^{R(0,x_a)}. \qquad (14.22)$$

It is convenient to solve for the avalanche current in terms of the macroscopic multiplication factors M_p and M_n rather than the ionization constants α_p and α_n. Starting with the continuity equations (10.8) and (10.9) in the form

AVALANCHE DEVICES AND CIRCUITS 301

$$\frac{1}{v_p}\frac{\partial J_p}{\partial t} + \frac{\partial J_p}{\partial x} = \alpha_n J_n + \alpha_p J_p \tag{14.23}$$

$$\frac{1}{v_n}\frac{\partial J_n}{\partial t} - \frac{\partial J_n}{\partial x} = \alpha_n J_n + \alpha_p J_p \tag{14.24}$$

Poisson's equation for the space charge induced electric field is

$$\epsilon \frac{\partial E_{sc}}{\partial x} = \frac{J_p}{v_p} - \frac{J_n}{v_n}, \tag{14.25}$$

and the spatial average conduction current is

$$J(t) = J_p(x,t) + J_n(x,t) + \epsilon \frac{\partial E_{sc}(x,t)}{\partial t}. \tag{14.26}$$

Kuvas and Lee[8] found an expression for the total current in the avalanche zone. The details of its derivation are found in Appendix A.

$$\left(\frac{\partial}{\partial t} + \frac{1}{M\tau}\right) J = \frac{J_s}{\tau} + \frac{\epsilon}{v_p + v_n} \left[\frac{v_n}{\tau_p} \frac{\partial E_{sc}(0,t)}{\partial t} + \frac{v_p}{\tau_n} \frac{\partial E_{sc}(x_a,t)}{\partial t} \right.$$
$$\left. - \frac{1}{\tau_n} \int_0^{x_a} \left(\alpha_n v_p + \alpha_p v_n - \frac{\partial}{\partial t}\right) \epsilon \frac{\partial E_{sc}}{\partial t} e^{R(x,x_a)} dx \right] \tag{14.27}$$

The intrinsic response times used in this expression are defined as follows.

$$\tau_n = \frac{1}{v_p + v_n} \int_0^{x_a} e^{R(x,x_a)} dx \tag{14.28}$$

$$\tau_p = \frac{1}{v_p + v_n} \int_0^{x_a} e^{-R(0,x)} dx \tag{14.29}$$

From these expressions, the ratio of the electron and hole response times is found to be the same as that found in (14.22).

$$\frac{\tau_n}{\tau_p} = \frac{M_p}{M_n} = e^{R(0,x_a)} \tag{14.30}$$

The quantity $M\tau$ used in (14.27) is defined as

$$M\tau \triangleq M_p \tau_p = M_n \tau_n. \tag{14.31}$$

Finally, the saturation current is defined as

$$\frac{J_s}{\tau} \triangleq \frac{J_{ps}}{\tau_p} + \frac{J_{ns}}{\tau_n}. \tag{14.32}$$

The expression (14.27) includes the effects of space charge. It neglects the effects of diffusion on the carrier velocities and the modulation of the depletion layer by a large AC signal.

Both Kuvas and Lee[8] and Decker[9] simplified (14.27) by using a quasistatic analysis. By this is meant that the ratio of the AC particle current to the total current is the same ratio as the DC particle current to the total current. That is, the ratios

$$r_p(x) \triangleq \frac{J_p}{J} = \frac{J_{p0}}{J_0} \tag{14.33}$$

$$r_n(x) \triangleq \frac{J_n}{J} = \frac{J_{n0}}{J_0} \tag{14.34}$$

are independent of time where J_{p0}, J_{n0}, and J_0 are the DC currents. The quasistatic approximation reduces (14.27) to the following expression for the avalanche current.[9]

$$\frac{\partial J}{\partial t} + \frac{J}{\kappa M \tau} = \frac{J_s}{\kappa \tau} \tag{14.35}$$

$$\kappa = \frac{v_p + v_n}{\tau_p v_p v_n} \int_0^{x_a} r_p(x)[1 - r_p(x)] e^{-R(0,x)} \, dx \tag{14.36}$$

The ratio r_p is obtained from (14.14).

$$r_p(x) = \int_0^{x_a} \alpha_n e^{R(x',x)} \, dx' + \frac{e^{R(0,x)}}{M_p} \tag{14.37}$$

The derivation of (14.35) and (14.36) may be found in Appendix B.

For Ge, Si, and GaAs the saturated drift velocities for both holes and electrons are approximately equal ($v_p = v_n = v_s$). For GaAs and GaP, the hole and electron ionization rates are also approximately equal ($\alpha = \alpha_p = \alpha_n$).[10] For the latter two it may be assumed that the ionization coefficient is independent of position within the avalanche zone, and that consequently $\alpha x_a = 1$. In addition the response time is $\tau_p = x_a/(2v_s)$, and the avalanche zone transit time is $\tau_a = x_a/v_s$. Using these approximations

$$\kappa \cong \frac{2}{3} - \frac{2}{M_p^2} \cong \frac{2}{3}, \qquad (14.38)$$

and the equation for the avalanche current can be written as

$$\frac{\partial J}{\partial t} = \frac{3J}{\tau_a}\left[\int_0^{x_a} \alpha \, dx - 1\right] + \frac{3J_s}{\tau_a}. \qquad (14.39)$$

The original derivation by Read contained the factor of 2 rather than 3 on the right side of (14.39). The discrepancy results from Read's assumption of constant current in the avalanche zone, while the above result was obtained without this approximation.[11]

The quasistatic model forms the basis for the large signal circuit model in Figure 14.4 derived by Gannett and Chua.[12] The capacitances C_a and C_d are simply the geometric capacitances across the avalanche and drift regions given by $C_a = \epsilon A/x_a$ and $C_d = \epsilon A/(W - x_a)$. The current source I_0 represents the DC bias while V_0 is the resulting DC voltage. This latter voltage does not affect the RF characteristics. The volt-ampere relationship of the nonlinear avalanche zone circuit element N is given by (14.35) in the following form.

$$\frac{dI_a(t)}{dt} + \frac{I_a(t) + I_0}{M\kappa\tau} = \frac{I_s}{\kappa\tau} \qquad (14.40)$$

The current I_e for the drift zone in the large signal model in Figure 14.4 may be found. Assuming the carriers continue to move at their saturated velocities in the drift zone, and no further ionization occurs, the total particle current is

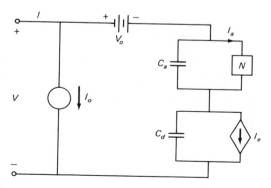

Figure 14.4. Large signal circuit model for the Read IMPATT diode *(Gannett and Chua[12] © 1978 IEEE.)*

$$I_D(x,t) = I_p(x,t) + I_{ns}. \qquad (14.41)$$

A solution for the hole current may be obtained from the continuity equation for hole current

$$q\frac{\partial p}{\partial t} + \frac{\partial J_p}{\partial x} = 0 \qquad (14.42)$$

and from (10.6) as reproduced below.

$$J_p = q\left(v_p p - D_p \frac{\partial p}{\partial x}\right) \qquad (14.43)$$

The expression (14.42) is differentiated with respect to x and (14.43) with respect to t. These two equations are combined by eliminating the term $\partial^2 p/\partial x \partial t$ between them. Finally the time derivative of p is replaced under (14.42) resulting in the following equation.

$$\frac{\partial^2 I_p}{\partial x^2} - \frac{v_p}{D_p}\frac{\partial I_p}{\partial x} - \frac{1}{D_p}\frac{\partial I_p}{\partial t} = 0 \qquad (14.44)$$

Since the AC part of the drift current is the same as the AC part of the hole current, (14.44) may be rewritten in the form

$$\frac{\partial^2 I_d(x,t)}{\partial x^2} - \frac{v}{D}\frac{\partial I_d(x,t)}{\partial x} - \frac{1}{D}\frac{\partial I_d(x,t)}{\partial t} = 0. \qquad (14.45)$$

The subscripts on v and D have been dropped, and the hole drift current is I_d. The boundary conditions are given by

$$I_d(x,t) = 0 \qquad t \leq 0^+ \qquad (14.46)$$
$$I_d(0^+,t) = I_a(t) \qquad (14.47)$$

where I_a is the AC part of the avalanche current injected into the drift zone. Solution of the equation is easily handled by reducing it to an ordinary differential equation by means of the Laplace transform.

$$\frac{d^2 i_d(x,s)}{dx^2} - \frac{v}{D}\frac{di_d(x,s)}{dx} - \frac{s}{D}i_d(x,s) = 0. \qquad (14.48)$$

The solution of this is

$$i_d(x,s) = i_a(s) \exp\left[\left(\frac{v}{2D} - \sqrt{\frac{v^2}{4D} - s}\right)x\right] \qquad (14.49)$$

and the inverse transform gives

$$I_d(x,t) = \int_0^t I_a(t - \lambda)g(x,\lambda)\, d\lambda \qquad (14.50)$$

where

$$g(x,\lambda) = \frac{x}{2\sqrt{D\pi\lambda^3}} \exp\left[-\frac{(x - v\lambda)}{4D\lambda}\right]. \qquad (14.51)$$

The total current I going through the drift region is independent of position.

$$I(t) = I_0 + \frac{1}{x_d}\int_0^{x_d} I_d(x,t)\, dx + C_d \frac{dV_D}{dt} \qquad (14.52)$$

$$\stackrel{\Delta}{=} I_0 + I_e + C_d \frac{dV_D}{dt}. \qquad (14.53)$$

The spatially averaged drift current is thus related to the required current source I_e in the large signal IMPATT diode model.

The large signal equivalent circuit has been employed by Bates and Khan.[13] They analytically predicted the oscillation frequency and second harmonic output power of an IMPATT diode placed in a waveguide or coaxial circuit. Their procedure was to find the conditions where the sum of the circuit and device impedance is zero. Thus the equivalent circuit in Figure 14.4 was found to be a useful and accurate analysis tool.

Alternate approaches to analysis of IMPATT diode circuits have been suggested. Gupta and Lomax[14] found a self-consistent solution by iterating between the time domain and frequency domain by use of the fast Fourier transform. The device is described in the time domain and the circuit in the frequency domain. Brazil and Scanlan[15] used a state variable approach to produce a system of first order differential equations describing a circuit with an IMPATT diode. All these procedures require lengthy computer programs to implement.

14.4 SMALL SIGNAL ANALYSIS OF THE READ DIODE

The large signal analysis considered above is helpful in accurately obtaining the effects of a given circuit on the diode operation, but a small signal analysis provides better insight into the input impedance of the diode. The theory given by Gilden and Hines[16] results in an expression for the diode input impedance and clearly displays the origin of the negative resistance. The approximation involves splitting up the total current and voltage into a large average or DC part and a small AC part. The assumptions used in obtaining (14.19) such as $v_p = v_n = v_s$, $\alpha_p = \alpha_n = \alpha$, and α independent of x are also incorporated here. It should be noted that while α has no spatial variation in the avalanche zone, it is strongly dependent on the applied electric field. The small signal assumptions for the avalanche zone are listed below.

$$J_a = J_0 + \tilde{J}_a e^{j\omega t} \qquad (14.54)$$
$$E_a = E_0 + \tilde{E}_a e^{j\omega t} \qquad (14.55)$$

The ionization coefficient may be expanded in a Taylor series about the electric field where the variation dE is the small AC part of (14.55).

$$\alpha = \alpha_0 + \frac{d\alpha}{dE} dE + \cdots$$
$$\cong \alpha_0 + \frac{d\alpha}{dE} \tilde{E}_a e^{j\omega t} \qquad (14.56)$$

Multiplying (14.56) through by x_a gives an approximation for the ionization integral in (14.39).

$$\alpha x_a = 1 + x_a \alpha' \tilde{E}_a e^{j\omega t} \qquad (14.57)$$

where $\alpha' = d\alpha/dE$ and $\alpha_0 x_a = 1$. When this and (14.54) are substituted into (14.39), and products of the small AC parts are neglected, a simple expression for the avalanche current is obtained.

$$\tilde{J}_a = \frac{3 J_0 \alpha' x_a \tilde{E}_a}{j \omega \tau_a} \qquad (14.58)$$

In addition there is also a displacement current caused by the physical capacitance of the avalanche region.

$$\tilde{J}_{ad} = j \omega \epsilon \tilde{E}_a \qquad (14.59)$$

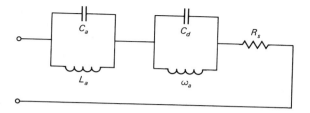

Figure 14.5. Small signal equivalent circuit for the Read IMPATT diode *(Gilden and Hines[16]* © *1966 IEEE.)*

The phase angle of the current in (14.58) is inductive while that of (14.59) is capacitive. Therefore the equivalent circuit for the avalanche zone is a simple parallel LC circuit shown in Figure 14.5, where

$$L_a = \frac{\tau_a}{3I_0\alpha'} \qquad (14.60)$$

$$C_a = \frac{\epsilon A}{x_a}. \qquad (14.61)$$

The resonant frequency is known as the avalanche frequency.

$$\omega_a = \sqrt{\frac{3J_0\alpha' x_a}{\tau_a \epsilon}} \qquad (14.62)$$

The impedance of the avalanche zone is the parallel combination of the L_a and C_a, which in terms of the avalanche frequency, is given by

$$Z_a = \{j\omega C_a[1 - (\omega_a/\omega)^2]\}^{-1}. \qquad (14.63)$$

The ratio of the AC avalanche conduction current to the total current $\tilde{J} = \tilde{J}_{ad} + \tilde{J}_a$ is found from (14.58) and (14.59).

$$\frac{\tilde{J}_a}{\tilde{J}} = \frac{1}{1 + \tilde{J}_{ad}/\tilde{J}_a}$$
$$= \frac{1}{1 - (\omega/\omega_a)^2} \qquad (14.64)$$

The current $\tilde{J}_a$ generated in the avalanche zone is assumed to enter the drift region without transit time delay and propagate through the drift region at a constant velocity v_s. Hence, the conduction current in the drift region is an unattenuated wave with a phase shift proportional to the length of the drift region.

$$\tilde{J}_c = \tilde{J}_a \exp(-j\omega x/v_s) \tag{14.65}$$

As in the avalanche zone, there is also a displacement current in the drift zone.

$$\tilde{J}_d = j\omega\epsilon\tilde{E}_d(x) \tag{14.66}$$

The total AC current is independent of position.

$$\tilde{J} = \tilde{J}_c(x) + \tilde{J}_d(x) \tag{14.67}$$

Solving (14.67) for the AC drift zone electric field gives

$$\tilde{E}_d = \frac{\tilde{\tau}}{j\omega\epsilon}\left[1 - \frac{\exp(-j\omega x/v_s)}{1 - (\omega/\omega_a)^2}\right], \tag{14.68}$$

which can be easily integrated over the length of the drift zone from $x = x_a$ to $x = W$.

$$v_d = \frac{\tilde{J}(W - x_a)}{j\omega\epsilon}\left[1 - \frac{1}{1 - (\omega/\omega_a)^2}\frac{1 - \exp(-j\theta_d)}{j\theta_d}\right] \tag{14.69}$$

In the previous expression $\theta_d = \omega(W - x_a)/v_s$, and the phase function $\exp(j\omega x_a/v_s) \cong 1$ since the avalanche transit time was assumed to be zero. Furthermore, a drift zone capacitance given by $\epsilon A/(W - x_a)$ can be identified with the physical length of the drift zone. Separating the real and imaginary parts of (14.69) yields

$$Z_d = \frac{1}{\omega C_d}\left[\frac{1}{1 - (\omega/\omega_a)^2}\frac{1 - \cos\theta_d}{\theta_d}\right] \tag{14.70}$$
$$- \frac{j}{\omega C_d}\left[1 - \frac{1}{1 - (\omega/\omega_a)^2}\frac{\sin\theta_d}{\theta_d}\right].$$

At low frequencies the resistive term approaches the value for the space charge resistance.

$$R_{sc} = \frac{(W - x_a)^2}{2\epsilon A v_s} \tag{14.71}$$

Gilden and Hines[16] pointed out that this resistive term can help insure uniform current density in the avalanche zone by providing a barrier to local excess current densities. Because of this, a large portion of the avalanche zone can be put to use without burning out the diode. Finally, the avalanche zone and the drift zone impedances can be combined with a third term, R_s, that represents the unavoidable inactive region needed in connecting the device to an external circuit. The total small signal impedance is*

$$Z = R_s + 2R_{sc}\left[\frac{1}{1-(\omega/\omega_a)^2}\frac{1-\cos\theta_d}{\theta_d^2}\right]$$
$$-\frac{j}{\omega C_d}\left[1 - \frac{\sin\theta_d}{\theta_d} + \frac{\frac{\sin\theta_d}{\theta_d} + \frac{x_a}{W-x_a}}{1-(\omega_a/\omega)^2}\right] \quad (14.72)$$

and is represented by the equivalent circuit shown in Figure 14.5. An estimate for x_a can be calculated from the convention that 95% of the avalanche is completed at x_a

$$\int_0^{x_a} \alpha(E)\,dx = 0.95, \quad (14.73)$$

where from Figure 14.1

$$E(x) = \begin{cases} E_c - \dfrac{q(N_1-N_s)x}{\epsilon} & 0 \le x \le b \\ E_c - \dfrac{q(N_1-N_2)b}{\epsilon} - \dfrac{q(N_2-N_s)x}{\epsilon} & b < x \le W \\ E_c - \dfrac{q(N_1-N_2)b}{\epsilon} - \dfrac{q(N_2-n^+)W}{\epsilon} - \dfrac{qn^+}{\epsilon}x & x > W. \end{cases} \quad (14.74)$$

The critical field E_c is that value of E where sustained avalanche is obtained, and $N_s = J_0/qv_s$ is the space charge caused by the DC current.

As an example, a typical Read IMPATT diode designed to operate at around 10 GHz might have the following parameters.

$\epsilon_r = 10.9$
$v_s = 8\ 10^6 - 1.14\ 10^4(T-25)$ cm/sec (ref[17,18])
$W = 3\ \mu\text{m}$

---

*There is a typographical error of a factor of 2 in Gilden and Hines.[16]

$$E_0 = 500 \text{ kV/cm}$$
$$T = 200°\text{K}$$
$$A = 6 \; 10^{-4} \text{ cm}^2$$
$$J_0 = 500 \text{ A/cm}^2$$

Based on these parameters, the small signal resistance and reactance are plotted in Figure 14.6. The small signal theory predicts negative resistance only above the avalanche frequency ω_a. Although the magnitude of the negative resistance is maximum near ω_a, the diode is generally used somewhat above ω_a where the magnitude of the reactance is smaller.

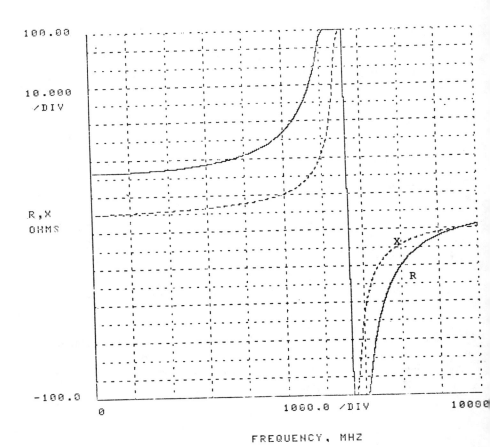

Figure 14.6. Small signal IMPATT diode impedance.

14.5 TEMPERATURE EFFECTS

IMPATT diodes have come to be used more and more in pulsed locked amplifiers and oscillators. In these pulsed applications the average power can be pushed beyond that achievable on a CW basis. Pulsed diodes, though, have thermal transients that must be considered as they affect the rise and fall time of the device and the frequency chirp during the on time. If a rise in the junction temperature is considered to modify only the ionization coefficient and the carrier saturation velocity, v_s and ω_a decrease while x_a and θ_d increase. The decrease in ω_a with temperature rise is exhibited as an onset of oscillation at lower bias levels, while a rise in x_a means more of the highly doped avalanche zone is being used. An increase in θ_d implies a longer transit angle in the drift zone so that the frequency of oscillation will decrease. These qualitative predictions have been confirmed experimentally.[19]

High power devices such as IMPATTs must be cooled through a carefully engineered heat sink. In order to maximize the heat transfer from the diode to the heat sink, toroidal-shaped devices and multiple chips on a single heat sink have been incorporated. A complete transient analysis of the thermal problem for pulsed devices has been carried out by Holway[20] and briefly discussed in chapter 7.

14.6 SPURIOUS OSCILLATIONS

IMPATT diodes, particularly the high efficiency types, are prone to spurious oscillations. These can cause premature diode burnout, low RF power output, and reduced efficiency. There are two primary sources of these spurious signals: bias circuit oscillations and subharmonics. Bias circuit oscillations can occur at very low frequencies to as high as a few hundred megahertz. The negative resistance responsible for these oscillations has been characterized by Brackett.[21] This arises when the diode is biased by a constant current source, and the DC voltage is lowered by the large signal rectification of the microwave signal. He found that circuit stability can be guaranteed if at the intersection of the circuit impedance Z_b and the negative of the diode impedance $-Z_d$, the frequency on the $-Z_d$ locus is lower than the frequency on the Z_b locus. As shown in Figure 14.7 this means the bias circuit impedance should be kept high for frequencies below some cutoff frequency of around 500 MHz.

The second source of spurious oscillations occurs at or near the subharmonic of the oscillation frequency. Hines[22] related this to a parametric interaction based on the IMPATT diode's nonlinear inductance. He found that avoidance of parametric oscillations requires the circuit to present a large inductive admittance to the diode terminal. This criterion was examined in subsequent papers and has been generally confirmed.[23] Thus a well-designed IMPATT

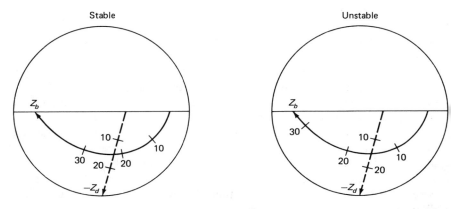

Figure 14.7. Circuit conditions for stable and unstable amplifiers where Z_b is the bias circuit impedance and Z_d is the diode impedance. The numbers indicate the frequency in MHz (*Bracket.*[21] Reprinted with permission from The Bell System Technical Journal. Copyright 1973.)

circuit must provide the appropriate impedance to the diode over a broad frequency range. The circuit should have a high impedance from DC to a few hundred megahertz, a large inductive admittance near the subharmonic frequency, and the appropriate low impedance at the operating frequency.

14.7 IMPATT NOISE

IMPATT diodes are relatively noisy because of the random hole-electron generation in the avalanche zone. Whereas spurious oscillations can be controlled by proper circuit design, very little can be done to make an IMPATT diode quiet. A locking signal or high Q cavity will help reduce FM noise. Hines[24] found an expression for the noise figure of an IMPATT diode amplifier by calculating the mean square noise current $\overline{I}_n^2$. This current is the sum of the idealized zero AC field current, the current induced by the AC field, and the displacement current. The noise figure is given by

$$F = 1 + \frac{\overline{I}_n^2 R_L}{GkT\,\Delta f} \qquad (14.75)$$

where R_L is the circuit load resistance, T is the diode junction temperature in °K, k is Boltzmann's constant, G is the amplifier gain, and Δf is the bandwidth. Using a high gain expression for the amplifier, Hines obtained a simple expression for the noise figure. When the procedure outlined in his paper is modified to include the revised value for ω_a in (14.62), and correction made for the typographical error of a factor of 2 in the original derivation, the noise figure expression becomes

$$F = 1 + \frac{q}{3kT\tau_x^2\omega^2\alpha'(1 - \omega_a^2/\omega^2)} \qquad (14.76)$$

The derivative of the ionization coefficient from (14.2) is $\alpha' = 2\alpha b^2/E^3$. Hines chose τ_x as approximately ½ the transit time through the avalanche zone, which for the 9 GHz diode model would be about $1.7\ 10^{-12}$ sec. At 9 GHz this expression predicts a noise figure of about 40 dB. This is a pessimistic prediction, and more sophisticated numerical models predict values closer to the experimentally observed values of approximately 30 dB. Nevertheless, this expression gives a functional relationship between diode parameters and indicates what factors control the noise figure.

14.8 EFFICIENCY

A simple calculation of the diode output power and efficiency can be made based on an assumed sinusoidal AC voltage and a squarewave current pulse (Figure 14.8). The output power is given by[26]

$$\begin{aligned}P &= \frac{1}{2\pi} \int_{\theta_1}^{\theta_2} I_{\max}V_{ac} \sin\theta\, d\theta \\ &= \frac{I_{\max}V_{ac}}{2\pi}(\cos\theta_1 - \cos\theta_2).\end{aligned} \qquad (14.77)$$

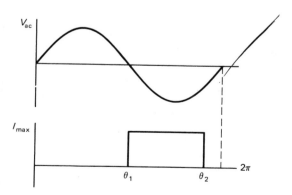

Figure 14.8. Waveform for maximum efficiency calculation.

The average or DC current

$$I_0 = \frac{I_{max}(\theta_2 - \theta_1)}{2\pi} \qquad (14.78)$$

is used in (14.77) to obtain the expression for the efficiency.

$$\eta = \left|\frac{P}{V_0 I_0}\right| = \left|-\frac{V_{ac}}{V_0}\frac{\cos\theta_2 - \cos\theta_1}{\theta_2 - \theta_1}\right| \qquad (14.79)$$

The efficiency is maximized when $\theta_1 = \pi$.

$$\eta = \frac{V_{ac}}{V_0}\frac{1 + \cos\theta_2}{\theta_2 - \pi} \qquad (14.80)$$

This term is in turn maximized when $\theta_2 = 1.742\pi$. If in addition $V_{ac}/V_0 = \frac{1}{2}$, the efficiency is 36%. Experimental efficiencies as high as 35.5% have been reported,[27] but this simple explanation is inadequate to explain the observed efficiency. Kuvas and Schroeder[28] explained the high efficiency by postulating the premature collection mode which is caused by the modulation of the space charge region width. The modulation is a result of the large RF signal. This allowed the drifting packet of carriers to be collected at a more optimum transit angle. Kuvas and Schroeder showed how this effect could explain the experimentally observed sudden increase in output power and efficiency with a gradual increase in bias current. Later Statz et al.[27] showed how efficiencies as high as 41% might be achieved by adding a moderate doping step. The distance between the step and the junction is proportional to the chosen transit time. They also found it necessary to increase the doping in the drift region and decrease the doping of the avalanche zone to some extent.

The analysis rules change for millimeter wave IMPATT diodes. Millimeter wave diodes are punched through at the operating voltages, have higher electric fields in the drift zone, and have higher charge packet densities. The observed high efficiencies and output power for GaAs diodes operating near 40 GHz could not be explained by the premature collection mode theory. Thoren[29,30] postulated that the fields in the drift region were sufficiently high to cause secondary avalanche. Minority carriers thus generated would propagate back to the primary avalanche zone where they could either enhance or quench the oscillation. This mechanism, named the delayed secondary avalanche mode was simulated on a computer using carrier velocities of 3.8 10^6 cm/sec for GaAs. This velocity is substantially lower than what has been normally assumed but was based on extrapolating data from low temperature studies. The simulation based on the delayed secondary avalanche mode and this velocity provide remarkable correlation with observed behavior of GaAs millimeter IMPATT diodes.

14.9 THE TRAPATT MODE

The TRAPATT mode (TRApped Plasma Avalanche Triggered Transit) was discovered by Prager, Chang, and Weisbrod,[31] and was found to provide microwave oscillations in approximately the 0.5 to 4 GHz range with efficiencies of 50%. However, the TRAPATT diode requires a complicated circuit and gives a higher noise level than the IMPATT diode. The circuit must provide a voltage pulse that is 50% to 100% higher than the critical voltage of the diode where avalanching begins. Consequently, carrier multiplication occurs rapidly and provides more carriers than the external current requires. The field in the diode then collapses leaving a plasma that drifts more slowly than the saturated drift velocity. Thus the TRAPATT diode operates at a lower frequency than its comparable IMPATT mode, and the high efficiency results from swings between a high voltage-low current state to a high current-low voltage state.

A useful circuit for the TRAPATT mode is a lowpass filter placed approximately $\lambda/2$ away from the diode. At harmonic frequencies the filter acts as a short circuit, so that the pulse of energy composed of the higher frequencies is reflected back into the diode at the proper time to initiate a new avalanche of carriers.

Although much research effort has gone into TRAPATT devices, they have presently fallen out of favor. This is due in part to their low average power capability as well as the complexity of the circuit design. Nevertheless, their high efficiency may yet prove valuable in some applications.

14.10 POWER-COMBINING CIRCUITS

The primary characteristic that distinguishes the IMPATT diode from other semiconductor microwave devices is its high power capability. It is therefore only natural to combine several IMPATT diodes in such a way that the total power is approximately equal to the sum of the powers of the individual diodes. Basically there are two types of combiners: 2-way and N-way. Of course 2-way combiners such as rat race couplers, hybrid couplers, or 2-way Wilkinson combiners[32] can be cascaded by using some form of corporate feed structure. The resulting N-way combiner where $N = 2^n$, and n is an integer, is cumbersome and becomes lossy for $N > 4$.

The alternative N-way combiner combines diodes in a single structure. This structure may be either a nonresonant type or a resonant type. In the nonresonant type, the power sources are isolated from one another so power-combining can only occur if it is excited by an external signal. On the other hand, in the resonant type of combiner, the power sources are coupled to one another so this circuit can be used as an oscillator without the use of an external locking signal. The nonresonant combiner is discussed briefly in the following section while a fuller discussion of a successful resonant type is given in the remaining section.

14.10.1 Nonresonant Power Combining

The Wilkinson combiner[32] shown in Figure 14.9 is a nonplanar circuit that provides high isolation between each of the input ports. This circuit consists of N input lines of characteristic impedance $\sqrt{N}Z_0$ and a resistive star network used to provide high isolation between each diode source. Cohn[33] extended the 2-way Wilkinson combiner by adding quarter wavelenth lines and resistors

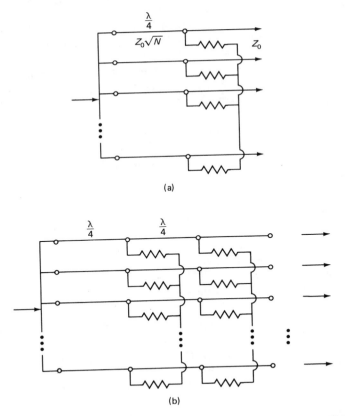

Figure 14.9. (a) Wilkinson power combiner[32] and (b) multisection power combiner *(Yee, Change, and Audeh[34] © 1960, 1970 IEEE.)*

between the lines to obtain broader bandwidth. Similarly, Yee, Change, and Audeh[34] added line lengths and resistive stars to broadband the N-way Wilkinson combiner. Since these resistors are ungrounded and therefore do not have good heat sinks, the power capability of this type of combiner is limited by the power capability of the resistors.

However, Gysel[35] showed by adding an extra quarter wavelength line to each of the N arms of the Wilkinson combiner, the stabilizing resistors could be brought out to ground (Figure 14.10). The only restriction on power capability is the breakdown voltage of the conductors since the loads can now be brought to the edge of the circuit and made as large as required. Although there is no closed form synthesis of this circuit, it may be analyzed by means of the even and odd mode excitation method that Taub and Fitzgerald[36] used on the Wilkinson combiner (Figures 14.11 and 14.12).

14.10.2 Resonant Cavity Combiner

Since the power sources in the resonant type of combiner are electrically coupled together, the devices used in this combiner must all be similar. The most successful type of resonant combiner is the cylindrical cavity combiner invented by Harp and Stover[37] that operates in the $TM_{mn\ell}$ mode. In practice, only TM_{010} and TM_{020} mode cavities are used since the TM_{010} mode is the fundamental mode of oscillation. The TM_{020} mode is used when the cylinder diameter needs to be enlarged in order to accommodate more diodes around the periphery. In this case some method must be employed to suppress the

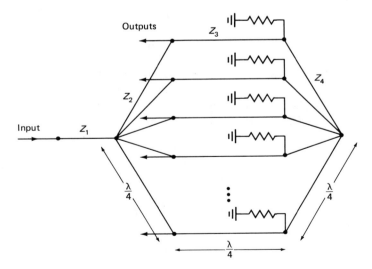

Figure 14.10. Gysel[35] power combiner with 1 input and N outputs.

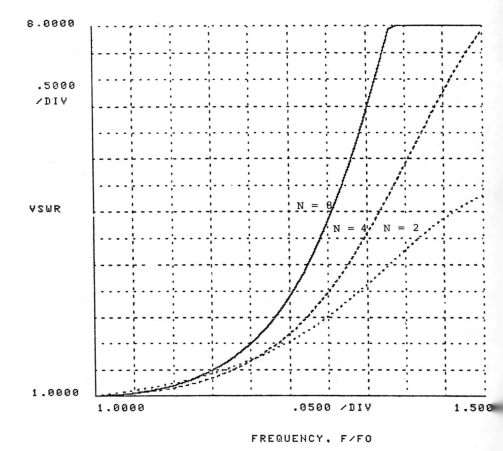

Figure 14.11. Input VSWR for Gysel power combiner with $N = 2$, 4, and 8 ports.

TM_{010} mode. The fields in a cylindrical cavity can be found by matching boundary conditions for Maxwell's equations. For the TM modes, this reduces to

$$H_z = 0 \tag{14.81}$$

$$\mathbf{E}_z = E_0 \hat{z} \tag{14.82}$$

$$\mathbf{E}_t = -\frac{j\beta}{k_c^2} \nabla_t E_z \tag{14.83}$$

$$\mathbf{H}_t = \frac{\omega \epsilon_o}{\beta} \hat{z} \times \mathbf{E}_t \tag{14.84}$$

$$\left(\frac{\omega}{c}\right)^2 = k_c^2 + \beta^2. \tag{14.85}$$

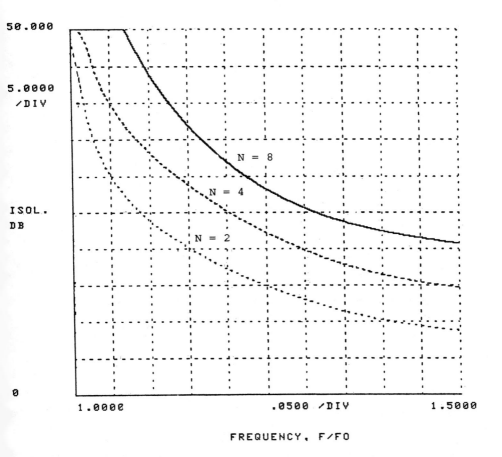

Figure 14.12. Isolation between ports for a Gysel power combiner with $N = 2, 4,$ and 8 ports.

In the above equations, β is the propagation constant, $k_c = x_{mn}/a$ is the wave number, and the subscript t refers to the radial and circumferential directions $\hat{r}$ and $\hat{\phi}$. For the $TM_{mn\ell}$ mode cavity, the resonant frequency is obtained from (14.85).

$$\omega = c \sqrt{(x_{mn}/a)^2 + (\ell\pi/n)^2} \qquad (14.86)$$

For the TM_{010} mode cavity of radius a, the nonzero fields are

$$E_z = E_0 J_0(x_{01} r/a) \qquad (14.87)$$

$$H_\phi = \frac{j\omega\epsilon_o a E_0}{x_{01}} J_1(x_{01} r/a) \qquad (14.88)$$

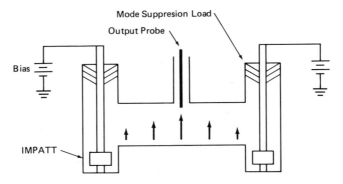

Figure 14.13. Cavity diode combiner.

where the $e^{j\omega t}$ time dependent term has been dropped, x_{01} is the root of the zero order Bessel function, and E_0 is an arbitrary excitation amplitude. The cavity combiner for two diodes is illustrated in Figure 14.13. The two equations above indicate the electric field is maximum at the center of the cavity where it can be coupled to the outside circuit via an electric field probe. Since the resonant frequency for the TM_{010} cavity is $\omega = cx_{01}/a$, (14.88) can be written as

$$H_\phi = \frac{jE_0}{\eta} J_1(x_{01}r/a) \qquad (14.89)$$

where $\eta = \sqrt{\mu_o/\epsilon_o}$ is the intrinsic impedance of a vacuum. The magnetic field is maximum at the outer walls of the cavity where it can couple with the coaxial magnetic field generated by the diode sources. A stabilizing load R_t is used in the opposite end of the coaxial line to dampen unwanted signals at out-of-band frequencies. The equivalent circuit for a two-diode combiner may be readily analyzed using standard loop analysis (Figure 14.14). Extending this to N identical diode sources results in an input impedance seen by the diode of

$$Z_{in} = R_g + R_t + \frac{N}{n_1^2[1/R_c + j(\omega C - 1/\omega L) + 1/(n_2^2 R_L)]} . \qquad (14.90)$$

The input and output coupling coefficients β_1 and β_2 are defined as the ratio of the input to source resistance.

$$\beta_1 \triangleq R_c/(n_1^2 R_g) \qquad (14.91)$$

$$\beta_2 \triangleq R_c/(n_2^2 R_L) \qquad (14.92)$$

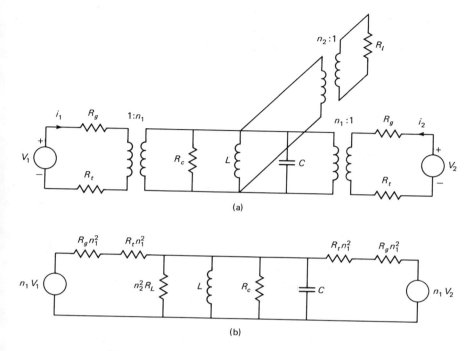

Figure 14.14. Equivalent circuit of a two-diode combiner.

Furthermore, if the unloaded cavity Q is $Q_0 = \omega_o R_c C_c$, then the impedance seen by the diode can be written in terms of the coupling coefficients.

$$Z_{in} = R_g + R_t + \frac{R_g N \beta_1}{1 + \beta_2 + jQ_o \left(\frac{\omega}{\omega_o} - \frac{\omega_o}{\omega}\right)} \quad (14.93)$$

Also, the loaded Q of the system is

$$Q_L = \frac{\omega_o C}{\frac{1}{R_c} + \frac{N}{n_1^2(R_g + R_t)} + \frac{1}{n_2^2 R_L}}$$

$$= \frac{Q_0}{1 + \beta_2 + \frac{\beta_1 N}{1 + R_t/R_c}}. \quad (14.94)$$

As the number of diodes N increases, the input-coupling coefficient β_1 must decrease or β_2 increase to provide the same input match to the diode.

The insertion loss between a diode and the load is the ratio of the available power to the power delivered to the load. The power in the load contributed by one diode is

$$P_L = \frac{|n_1 V|^2}{2 n_2^2 R_L}$$

$$= \frac{|I|^2 (n_1 R_g)^2}{n_2^2 R_L}$$

$$\times \frac{\left[\left(1 + \frac{R_t}{R_g}\right)(1 + \beta_2) + N\beta_1\right]^2 + \left[Q_0 \left(1 + \frac{R_t}{R_g}\right)\left(\frac{\omega}{\omega_o} - \frac{\omega_o}{\omega}\right)\right]^2}{(1 + \beta_2)^2 + Q_0^2 \left(\frac{\omega}{\omega_o} - \frac{\omega_o}{\omega}\right)^2} \tag{14.95}$$

Since the available power is $|I|^2 R_g / 8$, the insertion loss α^2 is

$$\alpha^2 = \frac{8 P_L}{|I|^2 R_g}. \tag{14.96}$$

At resonance, $R_t \cong 0$, and the insertion loss becomes

$$\alpha^2 \cong \frac{4\beta_2 (1 + \beta_2 + N\beta_1)^2}{\beta_1 (1 + \beta_2)^2}. \tag{14.97}$$

If the input impedance is held constant as N increases by keeping the product $N\beta_1$ constant, the insertion loss decreases as N increases.

This analysis provides good agreement with impedance measurements when N is small. For N approaching 16, the large number of coaxial lines entering the cavity alter the effective cavity diameter. Nevertheless, an approximation for the internal cavity capacitance and inductance can be obtained from the expression for the electric energy storage in a cavity with radius a and height h.

$$\mathcal{E}_e = \frac{n_1^2 C_c |V|^2}{4} = \frac{\epsilon}{4} \int_0^a \int_0^{2\pi} \int_0^h |E|^2 r \, dr \, d\phi \, dz \tag{14.98}$$

The electric field given by (14.87) is substituted into (14.98) where the cavity voltage is given by

$$V = hE_z. \qquad (14.99)$$

The evaluation of the integral is aided by the following identity for Bessel functions.

$$\int v J_\nu^2(cv) \, dv = \frac{v^2}{2} [J_\nu(cv) - J_{\nu-1}(cv) J_{\nu+1}(cv)] \qquad (14.100)$$

The equivalent lumped internal capacitance and inductance of the cavity is found in terms of the input transformer turns ratio n_1.

$$C = \frac{\epsilon_o \pi a^2}{n_1^2 h} \frac{J_1^2(x_{01})}{J_0^2(x_{01} r/a)} \qquad (14.101)$$

$$L = 1/(\omega_o^2 C) \qquad (14.102)$$

REFERENCES

1. W. T. Read, "A Proposed High-Frequency, Negative-Resistance Diode," *Bell System Technical Journal*, Vol. 37, pp. 401–446, March 1958.
2. R. L. Johnston, B. C. DeLoach, and B. G. Cohen, "A Silicon Diode Microwave Oscillator," *Bell System Technical Journal*, Vol. 44, pp. 369–372, February 1965.
3. C. A. Lee, R. L. Batdorf, W. Wiegman, and G. Kaminsky, "The Read Diode, an Avalanche Transit-Time, Negative-Resistance Oscillator," *Applied Physics Letters*, Vol. 6, p. 89, 1965.
4. B. C. DeLoach, "The IMPATT Story," *IEEE Trans. on Electron Devices*, Vol. ED-23, pp. 657–660, July 1976.
5. B. C. DeLoach, "Avalanche Transit-Time Microwave Diodes," *Microwave Semiconductor Devices and Their Circuit Applications*. ed. H. A. Watson. New York: McGraw-Hill, Chapter 5, 1969.
6. T. Misawa, "Negative Resistance in p-n Junctions Under Avalanche Breakdown Conditions, Parts I and II," *IEEE Trans. on Electron Devices*, Vol. ED-13, pp. 137–151, January 1966.
7. H. Holway Jr., S. R. Steele, and M. G. Alderstein, "Measurement of Electron and Hole Properties in p-type GaAs," *7th Bienn. Cornell Conference on Active Semiconductor Devices and Circuits*, pp. 198–208, August 1979.
8. R. Kuvas and C. A. Lee, "Quasistatic Approximation for Semiconductor Avalanches," *Journal of Applied Physics*, Vol. 41, pp. 1743–1755, March 15, 1970.
9. D. R. Decker, "IMPATT Diode Quasi-Static Large-Signal Model," *IEEE Trans. on Electron Devices*, Vol. ED-21, pp. 469–479, August 1974.
10. S. M. Sze, *Physics of Semiconductor Devices*. New York: Wiley, 1969.
11. S. O. Scanlan and T. J. Brazil, "Large-Signal Computer Simulation of IMPATT Diodes," *IEEE Trans. on Electron Devices*, Vol. ED-28, pp. 12–21, January 1981.
12. J. W. Gannett and L. O. Chua, "A Nonlinear Circuit Model for IMPATT Diodes," *IEEE Trans. on Circuits and Systems*, Vol. CAS-25, pp. 299–307, May 1978.
13. B. D. Bates and P. J. Khan, "Analysis of Waveguide IMPATT Oscillator Circuits," *1981 IEEE International Microwave Symposium Digest*, pp. 232–234, June 1981.
14. M. S. Gupta and R. J. Lomax, "A Self-Consistent Large-Signal Analysis of a Read-Type Diode Oscillator," *IEEE Trans. on Electron Devices*, Vol. ED-18, pp. 544–550, August 1971.
15. T. J. Brazil and S. O. Scanlan, "Self-Consistent Solutions for IMPATT Diode Networks," *IEEE Trans. on Microwave Theory and Techniques*, Vol. MTT-29, pp. 26–32, January 1981.

16. M. Gilden and M. E. Hines, "Electronic Tuning Effects in the Read Microwave Avalanche Diode," *IEEE Trans. on Electron Devices,* Vol. ED-13, pp. 169–175, January 1966.
17. Y. Takayama, "Effect of Temperature on Device Admittance of GaAs and Si IMPATT Diodes," *IEEE Trans. on Microwave Theory and Techniques,* Vol. MTT-23, pp. 673–680, August 1975.
18. W. R. Schroeder and G. I. Haddad, "Nonlinear Properties of IMPATT Devices," *Proc. of the IEEE,* Vol. 61, pp. 153–182, February 1973.
19. W. E. Schroeder and G. I. Haddad, "The Effect of Temperature on the Operation of an IMPATT Diode," *Proc. of the IEEE,* Vol. 59, pp. 1242–1244, August 1971.
20. L. H. Holway, "Heat Transport and Current Crowding in IMPATT Diodes," *IEEE Trans. on Electron Devices,* Vol. ED-23, pp. 1304–1312, December 1976.
21. C. H. Brackett, "The Elimination of Tuning-Induced Burnout and Bias-Circuit Oscillations in IMPATT Oscillators," *Bell System Technical Journal,* Vol. 52, pp. 271–306, March 1973.
22. M. E. Hines, "Large-Signal, Frequency Conversion, and Parametric Instabilities in IMPATT Diode Networks," *Proc. of the IEEE,* Vol. 60, pp. 1534–1548, December 1972.
23. D. F. Peterson, "Circuit Conditions to Prevent Second-Subharmonic Power Extraction in Periodically Driven IMPATT Diode Networks," *IEEE Trans. on Microwave Theory and Techniques,* Vol. MTT-22, pp. 784–790, August 1974.
24. M. E. Hines, "Noise Theory for the Read Type Avalanche Diode," *IEEE Trans. on Electron Devices,"* Vol. ED-13, pp. 158–163, January 1966.
25. M. S. Gupta, "Noise in Avalanche Transit-Time Devices," *Proc. of the IEEE,* Vol. 59, pp. 1674–1687, December 1971.
26. G. I. Haddad, P. T. Greiling, W. E. Schroeder, "Basic Principles and Properties of Avalanche Transit-Time Devices," *IEEE Trans. on Microwave Theory and Techniques,* Vol. MTT-18, pp. 752–772, November 1970.
27. H. Statz, H. A. Haus, and R. A. Pucel, "Large-Signal Dynamic Loss in Gallium Arsenide Read Avalanche Diodes," *IEEE Trans. on Electron Devices,* Vol. ED-25, pp. 22–33, January 1978.
28. R. L. Kuvas and W. E. Schroeder, "Premature Collection Mode in IMPATT Diodes," *IEEE Trans. on Electron Devices,* Vol. ED-22, pp. 549–558, August 1975.
29. G. R. Thoren, *The Effects of Delayed Avalanche Phenomena on the High Efficiency Operation of GaAs Millimeter Wave IMPATT Diodes.* Ithaca, New York: Cornell University Thesis, January 1981.
30. G. R. Thoren, "Delayed Secondary Avalanche Effects in Millimeter Wave GaAs IMPATT Diodes," *IEEE Electron Device Letters,* Vol. EDL-2, pp. 10–13, January 1981.
31. H. J. Prager, K. K. N. Chang, and S. Weisbrod, "High-Power, High-Efficiency Silicon Avalanche Diodes at Ultra High Frequencies," *Proc. of the IEEE,* Vol. 55, pp. 586–587, April 1967.
32. E. Wilkinson, "An N-Way Hybrid Power Divider," *IRE Trans. on Microwave Theory and Techniques,* Vol. MTT-8, pp. 116–118, January 1960.
33. S. B. Cohn, "A Class of Broadband Three Port TEM-Mode Hybrids," *IEEE Trans. on Microwave Theory and Techniques,* Vol. MTT-16, pp. 110–116, February 1968.
34. H. Y. Yee, F. Change, and N. F. Audeh, "N-Way TEM Mode Broadband Power Dividers," *IEEE Trans. on Microwave Theory and Techniques,* Vol. MTT-18, pp. 682–688, October 1970.
35. V. H. Gysel, "A New N-Way Power Divider/Combiner Suitable for High Power Applications," *1975 IEEE MTT-S International Microwave Symposium Digest,* p. 116, May 1975.
36. J. J. Taub and B. Fitzgerald, "A Note on N-Way Hybrid Power Dividers," *IEEE Trans. on Microwave Theory and Techniques,* Vol. MTT-12, pp. 260–261, March 1964.
37. R. S. Harp and H. L. Stover, "Power Combining of X-Band Circuits," *1973 IEEE International Solid State Conference Digest of Technical Papers,* p. 118, February 1973.

Chapter 15
Gunn Effect Devices

15.1 INTRODUCTION

In February of 1962, J. R. Gunn of IBM was investigating the dependence of the electric field strength on the carrier velocity in bulk semiconductor materials. While testing a DC biased bulk n-type GaAs material, he noticed what appeared to be noise on his oscilloscope screen. This "noise," he discovered, was a 4.5 GHz microwave oscillation. He found the frequency of oscillation to be equal to the transit time of the electrons through the material and to be largely independent of the external circuit. Furthermore, the oscillations were determined to be associated with the bulk semiconductor material and not with the contacts. He was convinced that indeed he had found a new mechanism for generating microwave signals. However, it was unclear how the oscillator worked. In his paper, Gunn stated, "The mechanism leading to the oscillations is not yet understood."[1] Further, he specifically rejected the theory given by Ridley and Watkins[2] because of the supposed incompatibility of the observed critical drift velocity with that required by the theory. In 1964, Kroemer[3] pointed out that the properties of the Gunn effect could be satisfactorily explained by the periodic nucleation and disappearance of space-charge domains originally discussed by Ridley.[4] With the added illumination given by Hilsum,[5] Kroemer observed that the theoretical electron energies were compatible with the observed velocities. In contrast with the early IMPATT diode work, the experimental discovery, in practice, preceded the theoretical explanations.

The theory given by Ridley and Watkins,[2] and Hilsum[5] is now known as the RWH mechanism. It showed how a voltage-controlled differential negative resistance could be obtained by exciting high mobility electrons from a lower conduction valley to an upper conduction valley where the electron mobility is low (see Section 15.3). By subjecting GaAs samples to hydrostatic pressure, direct confirmation of the RWH theory was provided by Hutson, et al.[6] These authors showed that when the energy separation between the upper valley and the central lower valley was reduced, the critical electric field for the onset of oscillations also decreased. This and other experiments confirmed the RWH theory for the origin of the microwave oscillations from a biased bulk sample of GaAs.

Among the semiconductors in which oscillations have been observed, GaAs has been the material most widely used. The term *Gunn Device* has been applied to bulk *n*-type GaAs semiconductors with a wide range of doping concentrations and physical lengths operating in several different modes. The major microwave applications of the Gunn device have been in low power oscillators in the 6 to 20 GHz range with average output power to about 200 mW. Recently, however, Gunn devices have been used in applications going to 90 GHz. They have also been found useful as broadband negative resistance amplifiers. The Gunn device has about 10 dB less noise figure and requires a lower bias voltage than the IMPATT diode. However, it generates less power and has lower efficiency than the IMPATT.

The following section presents a short description of GaAs material pertinent to Gunn devices. A qualitative description of the RWH mechanism and domain formation is found in sections 15.3 and 15.4. Section 15.5 contains a description of the various operating modes in which a Gunn device might be used. A more complete analysis of the dipole domain is given in Section 15.6. Since the negative resistance of a Gunn device can be described in terms of the device velocity-field characteristic, a simple analytical model for this curve is provided in Section 15.7. A short description of Gunn device noise is found in the following section. The final two sections are concerned with circuit design for a Gunn device. These sections describe an equivalent circuit for the device and provide a rigorous analysis of a waveguide circuit often used for Gunn devices.

15.2 PROPERTIES OF GaAs

The crystal structure of gallium arsenide is termed a zinc-blend structure after ZnS. It may be considered as two interpenetrating face-centered cubic structures with one sublattice of gallium and the other of arsenic. The primitive cell, as shown in Figure 15.1, is a face-centered cubic structure of four arsenic atoms with four gallium atoms located at the coordinates $(1/4,1/4,3/4)$, $(3/4,1/4,1/4)$, $(3/4,3/4,3/4)$, and $(1/4,3/4,1/4)$.

The simple band model for semiconductors presented in chapter 10 is insufficient for understanding the RWH mechanism, since attention must be paid to the way the band structure changes in the different crystallographic directions. For a free electron or for an electron bound to a hydrogen atom, the energy follows a parabolic curve when plotted against the momentum p.

$$\mathcal{E} = \frac{p^2}{2m} \qquad (15.1)$$

In the hydrogen atom, only certain discrete values of energy are allowable. In more complex crystals, the potential driving the carriers is very complicated.

GUNN EFFECT DEVICES 327

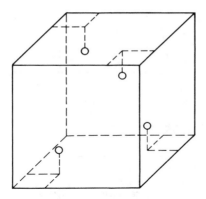

Figure 15.1. The cube represents a face-centered cubic arsenic cell with 4 gallium atoms marked inside the cell.

The only tractable relationship that can be expected is that the energy-momentum be periodic. The quantum-mechanical wave function can be written as

$$\psi(x,t) = A(k,x)e^{-j(kx-\mathcal{E}t/\hbar)}. \tag{15.2}$$

The symbol $\hbar$ is Planck's constant divided by 2π. The Bloch function, $A(k,x)$ is periodic in x with a period equal to the repetition of the unit cells in the crystal material. A particle moving in a periodic potential, which is provided by the fixed atoms in the crystal, travels with an average group velocity found by differentiating the exponent in (15.2).

$$v_g = \frac{1}{\hbar}\frac{d\mathcal{E}}{dk} \tag{15.3}$$

Defining the crystal momentum as $p = \hbar k$, an effective mass m^* for the moving particle in the crystal may be expressed as

$$\frac{dv_g}{dt} = \frac{d}{dt}\left(\frac{d\mathcal{E}}{dp}\right) = \frac{F}{m^*} \tag{15.4}$$

$$\frac{dp}{dt}\frac{d}{dp}\left(\frac{d\mathcal{E}}{dp}\right) = \frac{F}{m^*}. \tag{15.5}$$

Since the force is the time rate of change of momentum, the effective mass is

$$\frac{1}{m^*} = \frac{d^2\mathcal{E}}{dp^2}. \tag{15.6}$$

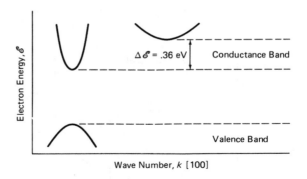

Figure 15.2. Energy-momentum diagram for GaAs.

The acceleration of an electron in a crystal is much different than in free space, and consequently it will appear to have a different mass $m^* \neq m_e$. The electron in a crystal will also be subject to an anisotropic field so that m^* is actually a tensor quantity, dependent on the direction in which the particle is traveling.

The crystal, GaAs, has an energy-momentum diagram like that shown in Figure 15.2. It is known as a direct band gap material since the minimum conduction band is directly above the maximum valence band. This property is the basis of the GaAs light-emitting diode. An additional feature of GaAs is important for Gunn devices. In the six crystallographic directions symmetrical to the [100] direction, there is a second conduction band of slightly higher energy than the direct band gap. This second or upper valley is the basis for the RWH mechanism in Gunn diodes. The energy gap between the valence band and the lower valley conduction band at room temperature is 1.43 eV, while the energy difference between the two conduction bands is only 0.36 eV. Furthermore, the electron mobility in the lower conduction band is approximately 50 times greater than the mobility in the upper band. In terms of the effective masses of the electrons, the upper and lower effective masses are $m_u^* = 0.072\, m_e$ and $m_l^* = 1.2\, m_e$, where m_e is the mass of a free electron. The desired material properties for the Gunn device can be summarized as follows:

1. There must be a relatively large forbidden energy gap between the valence band and the central, lower conduction band.
2. There should be an upper conduction band valley with a small minimum energy $\Delta \mathscr{E}$ that is slightly greater than the central valley energy, but still large compared to the thermal energy kT.
3. The lower valley should be relatively narrow, having a low density of available states.
4. The upper valley should be relatively broad, having a large density of states.

5. An n-type doping material should be added to provide a small donor population near the conduction band.

The materials GaAs, InP, GdTe, ZnSe, and InAs meet the above requirements and have been found to produce Gunn type oscillations. GaAs has had the most technological development and is by far the most widely used material in Gunn devices today.

15.3 THE RIDLEY-WATKINS-HILSUM MECHANISM

The explanation of the origin of the negative resistance observed in bulk n-type GaAs devices was the subject of the investigations of Ridley and Watkins,[2] and Hilsum[5]. At room temperature the small electron population lies entirely in the lower conduction valley since $\Delta \mathcal{E} < kT$. Electrons associated with the lower valley have low effective mass and high mobility. These electrons react to applied fields less than 3200 V/cm. As the electric field increases, the electron temperature rises to the point where some of the electrons start transferring to the upper valley where the mobility is low. Since density of available states in the upper valley is large relative to the lower valley, the number of electrons transferring to the upper valley is not limited by the density of states in the upper valley. For GaAs, 50% of the electrons have transferred out of the lower valley by the time the applied DC field has reached 8500 V/cm. This number continues to decrease slowly until the bias reaches about 300 kV/cm. Electrons in the lower valley have a saturation velocity of about 2×10^7 cm/sec. The proportion of electrons having high velocity decreases as the electric field increases between the values of 3.2 and 11 kV/cm. This mechanism is responsible for the differential negative resistance observed in Gunn device oscillators. Shockley[7] pointed out that, when boundary conditions are properly taken into

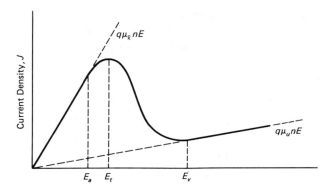

Figure 15.3. Current-electric field curve for a bulk GaAs Gunn diode.

account, the DC internal space charge injected into the cathode of the device must always be larger than the decrease in current caused by the negative electron mobility. However, steady-state solutions are not stable with respect to small AC fluctuations. Consequently, the observed decrease in current with rising electric field (Figure 15.3) is a result of a decreasing fraction of electrons in the high mobility conduction band relative to the low mobility conduction band.

15.4 DOMAIN FORMATION

The ohmic contacts at each end of the bulk n-type GaAs material have been found to be an important parameter in the design of devices. A perfect ohmic contact can inject or accept any number of electrons without voltage drop. They are often modeled as a very heavily doped n-type material. Gunn devices are thus fabricated on an epitaxial layer of $n^+\ n\ n^+$ GaAs. As a result, the active portion of the device is of length $L <$ the physical length of the semiconductor (Figure 15.4). The total carrier concentration n is the sum of the carriers in the lower and upper conduction bands.

$$n = n_l + n_u \qquad (15.7)$$

While the number of carriers n_l will decrease with applied field, the total number n will be independent of the field. When a time-varying voltage $V(t)$ is applied to the device, a current density results that is the sum of the drift, diffusion, and displacement current components. The one-dimensional model is justified because the device length L is much smaller than the cross sectional dimensions. Except near the ohmic contacts, where there is a sharp discontinuity in the doping concentration, the diffusion term may be neglected. The output current occurs in clumps of charge that begin at the cathode, propagate through the device, and enter the external circuit at the anode. The point where the charge clump begins to nucleate could, in principle, be anywhere in the sample. Normally, however, it nucleates at the cathode contact where there is a large doping discontinuity. The carrier densities below the threshold electric

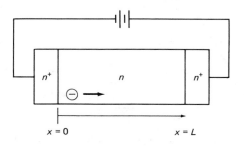

Figure 15.4. Bulk GaAs sample.

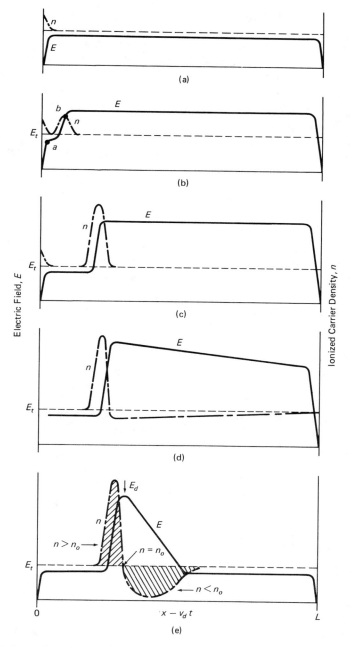

Figure 15.5. Formation of a dipole domain. (*Illustration reprinted from P. N. Butcher, et al., "A Simple Analysis of Stable Domain Propagation in the Gunn Effect,"* British Journal of Applied Physics, *Vol. 17, pp. 841–850, 1966.*)

field are shown in Figure 15.5a. Shortly after the electric field exceeds the threshold value, E_t, an accumulation of charge appears at the cathode. This accumulated charge is caused by the velocity of the high mobility electrons at point "a" exceeding the velocity of the low mobility electrons at point "b." The charge accumulation begins to grow and under the force of the electric field begins moving toward the anode. As this instability grows, a region of depleted electronic charge forms in front of the high electron charge density (Figure 15.5d). Finally a fully formed dipole domain is formed that propagates through the material at a field dependent drift velocity $v(E)$. It is generally understood that dipole domains occur in almost all operating modes for the Gunn device because of imperfect contacts and doping fluctuations.

When the domain reaches the anode, the electrons and holes rapidly recombine and produce a current pulse in the external circuit. The field behind the domain simultaneously begins to rise until it exceeds the threshold field E_T. At the doping notch in the cathode a new domain starts to form. The cycle repeats and in the time domain a series of current pulses are produced in the external circuit. The Fourier transform of the waveshape produces a fundamental frequency component associated with the time between pulses.

15.5 OPERATING MODES

Different operating modes of the Gunn diode are made possible by modifying the device to have a different nL product and/or by modifying the external circuit. These modes of operation are listed in Table 15.1. Dipole domains will form in the semiconductor when $nL > 10^{12}$ cm^{-2} and will not form when $nL < 10^{12}$ cm^{-2}. The impedance of the external circuit can be used to inhibit or quench these domains. Copeland[8] has conveniently summarized these modes as shown in Figure 15.6. In this diagram the frequency $\times$ length (fL product) is plotted against the device parameter nL. The vertical axis is thus dependent on the bias voltage and circuit impedance. The boundaries between various operating modes is approximate, since the device operation depends also on material properties, temperature, bias level, and external circuit impedance. What follows is a brief description of each of these operating modes.

15.5.1 Dipole Mode

The dipole mode described in the previous section is characterized by the formation of a charge dipole at the device cathode that propagates through the sample at the saturated drift velocity and is collected at the anode. In a low impedance circuit, the velocity of the dipole layer remains nearly independent of the applied voltage.[9] The transit time frequency is thus simply given by

$$f_o = \frac{v_d}{L} \tag{15.8}$$

Table 15.1. Gunn Diode Operating Modes.

MODE NAME	nL cm^{-2}	DOMAIN FORMED	CIRCUIT	FREQUENCY RANGE	COMMENTS
1 Dipole Domain Traveling Domain Transit Time Gunn	$>10^{12}$	yes	Low Impedance	$f_o = \dfrac{V}{L}$	Frequency is independent of bias and temperature.
2 Inhibited Delay	$>10^{12}$	yes	High Impedance, Resonant	$\tfrac{1}{2}f_o < f < f_o$	$V_{ac} > V_b - V_T$
3 Quenched	$>10^{12}$	yes	High Impedance, Resonant	$f_o < f < 2f_o$	$|V_{ac}| > V_s$
4 LSA	$>10^{12}$	no	Lowpass, Complicated	$2\,10^4 < \dfrac{n}{f} < 2\,10^5$	High Efficiency, Circuit sensitive.
5 Hybrid	$>10^{12}$	barely	Lowpass, Complicated	$f_o < f < 50f_o$ $f_o < f < 50f_o$	Domain forms just before it is cutoff. Between quenched and LSA.
6 Accumulation layer Negative conductance	$10^{11} < nL < 10^{12}$	no	amplifier		Low power. Gain of 30 to 40 dB possible.

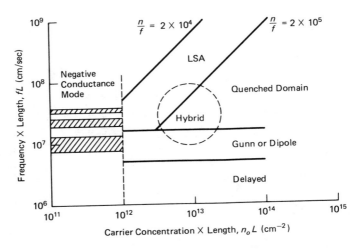

Figure 15.6. Chart of Gunn device operation showing approximate boundaries between modes.

where v_d is the domain velocity and L is the device length. For diodes operating in this mode, the frequency is fixed by the device length. For diodes operating in this mode, a 10 μm long sample would operate at $f_o = 10^7/10^{-3} = 10$ GHz. The threshold voltage would be simply $V_T = E_T L = 3.2$ V.

15.5.2 Inhibited Mode

With the circuit impedance is raised so that the AC voltage drops below the threshold voltage but remains above the sustaining voltage, the initiation of a new dipole domain may be delayed. As shown in Figure 15.7, the initiation of a new domain may be delayed until the total voltage exceeds the threshold voltage. However, once the dipole domain forms, it travels to the anode at the

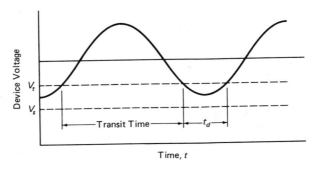

Figure 15.7. Inhibited domain mode.

saturation velocity. Consequently, the time between output pulses has been increased. In the limit, the delay may be as much as half the period of the AC voltage.

15.5.3 Quenched Domain Mode

If the AC voltage is large enough for the negative part of the cycle to drop below the sustaining voltage V_s, the domain may be quenched before it arrives at the anode (Figure 15.8). Consequently, the frequency of operation is raised since the effective transit time, t_e, has been shortened. In effect, the sample is made to look physically shorter. The criterion for the quenched domain mode is that the domain must be fully formed before it is extinguished. The maximum frequency is limited by the time required to develop a full domain. Both the quenched domain and inhibited modes can be considered as subsets of the dipole mode, since a fully formed dipole is formed in all three of these cases.

15.5.4 LSA Mode

The limited space charge accumulation (LSA) mode uses a diode like that described for the dipole mode where $nL > 10^{12}$ cm^{-2}, but its operation is completely determined by the circuit. The LSA mode is an extreme case of the quenched domain mode where only a small accumulation layer is formed before it is quenched by the AC voltage (Figure 15.5b). Consequently, a dipole is not allowed to form. During the short time that the accumulation layer is forming, the entire length of the device is subjected to an electric field greater than the threshold field E_T. The operating frequency range is limited by the rate of charge buildup and charge dissipation of the material. To prevent a high field domain from forming, the oscillation period for an LSA mode should

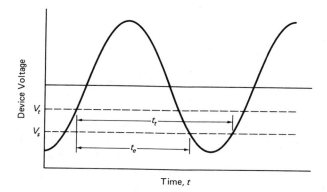

Figure 15.8. Quenched domain mode.

be shorter than the dielectric relaxation time or RC time constant of the material. The time required for a domain to form is

$$\tau_r = RC = \frac{L}{\sigma A} \cdot \frac{\epsilon A}{L} \qquad (15.9)$$

$$= \frac{\epsilon}{nq|\mu_n|}$$

where $|\mu_n|$ is the magnitude of the differential negative mobility. The low frequency limitation is governed by the rate τ_d of excess charge dissipation. This time constant has the same form as (15.9) except that the low mobility is used.

$$\tau_d = \frac{\epsilon}{nq\mu_\ell} \qquad (15.10)$$

Since $|\mu_n| \cong 100$ and $\mu_\ell \cong 8000$ cm^2/V−s, the time required to dissipate any accumulated charge is much shorter than the time required to accumulate the charge. Copeland[10] suggested that the frequency limitations for the LSA mode could be approximated by requiring $1/f < 3\tau_r$ and $1/f \gg \tau_d$. The latter restriction is used to avoid long periods of time where the bias is below threshold, which would lead to low conversion efficiency. Thus he found

$$10^3 \ll \frac{n}{f} < 2 \times 10^5 \text{ sec cm}^{-2} \qquad (15.11)$$

A more detailed calculation by Copeland[8] showed that

$$2 \times 10^4 < \frac{n}{f} < 2 \times 10^5 \text{ sec cm}^{-2} \qquad (15.12)$$

within the range of frequencies for the LSA mode.

A unique feature of this mode is that increasing the operating frequency increases the conversion efficiency while in most electronic devices the opposite is true. This behavior occurs because the distance the accumulation layer can travel decreases with a rise in frequency. Thus less energy is dissipated as heat and more charge is delivered to the load.

A second feature that distinguishes the LSA mode from other modes is its high efficiency. Since the entire LSA device is biased into the negative resistance region rather than a narrow region (as in the dipole mode), a larger percentage of the sample is useful for production of microwave energy. In contrast, when a domain is formed, the region outside the dipole remains passive and acts as a series resistance that reduces efficiency. Efficiencies in the 2% to 9% range are possible for the LSA mode.

A third feature is that the peak field in the LSA mode is less than the peak field in a dipole mode. Thus the operating voltage can be made considerably higher for the LSA mode before avalanching occurs. The input power and therefore the output power are higher. Levels of approximately 200 mW at around 10 GHz have been observed.

A fourth feature, as suggested previously, is the high operating frequency. In practice, an LSA mode device can operate between 1 and 50 times the transit frequency.

The major disadvantage of the LSA mode is its sensitivity to the circuit impedance. If the voltage-circuit conditions are not set just right, a dipole domain could form and create a local electric field high enough to destroy the device. Because of the difficulty in designing a circuit that would prevent inadvertent device burnout, the LSA mode is not used in very many applications today.

15.5.5 Hybrid Mode

This operating mode lies between the LSA and quenched domain mode. It is distinguished from the LSA mode as having slightly lower efficiency. The boundary for this operating mode is not well defined and is subject to some judgment. The device is not biased as high as it is in the LSA operation so if a domain is formed, it will not destroy the device as it would in the LSA mode. In practice, inhomogeneities in the doping cause high field domains to form, even in the LSA mode. Probably most results for the LSA mode are actually the hybrid mode

15.5.6 Accumulation Layer Mode

If the nL product of the device is less than 10^{12} cm^{-2}, then the device is too short or the available charge density too low for a dipole domain to form. Nevertheless, a negative resistance is developed at microwave frequencies; and this can be exploited in making high gain negative resistance amplifiers.[11] Hakki[12] has derived an analytical expression for the admittance of a lightly doped Gunn diode that incorporates electron diffusion effects. He assumes the existence of a uniform DC electric field inside the device, and he makes use of a small signal assumption. Amplifier gains as high as 30 to 40 dB have been observed. Since a high resistivity sample is used, the output power and efficiency are relatively low compared to the other modes. Typically the power levels are on the order of microwatts and the efficiency in the 1% range. Since competitive devices have been found to be superior in most low level amplifier applications, this subcritically doped device has not found many applications.

Perlman, Upadhyayula, and Marx[13] suggested that a negative resistance could be generated by a super critically doped diode operating in the dipole domain mode. The useful negative resistance would lie outside the oscillation frequency. The circuit is designed to present a short to the diode at the transit time frequency so that spurious oscillations are prevented from entering the amplifier. Perlman et al. achieved 6 dB linear gain over the frequency range of 8 to 12 GHz. The saturated added power was 250 mW and the maximum output power was 400 mW. Clearly, this amplifier is not limited to the smaller power levels seen in the negative resistance mode device.

15.6 LARGE SIGNAL DIPOLE DOMAIN ANALYSIS

A fully formed dipole domain, traveling at approximately 10^7 cm/sec will reach its saturated shape after traveling between 0.1 to 1 μm. This section will consider devices where the domain has been fully formed and propagates uniformly without changing shape. Oscillators using the domain mode can be tuned over one or two octaves. Electronic tuning with either a varactor diode or ferrite device can cover bandwidths up to 20% or from 4 to 10 GHz respectively.

The output current from a Gunn diode resulting from an applied voltage is the sum of the drift, diffusion and displacement currents.

$$J(t) = qn(x,t)v(E) - qD(E)\frac{\partial n(x,t)}{\partial x} + \epsilon \frac{\partial E(x,t)}{\partial t} \quad (15.13)$$

In this expression, n is the instantaneous carrier concentration, $v(E)$ is the field dependent carrier velocity, D is the diffusion constant, and ϵ the dielectric constant of the material. Poisson's equation can be written as

$$\frac{\partial E}{\partial x} = \frac{q}{\epsilon}(n - N_d) \quad (15.14)$$

where N_d is the ionized donor concentration. These two equations describe the current-voltage behavior of the Gunn device. By solving (15.14) for n and substituting into (15.13), an explicit expression for the RF current density is obtained.

$$J(t) = v(E)\left[qN_d + \epsilon\frac{\partial E}{\partial x}\right] - D(E)\left[q\frac{\partial N_d}{\partial x} + \epsilon\frac{\partial^2 E}{\partial x^2}\right] + \epsilon\frac{\partial E}{\partial t} \quad (15.15)$$

This equation has been solved numerically by Lakshminarayana and Partain.[14] A second approach known as the equal areas rule described by Butcher[15,16] provides good insight into the domain field shape. The critical conclusion from his work is that a stable domain must travel a velocity v_d exactly equal to the velocity v_o of the conduction electron current outside the domain. The field and carrier concentration outside the domain remains constant at $E = E_o$ and $n = n_o$ respectively. As shown in Figure 15.5(e), the dipole depletion layer occurs at the trailing edge; and the accumulation layer, on the leading edge of the domain. Somewhere inside the dipole the carrier concentration will equal n_o. At this point the field inside the domain will be maximum, $E = E_d$.

By setting $y = x - v_d t$, (15.13) and (15.14) can be solved inside the domain. With this change of variable, both the time derivative of E in (15.13) and the spatial derivative in (15.14) may be expressed in terms of the common variable y.

$$\frac{\partial E}{\partial x} = \frac{\partial E}{\partial y} \tag{15.16}$$

$$\frac{\partial E}{\partial t} = -v_d \frac{\partial E}{\partial y} \tag{15.17}$$

Outside the domain $n = n_o = N_d$ and the current consists entirely of conduction current $J = q n_o v(E_o) = q n_o v_o$. Consequently, (15.17) and (15.14) can be written in terms of y as

$$q n_o v_o = q n v(E) - q D(E) \frac{\partial n}{\partial y} - v_{d\epsilon} \frac{\partial E}{\partial y} \tag{15.18}$$

$$\frac{\partial E}{\partial y} = \frac{q}{\epsilon}(n - n_o). \tag{15.19}$$

Since,

$$\frac{\partial n}{\partial y} = \frac{\partial n}{\partial E} \frac{\partial E}{\partial y}, \tag{15.20}$$

equation (15.19) can be used to eliminate the y variable in (15.18) with the following result.

$$\frac{qD}{\epsilon}(n - n_o) \frac{\partial n}{\partial E} = n[v(E) - v_d] - n_o[v_o - v_d] \tag{15.21}$$

Dividing through the last equation by nn_o and integrating both sides yields

$$\int_{n_o}^{n} \left(\frac{1}{n_o} - \frac{1}{n'}\right) dn' = \frac{\epsilon}{q n_o D} \int_{E_o}^{E} \left\{ [v(E') - v_d] - \frac{n_o}{n(E)}[v_o - v_d] \right\} dE' \tag{15.22}$$

The left-hand side is easily evaluated to give the formal solution for the domain fields when the diffusion constant is assumed to be independent of electric field.

$$\frac{n}{n_o} - \ln\left(\frac{n}{n_o}\right) - 1 =$$
$$\frac{\epsilon}{qDn_o} \int_{E_o}^{E} \left\{ [v(E') - v_d] - \frac{n_o}{n(E')} [v_o - v_d] \right\} dE' \quad (15.23)$$

The assumption of a field independent diffusion constant is reasonably accurate except perhaps for those electrons near the outer edges of the domain.[17] When $E = E_d$, then $n = n_o$ and the left-hand side of (15.23) must vanish. As seen in Figure 15.9, the integration of (15.23) may be performed over two regions with identical results. The result is zero whether the integration is carried out from E_o to E_d over the accumulation region where $n(E) > n_o$ or over the depletion region where $n(E) < n_o$. The contribution to the integral from the first term is independent of n. If $v_o \neq v_d$, the contribution of the second term will be different when done over the accumulation region or over the depletion region. Since this is clearly impossible, $v_o = v_d$, and (15.23) is reduced to the following expression.

$$\int_{E_o}^{E_d} [v(E') - v_d] \, dE' = 0 \quad (15.24)$$

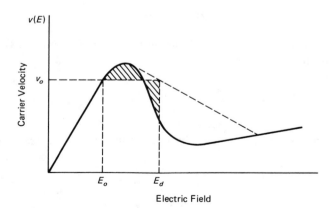

Figure 15.9. Equal areas rule states the areas above and below the characteristic are equal. *(Butcher[15] 1965 North-Holland Publishing Company, Amsterdam.)*

Thus the shaded areas in the two zones in Figure 15.9 must be equal. The dashed line represents the maximum field E_d vs. v_o curve, as determined by the equal areas rule. As the peak domain field E_d increases, $v_d = v_o$ decreases until the dashed line in Figure 15.9 intersects the static curve. If the applied voltage increased still further, the domain width, rather than E_d, would increase. The peak domain field E_d is related to the amount of charge depleted and accumulated within the dipole and to the value of the field outside the domain by the total applied voltage.

$$V = V_x + E_o L \qquad (15.25)$$

The excess domain voltage V_x is defined by Copeland[18] as the voltage drop across the dipole domain.

$$V_x = \int_0^L [E(x) - E_o]\, dx \qquad (15.26)$$

From Poisson's equation the width of the domain is

$$W = \frac{\epsilon}{qN_d}(E_d - E_o) \qquad (15.27)$$

which usually has a value in the 2 to 10 μm range. The excess voltage is then the triangular area under the curve in Figure 15.5.

$$V_x \cong \tfrac{1}{2}W(E_d - E_o) \qquad (15.28)$$
$$= \frac{\epsilon}{2qN_d}(E_d - E_o)^2$$

When this excess voltage reaches the anode, a pulse of current is delivered to the external circuit and a new dipole domain begins nucleating at the cathode. The oscillation frequency is approximately $f = v_d/L$.

15.7 CURRENT (VELOCITY)—FIELD CHARACTERISTIC

In the previous section, it was shown that the carrier velocity was dependent on the electric field so that over a range of electric field, the v-E slope was negative. In this section an analytical expression for the v-E curve will be given. Ruch and Kino[19] accurately measured the velocity-field characteristic of a relatively long 300 μm high quality sample of gallium arsenide. Electrons were injected into the device cathode with a 14 kV pulsed electron beam with rise and fall times approximately 0.1 ns long. The time required for the front of the pulse to reach the anode is a measure of the velocity of the dipole domain and can be measured as a function of the applied electric field.

Butcher and Fawcett[20] numerically integrated (15.21) incorporating a field dependent diffusion constant $D(E)$ and a carrier density $n(E)$. The diffusion constant is obtained from the Einstein relationship and is weighted between the lower and upper satellite valleys.

$$D(E) = \frac{n_\ell \mu_\ell T_\ell + n_u \mu_u T_u}{n_\ell + n_u} \frac{k}{q} \tag{15.29}$$

The results of their numerical calculation shows close agreement with experimental data. An approximate analytical expression for this curve can be found. The ratio of the carriers in the upper to lower valleys is proportional to a power of the electric field.[21]

$$\frac{n_u}{n_\ell} = \left(\frac{E}{E_o}\right)^\kappa \tag{15.30}$$

The electric field E_o is defined as the value of E where $n_u = n_l$, and κ is an as yet unknown constant. Two assumptions are made: (1) the mobilities are independent of the electric field, and (2) the local distribution of the electrons in the two valleys follows the electric field instantaneously. This latter assumption is valid for operating frequencies less than 10 GHz. When $n = n_l + n_u$, the expressions for the carrier densities in the two valleys can be found.

$$n_u = \frac{n}{1 + (E/E_o)^{-\kappa}} \tag{15.31}$$

$$n_\ell = n_u(E/E_o)^\kappa \tag{15.32}$$

Using a weighted average for the mean mobility

$$\bar{\mu} = \frac{\mu_\ell n_\ell + \mu_u n_u}{n}, \tag{15.33}$$

and the average velocity is

$$\bar{v} = \bar{\mu}E = \frac{E\mu_\ell[1 + (\mu_u/\mu_\ell)(E/E_o)^\kappa]}{1 + (E/E_o)^\kappa} \tag{15.34}$$

This simple expression may be used, in many cases, in place of the numerical calculation of the Boltzmann equation performed by Butcher and Fawcett.[20]

These authors obtained excellent correlation with the careful measurements of Ruch and Kino.[19] The velocity equation (15.34) can be made to approximate the experimental data if $E_o = 4.8$ kV/cm, $\kappa = 3.2$, $\mu_\ell = 8300$ and $\mu_u = 442$ cm^2/V-s. This correlation is shown in Figure 15.10 where (15.34) is plotted as a solid curve.

If interest is centered only on the negative resistance portion of the curve, then the expression

$$\frac{1}{v} = A + BE \qquad (15.35)$$

provides a closer approximation to the data[22], where $A = 2.5 \ 10^{-8}$ s/cm and $B = 0.64 \ 10^{-11}$ s/V.

McCumber and Chynoweth[9] used a temperature dependent model to find

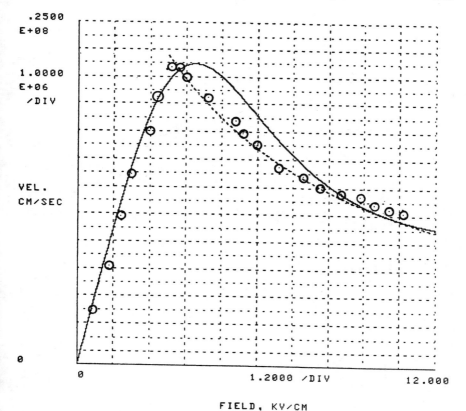

Figure 15.10. Measured *(Ruch and Kino[19])* and approximate static velocity-field characteristics where $\mu_\ell = 8300$ and $\mu_u = 442$ cm^2/V-sec.

an alternative expression for the velocity. Using the Maxwell-Boltzmann distribution function to define the relative populations of the lower and upper conduction valleys, they defined the energy of the carriers as

$$\mathcal{E} = \frac{3}{2} kT + \Delta\mathcal{E} \frac{N_u e^{-\Delta\mathcal{E}/kT}}{N_\ell + N_u e^{-\Delta\mathcal{E}/kT}} \quad (15.36)$$

where $\Delta\mathcal{E}$ is the energy separation between the lower and upper conduction valleys. From this and the energy transport equation, McCumber and Chynoweth found that when $kT > \Delta\mathcal{E}$,

$$\bar{v} = \mu_\ell E \frac{1 + \left(\left[1 + \frac{N_u}{N_\ell} + \frac{\Delta\mathcal{E}}{kT}\right]\frac{\mu_u}{\mu_\ell} - \left[1 + \frac{\Delta\mathcal{E}}{kT}\right]\right) e^{-\Delta\mathcal{E}/kT}}{1 + N_u/N_\ell e^{-\Delta\mathcal{E}/kT}} \quad (15.37)$$

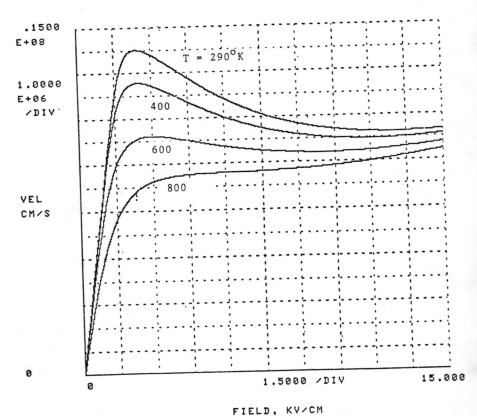

Figure 15.11. Temperature dependence of the velocity-field characteristic as given by McCumber and Chynoweth.[9]

In this equation the temperature is a function of the electric field, thermal relaxation time τ_t, the lattice temperature T_o, and the average carrier velocity. This relationship is given by

$$T = T_o + \frac{2}{3}\frac{\tau_t}{k} qE\bar{v}. \tag{15.38}$$

When the last two equations are solved simultaneously for a given ambient temperature T_o, the curves shown in Figure 15.11 are produced. Although this theory does not closely match the more recent experimental values, it does give a qualitative trend for the temperature dependence of the Gunn device. It shows that the magnitude of the negative resistance decreases with increasing lattice temperature until the negative resistance disappears above approximately 800°K.

15.8 NOISE IN THE GUNN DEVICE

The AM noise of a phase locked oscillating Gunn diode is comparable to that obtained in a kystron oscillator. However, it is characteristic of Gunn diodes to have a large FM noise component near the carrier frequency as well as a small AM noise power. Josenhans[23] compared the noise characteristics of a Gunn diode and an IMPATT diode. He found the Gunn diode had a 20 to 30 dB lower noise/carrier ratio than the IMPATT within 100 kHz of the carrier frequency. The FM noise, however, is much larger, and it is strongly dependent on the Q of the circuit. The rms frequency deviation in Hz in a bandwidth Δf is given by Edson[23] as

$$\Delta f_{\text{rms}} = \frac{f_o}{Q_{ex}}\frac{kT\,\Delta f}{P} \tag{15.39}$$

where P = power delivered to the load, and T = temperature of the effective noise generator in the diode. Experimental values for the Gunn diode follow this relationship qualitatively.

15.9 DEVICE EQUIVALENT CIRCUIT MODEL

A circuit designer would like to have a relatively simple equivalent circuit for the Gunn device. Ideally, the circuit would predict the RF output current when an RF voltage is applied to the terminals. The various known models have been reviewed by Pence[25] and placed in five categories: (1) physical models, (2) computer simulations, (3) analog models, (4) phenomenological models, and (5) empirical models. The physical model of Hobson[26] represents the diode as a series combination of a low-field and high-field equivalent circuit (Figure

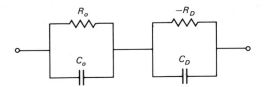

Figure 15.12. The physical model for the Gunn diode given by Hobson.[22]

15.12). Although the model does not explicitly include the effects of bias, it does provide a simple circuit topology. For a device using 1 ohm-cm GaAs where $N_d = 10^{15}$ cm^{-3}, 10 μm long and 100 μm in diameter, the element values are given as $R_o = 13$ ohms, $C_o = 0.08$ pF, $R_D = -500$ ohms, and $C_D = 0.30$ pF.

Based on known physical laws and known or assumed device and material parameters, computer simulations are most valuable in learning how the device operates. Because of the complexity of most computer simulations, they are not directly suitable for circuit design.

The analog model of Carroll and Giblin[27] is a refinement of the physical model. With the addition of ideal switches, additional lumped circuit elements, and ideal transistors, this model was used to predict current and voltage waveforms for various types of loads. By comparing the waveforms of the analog circuit with the actual device waveforms, the model is useful in determining the operating mode of a particular Gunn device.

The phenomenological models are characterized by modeling the observed voltage and current at the device terminals without undue concern for the internal physical laws of operation.

The model given by Robrock[28,29] is based on the physical model, but the equivalent circuit parameters are determined by a computer program that relates the RF current to a constant applied voltage. His model depicts the total voltage V as the sum of the low field voltage E_oL and the excess voltage V_x given by (15.25). Under quasistatic conditions, the current in the low field portion of the device is

$$I = qN_d A v(E_o) \tag{15.40}$$

where $v(E_o)$ is the electron velocity as obtained from the v vs. E curve, and A is the device cross sectional area. Since the voltage drop is simply $R_o I = E_o L$, an expression for the circuit resistance term is easily obtained (Figure 15.13).

$$R_o = \frac{E_o L}{qN_d v(E_o) A} \tag{15.41}$$

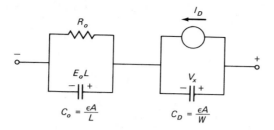

Figure 15.13. The lumped model for the Gunn diode given by Robrock.[24]

The high field region of the circuit model requires knowing V_x as a function of the field E_o outside the domain. This can be found by simultaneous solution of (15.25) and (15.26), repeated below for convenience.

$$V = V_x + E_o L \qquad (15.25)$$

$$V_x = \int_0^L [E(x) - E_o]\, dx \qquad (15.26)$$

Ordinarily the integral for the excess voltage is solved numerically and plotted. Intersection of this curve with the linear plot of (15.25) on a V_x versus E_o plot provides these two parameters for a given applied voltage V. Then (15.40) will give the value for domain current I.

The low field capacitance is found from the charge stored at the cathode and the anode. The resulting capacitance is simply

$$C_o = \frac{\epsilon A}{L}. \qquad (15.42)$$

For the high field domain, the displacement current is given by Robrock as

$$i_D = \frac{dQ_D}{dt} = \frac{dQ_D}{dV_x}\frac{dV_x}{dt}. \qquad (15.43)$$

From the divergence law $\nabla \cdot \mathbf{D} = \rho$, an approximate value for the domain width is given by

$$W = \frac{\epsilon}{qN_d}(E_d - E_o) \qquad (15.44)$$

where E_d is the maximum field of the domain. Also, the integral for the excess voltage (15.26) can be approximated by a triangular domain shape (Figure 15.5).

$$V_x = \frac{W}{2}(E_d - E_o) \tag{15.45}$$

By eliminating $(E_d - E_o)$ between these expressions, an approximate expression for the excess voltage is obtained.

$$V_x = \frac{qN_d}{2\epsilon} W^2 \tag{15.46}$$

The total charge stored in the domain depletion region is $Q_D = qN_d WA$ and by eliminating W with (15.46) the domain charge is found.

$$Q_D = qN_d \sqrt{\frac{V_x 2\epsilon}{qN_d}} A \tag{15.47}$$

The domain capacitance is then given by

$$C_D = \frac{dQ_D}{dV_x} = \sqrt{\frac{qN_d \epsilon}{2V_x}} A \tag{15.48}$$

or

$$C_D = \frac{\epsilon A}{W} \tag{15.49}$$

where W is seen to be proportional to the square root of V_x. The value for I_D in Figure 15.13 is obtained from the computer generated curve shown in Figure 15.14[30] and (15.40).

The equivalent circuit given by Robrock may be compared with that given by Hobson.[26] Hobson assumed the domain width was 2 μm which corresponds to an excess voltage from (15.46) of 2.81 V where $\epsilon_r = 12.9$. From Figure 15.14, $E_o = 1.8$ kV/cm, and from Figure 15.10 $v(E_o) = 1.4 \times 10^7$ cm/s. Then $R_o = 10.2$ ohms, $C_o = 0.077$ pF, $I = 176$ ma, and $C_D = 0.448$ pF. These values agree approximately with those given by Hobson. Robrock[29] showed that this model could be extended to account for the domain position as a function of time and dissolution process. However, for the purposes of circuit design, the approximate model above is adequate for an initial trial design.

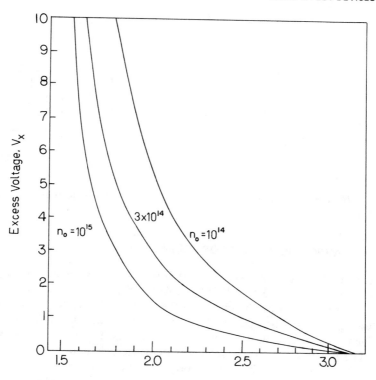

Outside Electric Field, (kV/cm)

Figure 15.14. The excess voltage in a domain vs. the electric field outside the domain. (Copeland[30] © 1966 IEEE.)

When operating in the LSA mode, no domain is formed, and the electric field across the device is approximately uniform. A simple equivalent circuit for this mode is the low voltage capacitance in parallel with a negative resistance (Figure 15.15).[31]

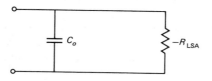

Figure 15.15. The equivalent circuit model for the LSA mode Gunn diode given by Harayan and Sterzer.[31]

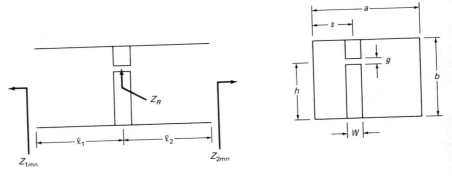

Figure 15.16. The equivalent circuit model for the LSA mode Gunn diode given by Harayan and Sterzer.[31]

15.10 WAVEGUIDE CIRCUIT FOR GUNN DIODES

A widely employed method of using the Gunn diode is to mount it on a post in a waveguide (Figure 15.16). It is possible to predict with high accuracy the impedance presented by the circuit to the diode package terminals using the analysis given by Eisenhart and Khan.[32] The dyadic Green's function is obtained for a rectangular waveguide by using the Lorentz reciprocity theorem. The reciprocity theorem relates the electric field and current density at two different locations. This method provides an expression for the impedance seen by a device placed on a post arbitrarily located in the waveguide. It is then possible to match the diode to the waveguide circuit by appropriate choice of waveguide and post dimensions. Furthermore, a Gunn device is usually placed in a waveguide with one arm shorted at a certain distance from the diode. Eisenhart's analysis can accommodate this type of circuit easily.

In general, the expression for the impedance Z_R seen by the diode involves the summation of an infinite number of waveguide modes, TE_{mn} and TM_{mn}. The loads on either side of the waveguide are described by the parameters τ_1 and τ_2 that are basically the input impedances of a terminated transmission line.

$$\tau_i = \frac{Z_{imn} + Z_{cmn} \tanh(\Gamma_{mn}\ell_i)}{Z_{cmn} + Z_{imn} \tanh(\Gamma_{mn}\ell_i)} \qquad i = 1,2 \qquad (15.50)$$

The impedances Z_{imn} are the terminating impedances in each arm, $i = 1, 2$; Γ_{mn} is the propagation constant; and Z_{cmn} is the characteristic impedance for the waveguide mode (m,n). These parameters are defined as follows.

$$\Gamma_{mn} = \sqrt{k_x^2 + k_y^2 + k_z^2} \qquad (15.51)$$

$$Z_{cmn} = \frac{j2\eta b}{ak} \frac{k^2 - k_y^2}{(2 - \delta_{n0})\Gamma_{mn}} \qquad (15.52)$$

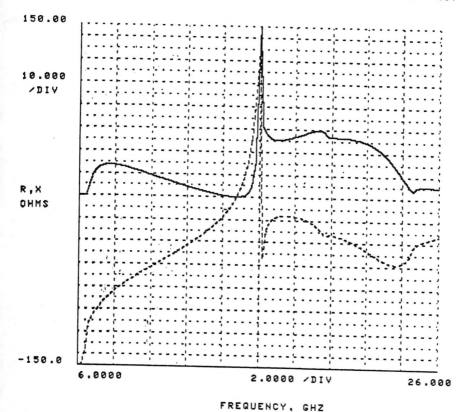

Figure 15.17. The impedance seen by a diode mounted in a post in a waveguide with inside dimensions of 0.4 × 0.9 inch. *(Eisenhart and Khan.[32])*

where

$$\delta_{n0} = \begin{cases} 1, & n = 0 \\ 0, & n \neq 0 \end{cases}$$

$$k_x = \frac{m\pi}{a}$$

$$k_y = \frac{n\pi}{b}$$

$$k = \omega/c$$

$$\eta = \sqrt{\mu_o/\epsilon_o}$$

The post-coupling factor and gap-coupling factor are

$$\kappa_{pm} = \sin k_x s \left(\frac{\sin \theta_m}{\theta_m}\right), \theta_m = \frac{m\pi W}{2a} \qquad (15.53)$$

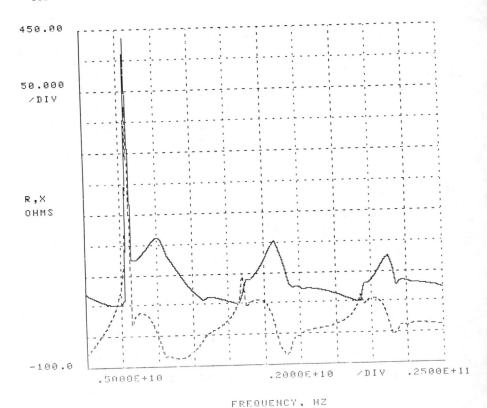

Figure 15.18. Broadband waveguide post impedance. *(Eisenhart and Khan.[32])*

$$\kappa_{gn} = \cos k_y h \left(\frac{\sin \phi_n}{\phi_n} \right), \phi_n = \frac{n\pi g}{2b} \quad (15.54)$$

The actual mathematics assumes the post may be replaced by an infinitesimally thin strip of width W. It has been found that a post of diameter d can be modeled with a strip if $W = 1.8\, d$. For the nth y-directed spatial harmonic, the impedance is the parallel combination of the waveguide impedances from the left and right sides.

$$Z_n = \left[\sum_{m=1}^{\infty} Z_{cmn} \left(\frac{\kappa_{pm}}{\kappa_{gn}} \right)^2 \tau_1 \right] \parallel \left[\sum_{m=1}^{\infty} Z_{cmn} \left(\frac{\kappa_{pm}}{\kappa_{gn}} \right)^2 \tau_2 \right] \quad (15.55)$$

The total admittance at the gap is the sum of these impedances.

$$Y_R = \sum_{n=0}^{\infty} 1/Z_n$$

The numerical value for the input impedance may be performed for a standard X-band waveguide with centered post and diode mounted on the surface of the waveguide ($x = a/2$, $b = 0$). The impedance presented to the diode from a waveguide matched in both directions is given in Figures 15.17 and 15.18. The quasi-periodicity in the impedance is evident in Figure 15.18. In truncating the infinite series, Eisenhart suggested the maximum $m = a/W$ and the maximum $n = b/g$. With this model for the waveguide, the previously given model for a Gunn device, and an equivalent circuit for the diode package, a complete description of a Gunn diode oscillator can be made. This was done by Eisenhart and Khan[33] where jumps in the oscillation frequency were accurately predicted using the above theory.

REFERENCES

1. J. B. Gunn, "Microwave Oscillations of Current III-V Semiconductors," *Solid State Communications*, Vol. 1, pp. 88–91, September 1963.
2. B. K. Ridley and T. B. Watkins, "The Possibility of Negative Resistance Effects in Semiconductors," *Proc. of the Physical Society*, Vol. 78, pp. 293–304, August 1961.
3. H. Kroemer, "Theory of the Gunn Effect," *Proc. of the IEEE*, Vol. 52, p. 1736, December 1964.
4. B. K. Ridley, "Specific Negative Resistance in Solids," *Proc. of the Physical Society*, Vol. 82, pp. 954–966, 1963.
5. C. Hilsum, "Transferred Electron Amplifiers and Oscillators," *Proc. of the IRE*, Vol. 50, pp. 185–189, February 1962.
6. A. R. Hutson, A. Jayaraman, A. G. Chynoweth, A. S. Corell, and W. L. Feldman, "Mechanism of the Gunn Effect from a Pressure Experiment," *Physical Review Letters*, Vol. 14, pp. 639–641, April 1965.
7. W. Shockley, "Negative Resistance Arising from Transit Time in Semiconductor Diodes," *Bell System Technical Journal*, Vol. 33, pp. 799–826, July 1954.
8. J. A. Copeland, "LSA Oscillator-Diode Theory," *Journal of Applied Physics*, Vol. 38, pp. 3096–3101, July 1967.
9. D. E. McCumber and A. G. Chynoweth, "Theory of Negative-Conductance Amplification and of Gunn Instabilities in 'Two-Valley' Semiconductors," *IEEE Trans. on Electron Devices*, Vol. ED-13, pp. 4–21, January 1966.
10. J. A. Copeland, "A New Mode of Operation for Bulk Negative Resistance Oscillators," *Proc. of the IEEE*, Vol. 54, pp. 1479–1480, October 1966.
11. H. W. Thim and M. R. Barber, "Microwave Amplification in GaAs Bulk Semiconductor," *IEEE Trans. on Electron Devices*, Vol. ED-13, pp. 110–114, January 1966.
12. B. W. Hakki, "Amplification in Two-valley Semiconductors," *Journal of Applied Physics*, Vol. 38, pp. 808–818, February 1967.
13. B. S. Perlman, C. L. Upadhyayula, and R. E. Marx, "Wideband Reflection-Type Electron Amplifiers," *IEEE Trans. on Microwave Theory and Techniques*, Vol. MTT-18, pp. 911–921, November 1970.
14. M. R. Lakshminarayana and L. D. Partain, "Numerical Simulation and Measurement of

Gunn Device Dynamic Microwave Characteristics," *IEEE Trans. on Electron Devices*, Vol. ED-27, pp. 546–552, March 1980.
15. P. N. Butcher, "Theory of Stable Domain Propagation in the Gunn Effect," *Physics Letters*, Vol. 19, pp. 546–547, December 1965.
16. P. N. Butcher, W. Fawcett, and C. Hilsum, "A Simple Analysis of Stable Domain Propagation in the Gunn Effect," *British Journal of Applied Physics*, Vol. 17, pp. 841–850, 1966.
17. I. B. Bott and W. Fawcett, "The Gunn Effect in Gallium Arsenide," *Advances in Microwaves*, ed. by Leo Young, New York: Academic Press, p. 251, 1968.
18. J. A. Copeland, "Stable Space-Charge Layers in Two-Valley Semiconductors," *Journal of Applied Physics*, Vol. 37, pp. 3602–3609, August 1966.
19. J. G. Ruch and G. S. Kino, "Measurement of the Velocity-Field Characteristic of Gallium Arsenide," *Applied Physics Letters*, Vol. 10, pp. 40–42, January 15, 1967.
20. P. N. Butcher and W. Fawcett, "Calculation of the Velocity-Field Characteristic for Gallium Arsenide," *Physics Letters*, Vol. 21, pp. 489–490, June 15, 1966.
21. H. Kroemer, "Detailed Theory of the Negative Conductance of Bulk Negative Mobility Amplifiers in the Limit of Zero Ion Density," *IEEE Trans. on Electron Devices*, Vol. ED-14, pp. 476–492, September 1967.
22. P. N. Robson, G. S. Kino, B. Fay, "Two-Port Microwave Amplification in Long Samples of Gallium Arsenide," *IEEE Trans. on Electron Devices*, Vol. ED-14, pp. 612–615, September 1967.
23. J. Josenhans, "Noise Spectra of Read Diode and Gunn Oscillators," *Proc. of the IEEE*, Vol. 54, pp. 1478–1479, October 1966.
24. W. A. Edson, "Noise in Oscillators," *Proc. of the IRE*, Vol. 48, pp. 1454–1466, August 1960.
25. I. W. Pence, *Impedance of Packaged Gunn Diodes and Parametric Circuit Applications*, Ph.D. Dissertation, Ann Arbor, Michigan: The University of Michigan, 1970.
26. G. S. Hobson, "Small-Signal Admittance of a Gunn Effect Device," *Electronics Letters*, Vol. 2, pp. 207–208, June 1966.
27. J. E. Carroll and R. A. Giblin, "A Low Frequency Analog for a Gunn Effect Oscillator," *IEEE Trans. on Electron Devices*, Vol. ED-14, pp. 640–656, October 1967.
28. R. B. Robrock II, "A Lumped Model for Characterizing Single and Multiple Domain Propagation in Bulk GaAs," *IEEE Trans. on Electron Devices*, Vol. ED-17, pp. 93–102, February 1970.
29. R. B. Robrock II, "Extension of the Lumped Bulk Device Model to Incorporate the Process of Domain Dissolutions," *IEEE Trans. on Electron Devices*, Vol. ED-17, pp. 103–107, February 1970.
30. J. A. Copeland, "Electrostatic Domains in Two-Valley Semiconductors," *IEEE Trans. on Electron Devices*, Vol. ED-13, pp. 189–192, January 1966.
31. S. Y. Harayan and F. Sterzer, "Transferred Electron Amplifiers and Oscillators," *IEEE Trans. on Microwave Theory and Techniques*, Vol. MTT-18, pp. 773–783, November 1970.
32. R. L. Eisenhart and P. J. Khan, "Theoretical and Experimental Analysis of a Waveguide Mounting Structure," *IEEE Trans. on Microwave Theory and Techniques*, Vol. MTT-19, pp. 706–719, August 1971.
33. R. L. Eisenhart and P. J. Khan, "Some Tuning Characteristics and Oscillation Conditions of a Waveguide-Mounted Transferred-Electron Diode Oscillator," *IEEE Trans. on Electron Devices*, Vol. ED-19, pp. 1050–1055, September 1972.

Chapter 16
Bipolar and Field Effect Transistors

16.1 INTRODUCTION

The two most commonly used microwave two-port semiconductor devices are the silicon bipolar transistor and the GaAs metal-semiconductor field effect transistor (MESFET). The commonly used symbols for these two devices are shown in Figure 16.1. The bipolar transistor is used predominately in applications where the operating frequency is below 4 to 5 GHz. Compared with the MESFET it can often provide adequate noise figure and gain at greatly reduced cost, with better reliability, and superior repeatability. These advantages are the result of the large expenditure of time and money spent on the development of the technology for low frequency transistors. Microwave bipolar transistors make full use of the developed technology of silicon processing.[1]

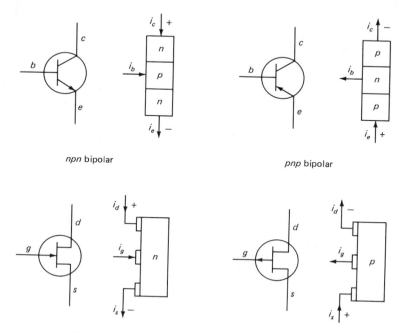

Figure 16.1. Symbols for the bipolar and FET transistor.

355

Above approximately 4 to 5 GHz, or when improved gain or noise figure is required, the GaAs MESFET is the preferred device. The considerably higher input impedance of the MESFET is often helpful in high frequency designs. The MESFET also has a negative temperature coefficient that prevents thermal runaway from occurring. Therefore, when the heat from thermal dissipation increases, the current through the device decreases. The MESFET can operate at higher frequencies and have smaller switching speeds than the bipolar transistor, because the latter suffers from minority carrier storage effects. Since the MESFET is a majority carrier device, it is not slowed down by this mechanism. The MESFET has a low voltage drop along the conduction channel, good heat dissipation properties, and better distortion characteristics than the bipolar transistor.

This chapter is concerned with the understanding of the operation of these two devices and their applications in typical microwave amplifier circuits. The following two sections are concerned with the physics of the two transistors. In section 16.4, amplifier gain and stability are derived for the two-port device, and the final section describes broadband design of an amplifier.

16.2 BIPOLAR TRANSISTORS

Since microwave bipolar transistors are scaled down versions of lower frequency devices, they may be understood in terms of the same basic theory of operation as their low frequency counterparts. More care, however, must be given to package design and minimizing device parasitic elements in order to achieve acceptable results. Most transistors today are made by the planar process. In this method an interdigital emitter-base geometry is diffused over the collector region (Figure 16.2). The lightly doped epitaxial collector region serves to provide high collector-base breakdown voltages while also providing a low capacitance.

The low frequency model is based on solving (10.6) through (10.9) for the minority carrier density in the base region. From this, the current density flow-

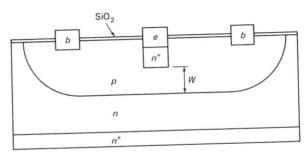

Figure 16.2. Profile of an *npn* bipolar transistor.

BIPOLAR AND FIELD EFFECT TRANSISTORS 357

ing through the device is determined. It may be assumed that the electric field is zero and that time derivatives are zero for the DC case. Then for the *npn* transistor

$$J_n = qD_n \frac{\partial n}{\partial x} \tag{16.1}$$

and

$$\frac{\partial J_n}{\partial x} = q \frac{n - n_p}{\tau_n}. \tag{16.2}$$

By eliminating the current density a simple differential equation for the minority carrier electron density in the base is obtained.

$$\frac{\partial^2 n}{\partial x^2} = \frac{n - n_p}{L_n^2} \tag{16.3}$$

The diffusion length $L_n = \sqrt{D_n \tau_n}$ is a measure of how far an electron can travel before it recombines with a hole. The general solution to (16.3) is

$$n = A \exp(-x/L_n) + B \exp(x/L_n) + n_p. \tag{16.4}$$

The boundary conditions at each of the two junctions, the base-emitter interface and the base-collector interface, are

$$n(0) = n_{pb} \exp(qV_{eb}/kT) \tag{16.5}$$
$$n(W) = n_{pb} \exp(qV_{bc}/kT) \tag{16.6}$$

where k is Boltzmann's constant and T is the absolute temperature of the transistor (Figure 16.3). Substituting these into (16.4) gives an expression for the electron density in the base region.

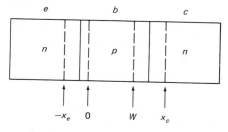

Figure 16.3. The undepleted base width W for an *npn* transistor.

$$n = n_{pb} + n_{pb} \frac{\sinh \frac{W-x}{L_n}}{\sinh \frac{W}{L_n}} [\exp(qV_{eb}/kT) - 1]$$

$$+ n_{pb} \frac{\sinh \frac{x}{L_n}}{\sinh \frac{W}{L_n}} [\exp(qV_{bc}/kT) - 1] \quad (16.7)$$

If the collector region is far away from the emitter, the above expression reduces to the diode equation and the emitter junction is unaffected by the collector. In such a case, collector current control by the base voltage becomes nil. A small additional term may be added to account for the hole density at the emitter-base interface. This value can be obtained from solving the dual of (16.3)

$$\frac{\partial^2 p}{\partial x^2} = \frac{p - p_n}{\tau_p} \quad (16.8)$$

for the boundary conditions at $x = -x_e$ and $x = -\infty$.

$$p(-x_e) = p_n[\exp(qV_{eb}/kT - 1)] \quad (16.9)$$
$$p(-\infty) = 0 \quad (16.10)$$

This results in

$$p = p_{ne}[\exp(qV_{eb}/kT) - 1] \exp[(x + x_e)/L_{pe}] \quad x \leq -x_e \quad (16.11)$$

and similarly for the collector-base junction

$$p = p_{nc}[\exp(qV_{bc}/kT) - 1] \exp[-(x - x_c)/L_{pc}] \quad x \geq x_c. \quad (16.12)$$

The electron current density is found by performing the indicated differentiation in (16.1); the hole current density is found by performing the differentiation indicated by the corresponding expression for the hole current.

$$J_p = -qD_p \frac{\partial p}{\partial x} \quad (16.13)$$

The total emitter and collector currents are found by summing the hole and electron currents.

$$J_e = J_n(x = 0) + J_p(x = -x_e)$$
$$= -n_{pb}\frac{qD_n}{L_n}\left\{\coth\left(\frac{W}{L_n}\right)\left[\exp\left(\frac{qV_{eb}}{kT}\right) - 1\right]\right.$$
$$\left. - \operatorname{csch}\left(\frac{W}{L_n}\right)\left[\exp\frac{qV_{bc}}{kT} - 1\right]\right\} \quad (16.14)$$
$$- \frac{qD_p p_{ne}}{L_{pe}}\left[\exp\left(\frac{qV_{eb}}{kT}\right) - 1\right]$$

$$J_c = J_n(x = W) + J_p(x = x_c)$$
$$= -n_{pb}\frac{qD_n}{L_n}\left\{\operatorname{csch}\left(\frac{W}{L_n}\right)\left[\exp\left(\frac{qV_{eb}}{kT}\right) - 1\right]\right.$$
$$\left. - \coth\left(\frac{W}{L_n}\right)\left[\exp\left(\frac{qV_{bc}}{kT}\right) - 1\right]\right\} \quad (16.15)$$
$$+ \frac{qD_p p_{nc}}{L_{pc}}\left[\exp\left(\frac{qV_{bc}}{kT}\right) - 1\right]$$

The extra subscript e or c on the diffusion length is necessary since the diffusion lengths in the emitter and collector regions may differ because of their different doping concentrations. Ordinarily the emitter-base junction is forward biased ($V_{eb} > 0$), the collector-base junction is reverse biased ($V_{bc} < 0$), and $W/L_n \ll 1$, so that

$$J_c = J_e = \frac{-qn_{pb}D_n}{W}[\exp(qV_{eb}/kT)]. \quad (16.16)$$

The emitter efficiency γ is defined as the ratio of the minority carrier emitter current to the total emitter current. Under normal bias conditions, the terms involving V_{bc} may be neglected.

$$\gamma = \frac{J_n(x = 0)}{J_n(x = 0) + J_p(x = -x_e)} \quad (16.17)$$
$$= \frac{1}{1 + \frac{D_p p_{ne} L_n}{D_n n_{pb} L_{pe}}\tanh\frac{W}{L_n}} \quad (16.18)$$

The ratio of the minority carriers reaching the collector to the number entering the base from the emitter is given by

$$\alpha_T = \frac{J_n(W)}{J_n(0)} \quad (16.19)$$
$$\cong \frac{1}{\cosh(W/L_n)} \quad (16.20)$$

Both γ and α_T are less than unity and describe the degree to which current must be supplied to the base. To ensure γ is close to one, the hole current must be much smaller than the electron current. This can be enhanced by requiring the electron density, n_{pb}, in the base to be much larger than the equilibrium hole density in the emitter. This is accomplished by making the impurity concentration in the emitter very high. Making α_T as large as possible requires making the base width W small.

The frequency dependent expression for γ and α_T are obtained by including the time dependent term of (10.9), and by assuming sinusoidal frequency dependence. Consequently, (16.2) is replaced by

$$\frac{\partial J_n}{\partial x} = \frac{qn(1 + j\omega\tau_n)}{\tau_n}. \qquad (16.21)$$

Replacing τ_n in (16.18) and (16.20) by

$$\tau_n^* = \frac{\tau_n}{1 + j\omega\tau_n} \qquad (16.22)$$

the AC solution of (16.3) can be found. Assuming $W/L_n \ll 1$, the cosh (W/L_n) term in α_T can be approximated by

$$\cosh(W/L_n) = 1 + \frac{W^2}{2D_n\tau_n} + \frac{jW^2\omega}{2D_n} \qquad (16.23)$$

$$\cong 1 + \frac{jW^2\omega}{2D_n}. \qquad (16.24)$$

With this approximation the magnitude of α_T and the transistor current gain $\alpha = \partial J_c/\partial J_e$ drops to approximately $1/\sqrt{2}$ of its low frequency value when

$$f = f_\alpha = \frac{2D_n}{W^2 2\pi}. \qquad (16.25)$$

This is known as the *alpha cutoff frequency*. High frequency operation is strongly dependent on the base width.

A frequency dependent model for the bipolar transistor is shown in Figure 16.4. The frequency at which the short circuit current gain $\beta = \partial J_c/\partial J_b$ reduces to 1 is known as the f_T cutoff frequency,

$$f_T = \frac{1}{2\pi\tau_{ec}}. \qquad (16.26)$$

BIPOLAR AND FIELD EFFECT TRANSISTORS 361

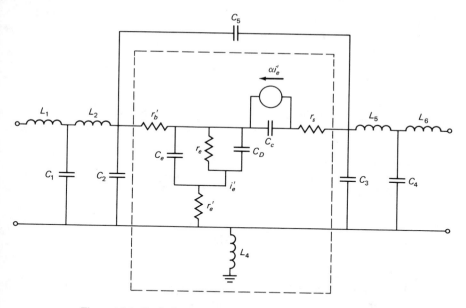

Figure 16.4. Equivalent circuit for microwave *npn* bipolar transistor.

The emitter to collector transit time τ_{ec} may be regarded as the sum of delay times for the signal to travel from the emitter to the collector.[2,3]

$$\tau_{ec} = t_e + t_b + t_d + t_c \qquad (16.27)$$

t_e = emitter-base junction capacity charging time
t_b = base transit time
t_d = collector depletion layer transit time
t_c = collector capacitance charging time

The first of these delay times is given by

$$t_e = r_e(C_e + C_c + C_x) \qquad (16.28)$$

where C_x includes all extraneous capacitances including those associated with the package. The space charge resistance of the emitter is obtained from the Shockley diode equation that is approximated by

$$r_e = \frac{kT}{qi_e}. \qquad (16.29)$$

The base transit time t_b can be found from the alpha cutoff frequency derived from the transport factor α_T.

$$t_b = \frac{1}{2\pi f_\alpha} = \frac{W^2}{2D_n} \qquad (16.30)$$

This expression is valid for a uniformly doped base. For the drift transistor where the doping density decreases in going from the emitter to the collector, this time delay may decrease by as much as a factor of 30.

The collector depletion layer transit time can be obtained from the scattering limited velocity v_s of the particles going through the base-collector junction. Cooke[2] found that this can be approximated by

$$t_d = \frac{x_c - W}{2v_s} \qquad (16.31)$$

where $v_s \cong 8 \times 10^4$ m/sec.

Finally in today's transistors, the time delay caused by the collector RC cutoff frequency $= 1/t_c$ may be considered negligible.

Most microwave transistors today are used in the grounded emitter configuration. Consequently, the most meaningful high frequency figure of merit is f_T. At high frequencies, the transistor power gain falls at the rate

$$\beta^2 \propto (f_T/f)^2 \qquad (16.32)$$

or about 6 dB/octave.

16.3 THE UNIPOLAR MESFET

To date the most successful microwave field effect transistors are the n-type GaAs metal semiconductor field effect transistors. These have the structure shown in Figure 16.5. The source, gate, and drain contacts must be made from metal, so there is always a metal-semiconductor junction at each of these contacts. For the gate, this junction forms a Schottky barrier while the other two are ohmic contacts. As shown in chapter 12, the Schottky barrier acts much like a pn junction. When the semiconductor and metal are brought into contact, the Fermi levels line up as required for energy equilibrium. Since $\mathcal{E}_{F \text{ semi}} > \mathcal{E}_{F \text{ metal}}$, the excess n-type carriers in the semiconductor spill over into the metal. This leaves a bound positive charge in the semiconductor and a negative charge in the metal. Thus, the energy bands in the semiconductor must bend. A depletion layer capacitance is formed under the gate given by

$$C = C_o[1 - (v - kT/q)/\phi]^{-1/2}. \qquad (12.6)$$

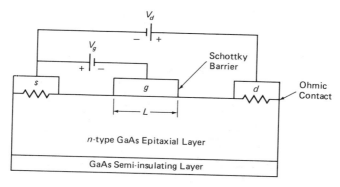

Figure 16.5. Structure for n-type MESFET.

The source and drain contacts form ohmic contacts. An ohmic contact is basically a very narrow Schottky barrier ~10 Å thick in which tunneling through the barrier can readily occur (Figure 16.6). In practice, it is made by substantially increasing the donor impurity concentration in the semiconductor near the contact. Thus, the contact provides very low resistance to currents flowing in either direction.

16.3.1 MESFET Operation

If two ohmic contacts are placed on either side of the semiconductor (Figure 16.7), electrons will migrate from the source through the n-type GaAs semiconductor channel to the drain. For silicon, the carrier velocity will saturate when the electric field reaches a certain point. For GaAs, the velocity will first increase with the electric field intensity, then decrease because of the low mobility satellite valleys. This is discussed in chapter 15 in conjunction with the Gunn device.

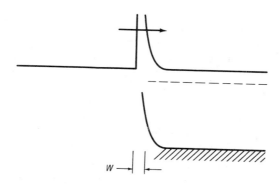

Figure 16.6. Ohmic contact at source and drain contacts.

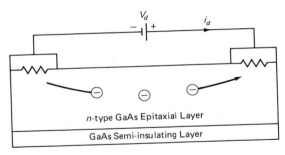

Figure 16.7. Two ohmic contacts on *n*-type semiconductor material.

The drain to source current is

$$I_d = Wqn(x)d(x)v(x) \tag{16.33}$$

where
W = gate width,
n = electron density,
v = electron drift velocity, and
d = conductive layer thickness.

For the device with no gate (Figure 16.7), the thickness *d* is the entire *n*-type epitaxial layer. However, when a gate electrode is added, a depletion layer may be formed in the semiconductor that effectively decreases the conductive layer thickness under the gate and thereby decreases the I_d. By varying the gate voltage the gate current I_d can be controlled.

Complication of this feature occurs in GaAs because of the non-monotonic character of the velocity-field characteristic. A numerical analysis by Himsworth[4] shows that, for a 3 μm gate length, the saturation velocity reaches a peak near the center of the gate, and another velocity peak occurs between the gate and the drain. This phenomenon is illustrated in Figure 16.8 where the first peak occurs at x_1. Since the channel thickness continues to decrease with increasing *x*, the electric field rises above the critical value. Electrons start transferring to the low mobility satellite valleys and result in a decrease in electron velocity. To compensate for this decrease of $v(x)$ in (16.33), $n(x)$ must increase in this region (Figure 16.8). Near the drain side of the gate at x_2 the channel thickness increases, the electric field decreases, the electron velocity rises toward a peak, and charge depletion ($\Delta n(x)$) takes place. The charge accumulation to the left of x_2 and the charge depletion to the right of x_2 form a stationary dipole in the channel.

Most microwave MESFETs have channel lengths on the order of 1 to 0.5 μm so that equilibrium conditions do not occur. In this case, as the electrons enter the high field region, they are accelerated to a peak velocity greater than

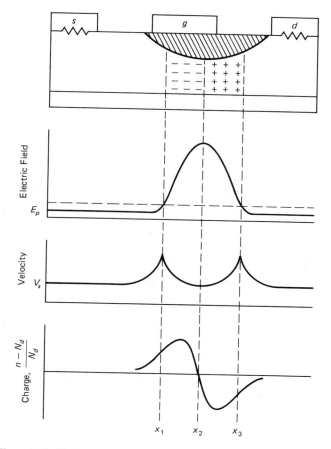

Figure 16.8. Field, velocity, and charge distribution in GaAs MESFET.

$2 \times$ the equilibrium value in about 0.6 μm. This velocity overshoot shifts the accumulation layer into the gap between the gate and the drain, shortens the electron transit time through the channel, and presumably improves the high frequency response of the MESFET.

16.3.2 Small Signal Model

The small signal common source equivalent circuit given by Liechti[5] is shown in Figure 16.9. This lumped model is useful up to about 12 GHz before transmission line elements would have to be added. The diagram with the model indicates the position and source of the various elements in the model. The intrinsic model is indicated within the dotted line. The total gate to channel capacitance is $C_{dg} + C_{gs}$, the capacitance of the dipole layer is represented by

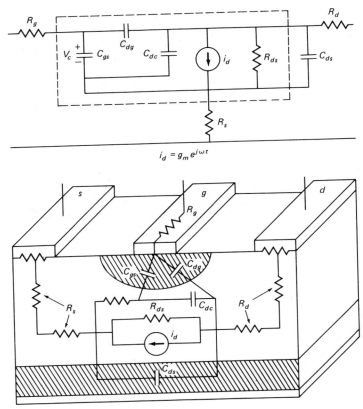

Figure 16.9. Equivalent circuit for a MESFET and the physical origin of the elements. *(Liechti[5] © 1976 IEEE.)*

C_{dc}, and the channel resistance is represented by R_i and R_{ds}. The voltage between the gate and source, V_c, controls the drain to source current given by $i_d = V_c g_m \exp(-j\omega t_o)$. The additional time delay t_o between 5 and 10 ps is associated with the carrier transit time when the electric field exceeds the critical value. Liechti gives the typical values for a MESFET (Table 16.1). This equivalent circuit when expressed in terms of S parameters will closely match measured S parameters.

16.3.3 Large Signal Analysis

For large signal applications such as in power amplifiers, oscillators, and multipliers, a small signal model or direct high power measurements are often used in the circuit design. If the small signal model is used, the assumption is made that there will be only minor changes in the transistor characteristics for a

Table 16.1. Hewlett-Packard Low
Noise-GaAs MESFET with 1 μm ×
500 μm Gate (Liechti[5]).

g_m = 53 mmhos	C_{ds} = 0.12 pF
t_o = 5 ps	R_g = 2.9 ohm
C_{gs} = 0.62 pF	R_d = 3 ohm
C_{dg} = 0.014 pF	R_s = 2 ohm
C_{dc} = 0.02 pF	V_{ds} = 5 volt
R_i = 2.6 ohm	V_{gs} = 0 volt
R_{ds} = 400 ohm	I_d = 70 ma

large signal application. A more exact procedure calls for a large signal simulation. Madjar and Rosenbaum[6] use analytical models for the carrier drift velocity and the electron density in the transition area between the depletion and conduction regions. Because of the use of these analytical expressions, their particular method provides reasonably accurate answers with only one numerical iteration. The speed advantage of their method makes possible device-circuit iteration studies with networks containing several devices.

16.3.4 Frequency Limitations

Just as the transistor cutoff frequency f_T was defined as the frequency where the short circuit current gain falls to unity, so the MESFET cutoff frequency is defined as that frequency where $\partial i_c / \partial i_g = 1$. From the model in Figure 16.9, the current gain is unity when the current through C_{gs} is equal to the source current, $g_m V_c$.

$$f_T = \frac{g_m}{2\pi C_{gs}} \quad (16.34)$$

However, owing to impedance transformations, power gain is achievable at frequencies above f_T. The frequency of maximum oscillation is given by[5]

$$f_{\max} = \frac{f_T}{2\sqrt{r_1 + f_T \tau_3}} \quad (16.35)$$

where the input to output resistance ratio is

$$r_1 = \frac{R_g + R_i + R_s}{R_{ds}} \quad (16.36)$$

and the time constant is

$$\tau_3 = 2\pi R_g C_{dg}. \tag{16.37}$$

The unilateral power gain can be found to be approximately[7]

$$G_u = (f_{max}/f)^2. \tag{16.38}$$

Like the bipolar microwave transistor, it has a gain slope of -6 dB/octave as the frequency increases. It is apparent that microwave MESFETs must have large f_T and low r_1 and τ_3. This can be achieved by minimizing C_{gs}, C_{dg}, R_g, and R_s, while at the same time optimizing the resistance ratio R_{ds}/R_i. In terms of device geometry, decreasing the gate length decreases C_{gs} and increases g_m. Thus, for very short gate lengths (~ 1 μm or less), f_T is inversely proportional to the gate length.[7] For the gate to exert adequate control over the channel current, it has been found that the gate length dimension L is limited by the channel thickness d so that $L/d > 1$. This, in turn, is limited by the maximum practical impurity doping concentration in the channel. Theoretically a minimum gate length $L = 0.1$ μm would have an associated $f_T = 100$ GHz.[8]

16.3.5 An Analytical Model for the MESFET

Computer aided design and measurement techniques are approaching the point where amplifier design is becoming "automatic." But, intuitive understanding of the device operation and the best design approach are done by an analytical model of the device that approximates the device physics. In this section an expression for the drain current as a function of the applied bias voltages will be found. From this function, an analytical model for the MESFET is derived.

The predominate control of the drain current is attained by the amount of voltage applied to the gate electrode. As the gate voltage V_g becomes more negative a larger portion of the region under the channel becomes depleted of charge carriers, thereby further restricting the current flow in the channel. An expression for the drain current based on a uniformly doped n region and the gradual channel approximation can be found by solving Poisson's equation. The model for the gradual channel approximation is shown in Figure 16.10 and can be expressed mathematically as

$$\left|\frac{\partial \psi}{\partial y}\right| \ll \left|\frac{\partial \psi}{\partial x}\right| \tag{16.39}$$

where ψ is the potential in the channel, and x, y are the longitudinal and vertical directions in the channel respectively. From (10.16) the depletion region height h is given by

$$h = \left[\frac{2\epsilon(\phi + V(x) - V_g)}{qN_d}\right]^{1/2} \tag{16.40}$$

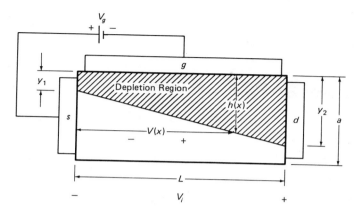

Figure 16.10. Ideal FET model for longitudinal current flow.

where the applied drain voltage is $-V(x)$ and the applied gate voltage is $-V_g$. The depths of the depletion layers at the two extreme ends of the gate region are

$$y_1 = \left[\frac{2\epsilon(\phi - V_g)}{qN_d}\right] \qquad x = 0 \qquad (16.41)$$

$$y_2 = \left[\frac{2\epsilon(\phi - V_g + V_i)}{qN_d}\right] \qquad x = L. \qquad (16.42)$$

The total voltage is known as the pinch off voltage V_p when the conductive channel is reduced to zero at $y_2 = a$. From (16.42) the pinch off voltage is

$$V_p = \frac{qN_d a^2}{2\epsilon}. \qquad (16.43)$$

The conduction current density in the channel is found from Ohm's law to be

$$I_d = -q\mu N_d E A \qquad (16.44)$$
$$= q\mu N_d \frac{dV}{dx}(a - h)W.$$

An expression for dh/dV is obtained from (16.40).

$$dV = \frac{qN_d h}{\epsilon} dh \qquad (16.45)$$

This is substituted into (16.44) and integrated along the channel length from $x = 0$ to $x = L$.

$$I_d = \frac{\mu(qN_d)^2 W}{\epsilon L} \int_{y_1}^{y_2} (a-h)h\,dh \qquad (16.46)$$

Integrating this expression gives

$$I_d = g_o \frac{(qN_d)}{6\epsilon}\left[3(y_2^2 - y_1^2) - \frac{2}{a}(y_2^3 - y_1^3)\right] \qquad (16.47)$$

where

$$g_o \triangleq \frac{q\mu N_d aW}{L}. \qquad (16.48)$$

Finally, substituting the values for y_1 and y_2 from (16.41) and (16.42) yields the fundamental channel current expression for field effect transistors.

$$I_d = g_o\left[V_i - \frac{2[(\phi - V_g + V_i)^{3/2} - (\phi - V_g)^{3/2}]}{3\sqrt{V_p}}\right] \qquad (16.49)$$

This is valid when $V_i \leq V_s$ where V_s is the saturation voltage. Under current saturation where $V_i = V_s$, the transconductance is given by

$$g_m = \left.\frac{\partial I_d}{\partial V_g}\right|_{V_i=V_s} = g_o\left[\frac{(V_s + \phi - V_g)^{1/2} - (\phi - V_g)^{1/2}}{\sqrt{V_p}}\right] \qquad (16.50)$$
$$V_i = V_g$$

Shur[9] pointed out that for typical MESFETs, where the gate length $L \cong 1$ μm, the saturation voltage is

$$V_s \ll \phi - V_g. \qquad (16.51)$$

Substituting this inequality into (16.50) yields

$$g_m \cong \left[\frac{qN_d \epsilon}{2(\phi - V_g)}\right]^{1/2} v_s W \qquad (16.52)$$

where v_s is the saturation velocity and W is the gate width.

The capacitances C_{dg} and C_{gs} are obtained from differentiation of the total charge under the gate in the linear region where $V_i \leq V_s$. The charge is obtained by integrating the depleted charge density over the depleted region. Assuming $V(x) = E_i x$, the depletion layer thickness from (16.40) is

$$h(x) = \left[\frac{2\epsilon}{qN_d}(\phi + E_i x - V_g)\right]^{1/2}. \tag{16.53}$$

The charge under the gate is then found from

$$Q = qN_d W \int_0^L h(x)\, dx \tag{16.54}$$

$$= \frac{2WL\sqrt{2qN_d\epsilon}}{3V_i}[(\phi + V_i - V_g)^{3/2} - (\phi - V_g)^{3/2}]. \tag{16.55}$$

The drain to gate and gate to source capacitances can now be found by straightforward differentiation.

$$C_{dg} = \frac{\partial Q}{\partial V_i}\bigg|_{V_g=\text{const}}$$

$$= \frac{2WL\sqrt{2qN_d\epsilon}}{3V_i^2}\left[\frac{3V_i}{2}(\phi + V_i - V_g)^{1/2}\right. \tag{16.56}$$

$$\left. + (\phi - V_g)^{3/2} - (\phi + V_i - V_g)^{3/2}\right]$$

$$C_{gs} = \frac{\partial Q}{\partial V_g}\bigg|_{V_i-V_g=\text{const}}$$

$$= \frac{2WL\sqrt{2qN_d\epsilon}}{3V_i^2}\left[(\phi + V_i - V_g)^{3/2}\right. \tag{16.57}$$

$$\left. - (\phi - V_g)^{3/2} - \frac{3}{2}V_i(\phi - V_g)^{1/2}\right]$$

Estimates for the cutoff frequency f_T given in (16.34) can be made with these expressions. Shur[9] has shown that $f_T \cong 25.5$ GHz for a 1 μm gate length device and has shown that this agrees well with numerical simulations. The values for the saturation channel current are within 1% and g_m within 13% of measured values.

16.3.6 Noise Behavior

The most complete analysis of noise for field effect transistors to date seems to have been done by Pucel, Haus, and Statz.[10,11] The equivalent circuit for their noise analysis is shown in Figure 16.11. The values for the noise generators in terms of the coefficients R and P are

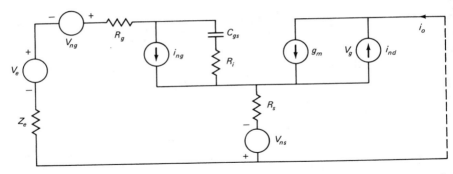

Figure 16.11. Equivalent noise circuit for the MESFET (*from "Signal and Noise Properties of Gallium Arsenide Microwave Field-Effect Transistors," by R. A. Pucel, H. A. Haus, and H. Statz, in* Advances in Electronics and Electron Physics, *Vol. 38, ed by L. Marten, 1975, by permission of Academic Press.*)[11]

$$\overline{i_{ng}2} = [4kT_0 \, \Delta f \omega^2 C_{gs}/g_m] R \qquad (16.58)$$
$$\overline{i_{nd}2} = [4kT_0 \, \Delta f g_m] P \qquad (16.59)$$
$$\overline{v_{ns}2} = 4kT_0 \, \Delta f R_s \qquad (16.60)$$
$$\overline{v_{ne}2} = 4kT_0 R_e \, \Delta f \qquad (16.61)$$

where k is Boltzmann's constant, Δf is the bandwidth, and T_0 is the absolute temperature of the device. The noise figure is expressed in the form

$$F = 1 + \frac{\left| \overline{i_{go} + i_{do} + i_{so} + i_{eo}} \right|^2}{\left| \overline{i_{so}} \right|^2} \qquad (16.62)$$

where i_{go}, i_{do}, i_{so}, and i_{eo} are the noise current components in the shorted drain circuit resulting from the noise sources (16.58) − (16.61). To evaluate (16.62), the coefficients R, P and a correlation coefficient C must be numerically calculated. The complexity of these expressions promoted Fukui[12] to obtain an empirically based but simplified expression for the minimum noise figure in terms of the coefficient $K_f \cong 2.5$, the gate metallization resistance R_g, and the source resistance R_s.

$$F_{min} = 1 + K_f \sqrt{g_m(R_g + R_s)} f/f_T \qquad (16.63)$$

Fukui noted that (16.63) is independent of gate bias and may be considered a special case of the more general form (16.62). He reported that the accuracy of predicting the minimum noise figure using (16.63) was within 0.67% for the five samples he measured where the parameters L, g_m, R_g, and R_s were all measured independently. He also found empirically that $g_m \propto W_T(N_d/aL)^{1/3}$ where W_T is the total device width. Since $f_T \propto 1/L$, the optimum noise figure may be expected to decrease with decreasing gate length approximately as

$$F \propto L^{5/6}. \qquad (16.64)$$

16.3.7 MESFET Power Limitations

The maximum output for a single cell of a MESFET is limited by the maximum allowable drain voltage and current swings, or more precisely, by the difference between the drain breakdown voltage and the maximum channel current under forward gate bias. However, a compromise must be reached in choosing the channel doping concentration N_d, since increasing N_d increases channel current but decreases breakdown voltage. Increasing the device power level beyond these limitations requires paralleling several MESFET cells so that the power from each cell adds to the total power. As soon as multiple cells are combined, there arises the fundamental problem of trying to connect three wires on a two-dimensional surface without crossovers. There are two solutions to this problem. The first approach is to make a conductance path through the device to the ground plane. This approach, called the *source via hole,* provides a low inductance current path from the source of each cell to ground. An alternative technology is called the *self-aligned flip chip* method. In this method, gold posts are plated on to the source pads so that the source pads are built up higher than the gate and drain electrodes. The chip is then flipped upside down and bonded to the copper pedestal. This method also provides a very low inductance path to ground and provides a good thermal heat sink for the device as well. These two approaches are illustrated in Figures 16.12 and 16.13.

16.4 CIRCUIT ANALYSIS OF TWO-PORT DEVICES

The circuit designer is primarily concerned with providing the optimum impedance to the input and output of a transistor in order to give the desired noise,

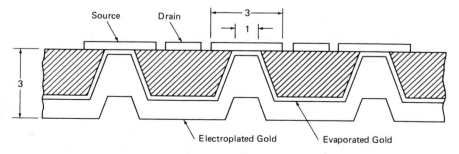

Figure 16.12. Cross sectional view of via hole source connections with dimensions given in mils.

Figure 16.13. Photograph of a flip chip transistor. *(Used with the permission of Microwave Semiconductor Corporation, Somerset, NJ.)*

gain or linearity from the amplifier. Because of the internal feedback of a transistor, it is possible for the amplifier to oscillate. Consequently, the circuit design must consider both the desired gain characteristic and the conditions under which the amplifier will be unstable. This section contains the derivation of the transducer power gain and the transistor stability factor in terms of S parameters.

16.4.1 Power Gain

The characteristics of either bipolar or field effect transistors are now usually given in the form of the readily measurable S parameters. The transducer power gain given by

$$G = \frac{\text{Power delivered to the load}}{\text{Power available from the source}} = \frac{P_L}{P_a} \qquad (16.65)$$

is a function of the device S parameters as well as both the load and source reflection coefficients. The power delivered to the load is the incident power minus the reflected power (Figure 16.14).

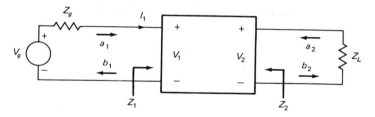

Figure 16.14. Two-port representation of a transistor.

$$P_L = \tfrac{1}{2}|b_2|^2(1 - |\Gamma_L|^2) \tag{16.66}$$

The available power from the source is

$$P_a = \frac{|V_g|^2}{8\operatorname{Re}(Z_g)} \tag{16.67}$$

where V_g represents the peak voltage rather than the rms value that is sometimes used. To calculate the available power in terms of the wave function b_1, the output of the voltage source is matched. This means the reflected wave b_1 is zero.

$$b_1 = \frac{V_1 - Z_1^* I_1}{2\sqrt{\operatorname{Re}(Z_1)}} = 0 \tag{16.68}$$

This expression implies that

$$V_1 = I_1 Z_1^* \tag{16.69}$$

Substituting this into the expression for the incident wave yields,

$$a_1 = \frac{v_1 + Z_1 I_1}{2\sqrt{\operatorname{Re}(Z_1)}}$$
$$= I_1 \sqrt{\operatorname{Re}(Z_1)}. \tag{16.70}$$

The current may be replaced by the input voltage to impedance ratio

$$a_1 = \frac{V_g \sqrt{\operatorname{Re}(Z_1)}}{Z_g + Z_1}. \tag{16.71}$$

The available power given in (16.67) can be put into a form containing the generator reflection coefficient and a_1. First (16.67) can be rewritten in the form

$$P_a = \frac{|V_g|^2 \operatorname{Re}(Z_1)}{2|Z_g + Z_1|^2(1 - |\Gamma_g|^2)} \qquad (16.72)$$

$$= \frac{|a_1|^2}{2(1 - |\Gamma_g|^2)} \qquad (16.73)$$

where

$$\Gamma_g = \frac{Z_g - Z_1^*}{Z_g + Z_1}. \qquad (16.74)$$

The transducer power gain is therefore

$$G = \frac{|b_2|^2}{|a_1|^2}(1 - |\Gamma_g|^2)(1 - |\Gamma_L|^2). \qquad (16.75)$$

Finding the ratio $|b_2|/|a_1|$ can be done by using the flow graph reduction technique found in chapter 1. The steps are outlined in Figure 16.15. The relationship between a_1 and b_2 can be written down by inspection of Figure 16.15(e), since the lower loop is a self-loop at node e_1'.

$$\frac{b_2}{a_1} = \frac{S_{21}}{(1 - S_{11}\Gamma_g)(1 - S_{22}\Gamma_L) - \Gamma_L\Gamma_g S_{12}S_{21}} \qquad (16.76)$$

The transducer power gain from (16.75) and (16.76) is

$$G = \frac{|S_{21}|^2(1 - |\Gamma_g|^2)(1 - |\Gamma_L|^2)}{|(1 - S_{11}\Gamma_g)(1 - S_{22}\Gamma_L) - S_{12}S_{21}\Gamma_L\Gamma_g|^2} \qquad (16.77)$$

This is the general expression for the transducer power gain with arbitrary load and generator impedances. The unilateral transducer power gain is obtained by setting $S_{12} = 0$. The maximum unilateral gain occurs when $\Gamma_g = S_{11}^*$ and $\Gamma_L = S_{22}^*$.

$$G_{u,\max} = \frac{|S_{21}|^2}{(1 - |S_{11}|^2)(1 - |S_{22}|^2)} \qquad (16.78)$$

When $S_{12} \neq 0$, the maximum gain occurs where the input and output ports are conjugately matched. To find the maximum non-unilateral gain, the flow graph reduction technique may again be used to find the input reflection coefficient for the two-port device terminated by a load with reflection coefficient

BIPOLAR AND FIELD EFFECT TRANSISTORS 377

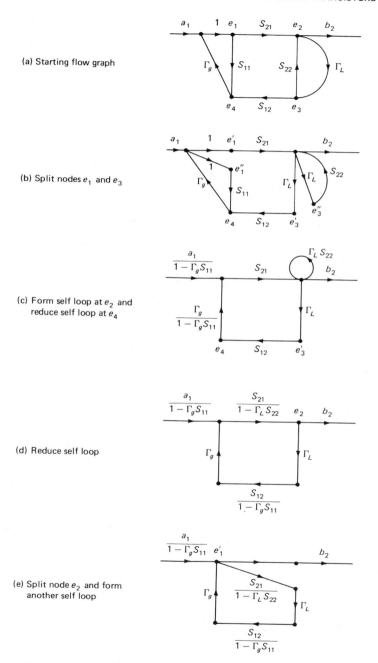

Figure 16.15. Reduction of the two-port flow graph for the transducer power gain.

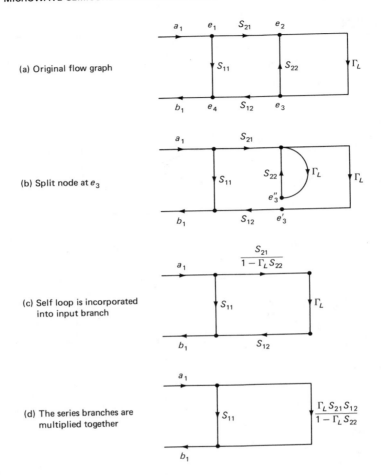

Figure 16.16. Flow graph for input reflection coefficient.

Γ_L. The overall reflection coefficient is obtained by adding the two parallel branches in Figure 16.16(d).

$$\Gamma_i = S_{11} + \frac{\Gamma_L S_{12} S_{21}}{1 - \Gamma_L S_{22}} \qquad (16.79)$$

Similarly the output coefficient is

$$\Gamma_o = S_{22} + \frac{\Gamma_g S_{12} S_{21}}{1 - \Gamma_g S_{11}}. \qquad (16.80)$$

The condition for matched input and output ports is $\Gamma_i = \Gamma_L^*$. Simultaneous solution of (16.79) and (16.80) provides expressions for Γ_g and Γ_L. The condition that the magnitude of these reflection coefficients be less than unity yields the following relation.

$$K = \frac{1 - |S_{11}|^2 - |S_{22}|^2 + |\Delta|^2}{2S_{12}S_{21}} > 1 \quad (16.81)$$

$$\Delta = S_{11}S_{22} - S_{12}S_{21} \quad (16.82)$$

The maximum non-unilateral gain can be expressed as[13]

$$G = \frac{|S_{21}|}{|S_{12}|}(K \pm \sqrt{K^2 - 1}). \quad (16.83)$$

The $+$ sign is used when $B_1 < 0$ and the $-$ sign when $B_1 > 0$, where

$$B_1 \triangleq 1 + |S_{11}|^2 - |S_{22}|^2 - |\Delta|^2. \quad (16.84)$$

16.4.2 Amplifier Stability

Rollett[14] found an invariant stability factor that characterizes the degree of conditional or unconditional stability of an amplifying device. Since Rollett's original paper, several forms for this stability factor have been found. In 1976 Woods[15] compared these various forms and indicated where the derivations of the different authors were equivalent. From this comparison, he proposed a form that was both simple and complete. In terms of the S parameters, the criterion for stable operation of an amplifier with passive source and load terminations is given by

$$K = \frac{1 - |S_{11}|^2 - |S_{22}|^2 + |\Delta|^2}{2|S_{12}S_{21}|} > 1 \quad (16.85)$$

$$|\Delta| < 1 \quad (16.86)$$

where Δ is the determinate of the S matrix given by (16.82). The equivalence of (16.85) and (16.81) means that simultaneous matching of a two-port device is possible if the circuit is conditionally stable. However, stability is not guaranteed by matching the input and output circuit ports.

In deriving these expressions, use is made of the condition that the input and output reflection coefficients must have a magnitude less than unity. Thus a two-port circuit is stable for all passive load and source impedances $|\Gamma_L| < 1$ and $|\Gamma_g| < 1$ if

$$\left| S_{11} + \frac{\Gamma_L S_{12} S_{21}}{1 - \Gamma_L S_{22}} \right| < 1 \qquad (16.87)$$

$$\left| S_{22} + \frac{\Gamma_g S_{12} S_{21}}{1 - \Gamma_g S_{11}} \right| < 1. \qquad (16.88)$$

Equation (16.87) can be re-expressed in terms of the determinate Δ. This is done by multiplying the numerator and denominator of the second term in (16.87) by S_{22} and adding 0 to the numerator in the form $S_{12}S_{21} - S_{12}S_{21}$.

$$\left| \frac{1}{S_{22}} \left(\Delta + \frac{S_{12}S_{21}}{1 - \Gamma_L S_{22}} \right) \right| < 1 \qquad (16.89)$$

The quantity $1 - S_{22}\Gamma_L$ becomes $1 - |S_{22}|e^{j\theta}$ when $|\Gamma_L| = 1$. Ha[16] pointed out that this is a circle with its center at 1 and with a radius of $|S_{22}|$ (Figure 16.17). Thus the quantity

$$\frac{1}{1 - |S_{22}|e^{j\theta}} \qquad (16.90)$$

is a circle centered at

$$\frac{1}{2} \left[\frac{1}{1 + |S_{22}|} + \frac{1}{1 - |S_{22}|} \right] = \frac{1}{1 - |S_{22}|^2} \qquad (16.91)$$

with radius

$$\frac{1}{2} \left[\frac{1}{1 - |S_{22}|} - \frac{1}{1 + |S_{22}|} \right] = \frac{|S_{22}|}{1 - |S_{22}|^2}. \qquad (16.92)$$

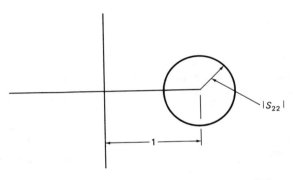

Figure 16.17. The circle with center at 1 and radius $|S_{22}|$.

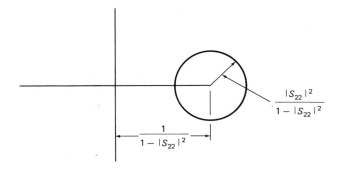

Figure 16.18. Diagram of the circle described by (16.91) and (16.92).

Therefore the inequality (16.89) may be written in terms of the maximum load reflection coefficient $|\Gamma_L| \leq 1$ with the phase angle θ (Figure 16.18).

$$\left| \frac{1}{S_{22}} \left(\Delta + \frac{S_{12}S_{21}}{1 - |S_{22}|^2} + \frac{S_{12}S_{21}|S_{22}|e^{j\theta}}{1 - |S_{22}|^2} \right) \right| < 1 \qquad (16.93)$$

Since this must be true for all θ, the inequality may be written as

$$\frac{1}{|S_{22}|} \left| \Delta + \frac{S_{12}S_{21}}{1 - |S_{22}|^2} \right| + \frac{|S_{12}S_{21}|}{1 - |S_{22}|^2} < 1 \qquad (16.94)$$

or

$$0 < \frac{1}{|S_{22}|} \left| \Delta + \frac{S_{12}S_{21}}{1 - |S_{22}|^2} \right| < 1 - \frac{|S_{12}S_{21}|}{1 - |S_{22}|^2}. \qquad (16.95)$$

Since the left-hand side of the inequality is a positive number, the right-hand side must also be positive.

$$0 < 1 - |S_{22}|^2 - |S_{12}S_{21}|. \qquad (16.96)$$

Starting with (16.88) a similar expression for S_{11} can be found.

$$0 < 1 - |S_{11}|^2 - |S_{12}S_{21}| \qquad (16.97)$$

Adding these last two inequalities together gives

$$|S_{12}S_{21}| < 1 - \tfrac{1}{2}(|S_{11}|^2 + |S_{22}|^2). \qquad (16.98)$$

The absolute value of the S matrix determinant is smaller than the sum of the absolute values of the components. Using this and (16.98),

$$|\Delta| < |S_{11}S_{22}| + |S_{12}S_{21}|$$
$$< |S_{11}S_{22}| + 1 - \tfrac{1}{2}(|S_{11}|^2 + |S_{22}|^2) \quad (16.99)$$

or

$$|\Delta| < 1 - \tfrac{1}{2}(|S_{11}| + |S_{22}|)^2 < 1. \quad (16.100)$$

This satisfies the second of the two stability criteria.

To find the value for K, the two inequalities (16.96) and (16.97) are multiplied together.

$$1 - |S_{11}|^2 - |S_{22}|^2 - 2|S_{12}S_{21}| + \chi > 0 \quad (16.101)$$

where

$$\chi = |S_{11}S_{22}|^2 + |S_{12}S_{21}|^2 - |S_{12}S_{21}|(|S_{11}|^2 + |S_{22}|^2). \quad (16.102)$$

From the self-evident inequality,

$$|S_{12}S_{21}|(|S_{11}| - |S_{22}|)^2 \geq 0, \quad (16.103)$$

the following expression is obtained.

$$|S_{12}S_{21}|(|S_{11}|^2 + |S_{22}|^2) > 2|S_{11}S_{22}S_{12}S_{21}|$$
$$> 2\mathrm{Re}(S_{11}S_{22}S_{12}^*S_{21}^*) \quad (16.104)$$

This can be used to show that χ in (16.102) is smaller than $|\Delta|^2$.

$$\chi < |S_{11}S_{22}|^2 + |S_{12}S_{21}|^2 - 2\mathrm{Re}(S_{11}S_{22}S_{12}^*S_{21}^*) = |\Delta|^2 \quad (16.105)$$

Consequently, (16.101) can be rewritten as

$$1 - |S_{11}|^2 - |S_{22}|^2 - 2|S_{12}S_{21}| + |\Delta|^2 > 0 \quad (16.106)$$

which is the value for K given by (16.85).

Transistor S parameters satisfying these stability criteria will provide stable amplifier operation for any passive load and generator impedances. If the stability criteria are not met, the device may still be conditionally stable; i.e., stable operation may be possible for a certain restricted range of Γ_g and Γ_L. One procedure for finding these stable regions would be simply to test the inequalities (16.87) and (16.88) for the particular load and source impedances. An alternative graphical approach is to find the circles of instability of load and generator reflection coefficients on a Smith chart. The circuit is unstable for impedances inside these generator and load stability circles. This can be shown by first observing that (16.79) can be solved for Γ_L in terms of the input reflection coefficient Γ_i.

$$\Gamma_L = \frac{S_{11} - \Gamma_i}{\Delta - S_{22}\Gamma_i} = \frac{1}{\Delta S_{22}}\left[\Delta + \frac{S_{12}S_{21}}{1 - \Gamma_i S_{22}\Delta^{-1}}\right] \quad (16.107)$$

As before with regard to (16.89), the quantity $1 - \Gamma_i S_{22}\Delta^{-1}$ is a circle centered at 1 with radius $|\Gamma_i S_{22}\Delta^{-1}|$. For $|\Gamma_i| < 1$ the reciprocal of this expression is a circle with center at

$$\frac{1}{2}\left[\frac{1}{1 + |\Delta^{-1}S_{22}|} + \frac{1}{1 - |\Delta^{-1}S_{22}|}\right] = \frac{1}{1 - |\Delta^{-1}S_{22}|^2} \quad (16.108)$$

and radius

$$\frac{1}{2}\left[\frac{1}{1 - |\Delta^{-1}S_{22}|} - \frac{1}{1 + |\Delta^{-1}S_{22}|}\right] = \frac{\Delta^{-1}S_{22}}{1 + |\Delta^{-1}S_{22}|}. \quad (16.109)$$

Since $|\Gamma_i| < 1$, the region of stability will be all points on the Smith chart outside this circle. From (16.107) the center of the load impedance circle is

$$\frac{1}{\Delta S_{22}}\left|\Delta + \frac{S_{12}S_{21}}{1 - |\Delta^{-1}S_{22}|^2}\right| \quad (16.110)$$

with radius

$$\frac{1}{|\Delta S_{22}|}\left|\frac{S_{12}S_{21}\,\Delta^{-1}S_{22}}{1 - |\Delta^{-1}S_{22}|^2}\right|. \quad (16.111)$$

When this circle is plotted on a Smith chart, load impedances that give unstable operation lie inside the circle described by (16.110) and (16.111). The condition of instability thus becomes immediately evident. A similar circle may also be plotted for the generator impedance.

16.5 AMPLIFIER DESIGN

Transistors can be optimized to provide low noise, high gain, or high power amplification of a signal. These three requirements are not normally compatible within a single amplifier stage. However, these requirements can often be best handled by using at least three separate stages in which the first stage is optimized for low noise, the second for high linear gain, and the third for high power. Impedance-matching the second and third types of transistor stages requires the measurement of the transistor S parameters: the high power device under high power conditions. The optimum reflection coefficient for a low noise amplifier is found experimentally under the appropriate low drain current bias level. This level is typically 0.1 to 0.2 I_{DSS}, where I_{DSS} is the saturation drain current at zero gate bias. The noise figure for a two-port circuit is[17,18]

$$F = F_m + \frac{R_n}{G_g} [(G_g - G_m)^2 + (B_g - B_m)^2] \qquad (16.112)$$

where F_m is the minimum noise figure, R_n is the equivalent thermal noise resistance, and $Y_m = G_m + jB_m$ is the optimum noise admittance. Optimum noise performance is obtained when $Y_g = Y_m$. Whatever the requirements for the amplifier, once the appropriate generator and load reflection coefficients are known, a circuit may be synthesized to provide the required match. If the objective is to design an amplifier chain with a minimum noise, the noise figure and gain of each stage must be considered. If each is identical, then the noise measure M defined by Haus[19] may be used profitably.

$$M = \frac{G(F - 1)}{G - 1} \qquad (16.113)$$

In this expression G is the gain and F is the noise figure of each stage of an amplifier chain with an infinite number of identical stages.

16.5.1 Analytical Matching Circuit Design

For synthesis purposes, the equivalent circuit for the bipolar transistor in Figure 16.4 or the MESFET in Figure 16.9 may be reduced to the unilateral model in Figure 16.19. This greatly simplifies the analytical design of input and output matching circuits since each of them can be designed independently of the others. For narrowband designs, the desired performance may be obtained by simply tuning out the device reactance at the center frequency and transforming the resulting real impedance to 50 ohms. Fine empirical tuning is often required because of the introduction of stray parasitic reactances that were not part of the measurement of the device, inaccurately measured device S parameters, and inaccuracy of the unilateral assumption.

For moderate bandwidth designs up to about 40% bandwidth, impedance and admittance inverters may be used for the input and output matching circuits.[20,21] The reactive load is incorporated into the first impedance inverter. The small inductances are realized by short bonding wires, and the series

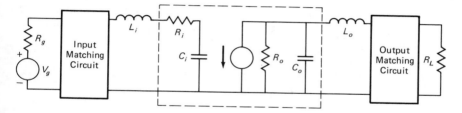

Figure 16.19. Unilateral microwave transistor model with package lead inductances and input and output matching networks.

capacitances are realized by an overlay or a gap in the transmission line. The gain slope of the transistor is compensated for by a series L, shunt R circuit on the output of the device.

To realize the full bandwidth capability of the microwave transistor, a third circuit design technique is needed. This technique requires compensating for the 6 dB/octave gain degradation with frequency of the previously discussed transistor models. To achieve flat gain across the desired frequency band, an inverse gain slope of the matching circuit is needed. The analytical approach will make the input matching circuit have a 6 dB/octave gain slope and the corresponding output circuit have a flat gain response. In Liechti and Tillman's[20] design, the gain slope compensation was done by the output circuit. Young and Scanlan[22] suggested that optimum results may be obtained by using some gain compensation in both the input and output circuits.

The two most widely used approximation functions for filter design are the Butterworth and Chebyshev polynomials. These provide a maximally flat or equal ripple passband response respectively. Of the two, the Chebyshev function offers better phase response and sharper bandpass skirts than the Butterworth function at a cost of small ripples in the passband. The transducer power gain or its reciprocal, the power loss ratio, is given by

$$G = \frac{1}{P_{LR}} = 1 - |\Gamma|^2 \tag{16.114}$$

where Γ is the circuit input reflection coefficient. The flat gain expression with the small Chebyshev ripple is

$$G = \frac{K}{1 + \epsilon^2 T_n^2(\omega')} \tag{16.115}$$

and the 6 dB/octave gain slope expression is

$$G = \frac{K(\omega/\omega_2)^2}{1 + \epsilon^2 T_n^2(\omega')} \tag{16.116}$$

for the frequency range ω_1 to ω_2. In these expressions, K is the passband gain, and ϵ is the passband ripple factor. The value $K \leq 1$ is chosen to insure that G in (16.116) is less than unity for all frequencies. Furthermore,

$$\omega' = \frac{1}{w}\left(\frac{\omega}{\omega_o} - \frac{\omega_o}{\omega}\right) \quad (16.117)$$

where the fractional bandwidth is

$$w = \frac{\omega_2 - \omega_1}{\omega_o} \quad (16.118)$$

and the mean frequency is

$$\omega_o = \sqrt{\omega_1 \omega_2}. \quad (16.119)$$

The Chebyshev polynomial used in (16.115) and (16.116) is

$$T_n(\omega') = \cos[n \text{ Arccos }(\omega')]. \quad (16.120)$$

Once the choice for the ripple factor and bandwidth has been made, the desired reflection coefficient $|\Gamma|^2$ is easily obtained as a ratio of two polynomials in s^2 of order $2n$.

$$\Gamma(s)\Gamma(-s) = |\Gamma(s)|^2 = 1 - G(s) \quad (16.121)$$

Since the denominator of $\Gamma(s)$ must be a Hurwitz polynomial, the left-hand plane roots of the numerator of $|\Gamma(s)|^2$ are usually employed to insure a minimal phase function. Usually, a root solver computer program must be used to actually find these roots. Once $\Gamma(s)$ is known, the input impedance function to be synthesized is

$$Z(s) = R_i \frac{1 + \Gamma(s)}{1 - \Gamma(s)} \quad (16.122)$$

where R_i is the transistor input resistance. As seen in Figure 16.19 the input of the transistor is a series LC circuit, so an inductor and a capacitor are extracted from this impedance. This leaves

$$Z_2(s) = Z(s) - L_1 s - 1/(C_1 s). \quad (16.123)$$

If $L_1 < L_i$ and $C_1 > C_i$, then the transistor elements may be incorporated into the first two elements of the bandpass-matching network. This procedure has

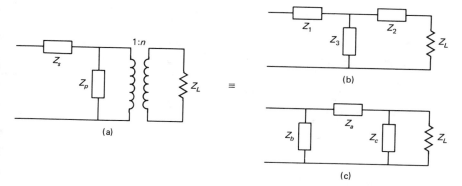

Figure 16.20. Equivalent T and π circuits for the ideal transformer and L circuit.

reduced the order of the numerator polynomial by 2 and the denominator by 1. In similar fashion, extracting admittance poles at $s = 0$ and $s = \infty$ from the reciprocal of $Z_2(s)$ yields a shunt LC section.

$$Y_4(s) = 1/Z_2(s) - 1/(L_2 s) - C_2 s \qquad (16.124)$$

This process continues until the required bandpass network is completed. It produces no resistive terms except for the terminating resistance. This is done by choosing K in (16.115) or (16.116) so that the maximum value of G is less than or equal to one over the entire frequency range. The addition of an ideal impedance transformer provides the necessary impedance matching between the transistor input resistance R_i and the 50 ohm source impedance.

Although a quarter wave stepped impedance transformer might easily be constructed, it is sometimes convenient to replace the transformer with an equivalent T circuit,[23] especially if the circuit is built to operate at 4 GHz or less. As indicated in Figure 16.20(a) and 16.20(b), the impedance matrix for each circuit must be the same.

$$\begin{bmatrix} Z_1 + Z_2 & Z_2 \\ Z_2 & Z_2 + Z_3 \end{bmatrix} = \begin{bmatrix} Z_s + Z_p & nZ_p \\ nZ_p & n^2 Z_p \end{bmatrix} \qquad (16.125)$$

In this expression Z_s and Z_p must both be either inductors or capacitors. Solving for the T circuit parameters yields

$$Z_1 = Z_s + (1 - n) Z_p, \qquad (16.126)$$
$$Z_2 = n Z_p, \qquad (16.127)$$
$$Z_3 = n(n - 1) Z_p. \qquad (16.128)$$

Realizability obviously requires

$$0 < n < 1 + Z_s/Z_p. \tag{16.129}$$

The alternative π equivalent circuit is also easily obtained by equating the y parameters of the two circuits in Figure 16.20(a) and 16.20(c).

$$\begin{bmatrix} \dfrac{1}{Z_s} & \dfrac{-1}{nZ_s} \\ -\dfrac{1}{nZ_s} & \dfrac{1}{n^2}\left(\dfrac{1}{Z_s} + \dfrac{1}{Z_p}\right) \end{bmatrix} = \begin{bmatrix} \dfrac{1}{Z_a} + \dfrac{1}{Z_b} & \dfrac{-1}{Z_a} \\ -\dfrac{1}{Z_a} & \dfrac{1}{Z_a} + \dfrac{1}{Z_c} \end{bmatrix} \tag{16.130}$$

Solving for the π circuit parameters yields

$$Z_a = nZ_s, \tag{16.131}$$

$$Z_b = \frac{nZ_s}{n-1}, \tag{16.132}$$

$$Z_c = \frac{n^2 Z_p Z_s}{Z_s + (1-n)Z_p}. \tag{16.133}$$

The output-matching network for the transistor must be handled differently since the reactive elements of the output of the transistor form the first two elements of a lowpass filter (shunt C, series L). Hence, the lowpass gain expression

$$G = \frac{K}{1 + \epsilon^2 T_n^2(\omega/\omega_c)} \tag{16.134}$$

for the flat gain approximation function, or

$$G = \frac{K(\omega/\omega_c)^2}{1 + \epsilon^2 T_n^2(\omega/\omega_c)} \tag{16.135}$$

for the 6 dB/octave slope approximation are used rather than the bandpass functions (16.115) and (16.116). In (16.134) and (16.135) the parameter ω_c is the lowpass cutoff frequency and ϵ is the ripple factor in the passband. The circuit elements giving the response for (16.134) may be found from a simple recursive algorithm (see chapter 2) or from various tabulated sources.[21] The sloped response (16.135) must be synthesized by the same methods employed for the input sloped circuit. The lowpass structure of the transistor in Figure 16.21(a) is readily accommodated by the lowpass matching circuit. A bandpass response may still be obtained by employing the lowpass to bandpass frequency transformation (16.117). This transformation replaces shunt capacitors with a

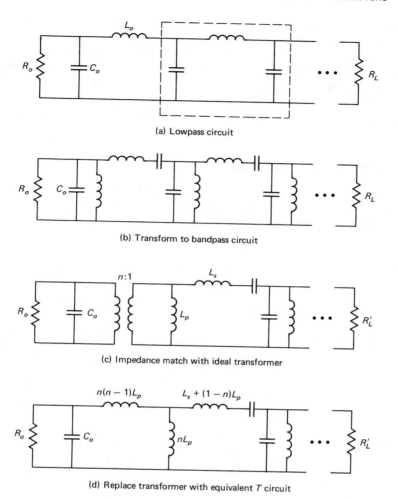

Figure 16.21. Output-matching network design.

shunt LC circuit and replaces series inductors with a series LC circuit (Figure 16.21(b)). Finally, an ideal impedance transformer is introduced to match the load R_L to the transistor output resistance R_o. Since the series output inductance of the transistor cannot be included as part of the external circuit, it is convenient to convert the inductive L circuit to an inductive T circuit (Figure 16.21(d)). This is done by transforming the ideal transformer through the circuit to a position adjacent to the shunt inductance.

When more than two amplifiers are cascaded together, each amplifier may be separated from the others by an isolator. This prevents unwanted feedback to an early stage from a later stage and insures that the input and output loads can be made to be 50 ohms. In cases where the isolators take up too much room or where they are found to restirct the amplifier bandwidth, they may be replaced by an interstage network. The design of an interstage network follows the same techniques outlined above for the input and output circuits. The impedance matching is between the input impedance of one transistor and the output impedance of the previous transistor rather than between two 50 ohm loads.

16.5.2 Numerical Design Example

The design procedure outlined above may be clarified by a numerical example. An input-matching network is needed to compensate the transistor gain slope of -6 dB/octave. The desired bandwidth extends from 6 to 12 GHz with a

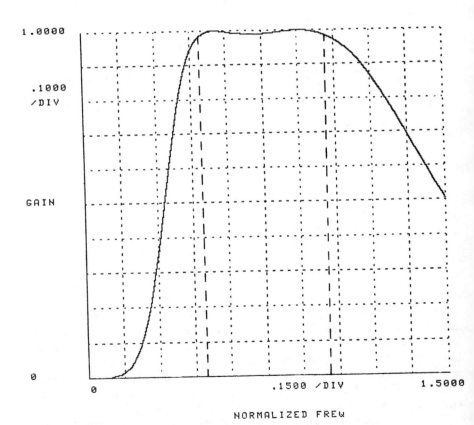

Figure 16.22. Chebyshev approximation function ($n = 2$) for constant gain from $\omega = 0.5$ to $\omega = 1$.

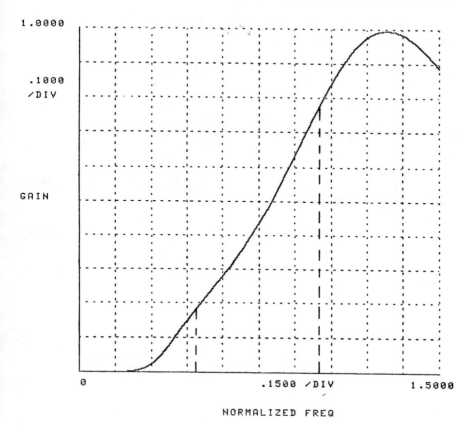

Figure 16.23. Chebyshev approximation function ($n = 2$) for 6 dB/octave gain slope from $\omega = 0.5$ to $\omega = 1$.

passband ripple of 0.1 dB. For these parameters and $n = 2$ it may be shown numerically that (16.116) reaches a maximum at $\omega = 1.28\omega_2$, and the gain remains less than unity if $K = 0.78$. If the frequencies are normalized so that $\omega_1 = 0.5$ and $\omega_2 = 1$, then the expected frequency response from (16.115) and (16.116) are plotted in Figures 16.22 and 16.23 respectively. For $s = j\omega$ the gain from (16.116), calculated to seven places, is

$$G(s) = \frac{-1.219s^6}{s^8 + 2.250s^6 + 3.328s^4 + 0.563s^2 + 0.0625}. \quad (16.136)$$

The magnitude squared of the reflection coefficient is obtained from (16.121).

$$|\Gamma(s)|^2 = \frac{s^8 + 3.469s^6 + 3.328s^4 + 0.563s^2 + 0.0625}{s^8 + 2.250s^6 + 3.328s^4 + 0.563s^2 + 0.0625} \quad (16.137)$$

The reflection coefficient is obtained from the left-hand plane roots of the numerator and denominator.

$$\Gamma(s) = \frac{(s + 0.176 + j.347)(s + 0.176 - j0.347)}{(s + 0.170 + j.342)(s + 0.170 - j0.342)} \\ \times \frac{(s + 0.039 + j1.283)(s + 0.039 - j1.283)}{(s + 0.582 + j1.173)(s + 0.582 - j1.173)} \\ = \frac{s^4 + 0.431s^3 + 1.827s^2 + 0.593s + 0.250}{s^4 + 1.504s^3 + 2.256s^2 + 0.752s + 0.250}$$ (16.138)

If for now the input resistance R_i is assumed to be unity, the input impedance is found from (16.122).

$$Z(s) = \frac{2s^4 + 1.935s^3 + 4.084s^2 + 1.347s + 0.5}{1.073s^3 + 0.429s^2 + 0.159s}$$ (16.139)

The network synthesis procedure begins by removing an impedance pole at $s = 0$.

$$Z_1(s) = Z(s) - \frac{1}{C_1 s} \\ = \frac{2s^4 + 1.936s^3 + 0.720s^2 + 0.000085s}{1.073s^3 + 0.429s^2 + 0.159s}$$ (16.140)

when

$$C_1 = 0.318774$$ (16.141)

The coefficient of s in the numerator is zero if K is chosen to give a maximum gain of exactly one. Otherwise a lossy resistor has to be included in the matching network. In the present case, it is small enough to be neglected. After reducing the order of the polynomials in both the numerator and denominator by one, an impedance pole at $s = \infty$ is extracted. This represents, at least in part, the series package inductance (Figure 16.24(b)) and acts to eliminate the coefficient of the highest power in s of the numerator. The remaining impedance is

$$Z_2(s) = Z_1(s) - L_1 s \\ = \frac{1.136s^2 + 0.423s}{1.073s^2 + 0.429s + 0.159}$$ (16.142)

when

BIPOLAR AND FIELD EFFECT TRANSISTORS   393

$$L_1 = 1.864055 \tag{16.143}$$

An admittance pole at $s = 0$ is then extracted which eliminates the constant term in the numerator of $Y_2(s) = 1/Z_2(s)$ (Figure 16.24(c)).

$$\begin{aligned} Y_3(s) &= Y_2(s) - 1/(L_2 s) \\ &= \frac{1.073s + 0.000232}{1.136s + 0.423} \end{aligned} \tag{16.144}$$

when

$$L_2 = 2.649884 \tag{16.145}$$

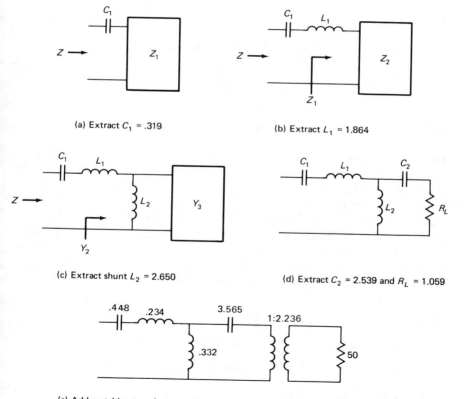

(a) Extract $C_1 = .319$

(b) Extract $L_1 = 1.864$

(c) Extract shunt $L_2 = 2.650$

(d) Extract $C_2 = 2.539$ and $R_L = 1.059$

(e) Add matching transformer and un-normalize to give above values in pF and nH

Figure 16.24. Synthesis steps for transistor input-matching network.

Finally, if the small constant term in Y_3 is neglected, the remaining impedance is found by inspection.

$$Z_3(s) = R_L + 1/(C_2 s) \qquad (16.146)$$
$$= 1.05869 + \frac{1}{2.538746s}$$

The circuit may now be unnormalized to the 6 to 12 GHz frequency range and to the typical transistor input resistance of 10 ohms. The circuit elements are transformed by the following relations.

$$R \to R_i R \qquad (16.147)$$
$$L \to \frac{R_i L}{\omega_2} \qquad (16.148)$$
$$C \to \frac{C}{R_i \omega_2} \qquad (16.149)$$

The resulting circuit is shown in Figure 16.24(e). Impedance matching in this case must be done with an impedance transformer since the realizability criterion (16.129) for conversion to an equivalent T circuit is not met for this circuit.

16.5.3 Distributed Circuit Design

The synthesis procedure described in the previous section produced a lumped parameter network which, with chip capacitors and spiral inductors, can be applied to amplifier design up to approximately 4 GHz. Higher frequency amplifiers should make use of the techniques described in chapter 2. These techniques included transforming lumped circuit elements in the s plane to distributed circuit elements in the S plane, extracting of unit elements and S plane elements from the immitance function, and using impedance/admittance inverters. The specific technique employed is governed by the design requirements and the realizability of the circuit parameters in the chosen media.

REFERENCES

1. C. P. Snapp, "Bipolars Quietly Dominate," *Microwave Systems News*, Vol. 9, pp. 45–67, November 1979.
2. H. F. Cooke, "Microwave Transistors: Theory and Design," *Proc. of the IEEE*, Vol. 59, pp. 1163–1181, August 1971.
3. R. L. Pritchard, *Electrical Characteristics of Transistors*. New York: McGraw-Hill, 1967.
4. B. Himsworth, "A Two-Dimensional Analysis of Gallium-Arsenide Junction Field-Effect Transistors with Long and Short Channels," *Solid-State Electronics*, Vol. 15, pp. 1353–1361, December 1972.

5. C. A. Liechti, "Microwave Field-Effect Transistors—1976," *IEEE Trans. on Microwave Theory and Techniques*, Vol. MTT-24, pp. 279–300, June 1976.
6. A Madjar and F. J. Rosenbaum, "A Large-Signal Model for the GaAs MESFET," *IEEE Trans. on Microwave Theory and Techniques*, Vol. MTT-29, pp. 781–788, August 1981.
7. P. Hower and G. Bechtel, "Current Saturation and Small-Signal Characteristics of GaAs Field Effect Transistors," *IEE Trans. on Electron Devices*, Vol. ED-20, pp. 213–220, March 1973.
8. S. M. Sze, *Physics of Semiconductors*. New York: John Wiley, 1981.
9. M. S. Shur, "Analytical Model of GaAs MESFET's," *IEEE Trans. on Electron Devices*, Vol. ED-25, pp. 612–618, June 1978.
10. H. Statz, H. Haus, and R. Pucel, "Noise Characteristics of Gallium Arsenide Field-Effect Transistors," *IEEE Trans. on Electron Devices*, Vol. ED-21, pp. 549–562, September 1974.
11. R. A. Pucel, H. A. Haus, and H. Statz, "Signal and Noise Properties of Gallium Arsenide Microwave Field-Effect Transistors," in *Advances in Electronics and Electron Physics*. Vol. 38, New York: Academic Press, pp. 195–265, 1975.
12. H. Fukui, "Optimal Noise Figure of Microwave GaAs MESFET's," *IEEE Trans. on Electron Devices*, Vol. ED-16, pp. 1032–1037, July 1979.
13. G. E. Bodway, "Two Port Power Flow Analysis Using Generalized Scattering Parameters," *The Microwave Journal*, Vol. 10, pp. 61–69, May 1967.
14. J. M. Rollett, "Stability and Power-Gain Invariants of Linear Twoports," *IRE Trans. on Circuit Theory*, Vol. CT-9, pp. 29–32, March 1962.
15. D. Woods, "Reappraisal of the Unconditional Stability Criteria for Active 2-port Networks in terms of S Parameters," *IEEE Trans. on Circuits and Systems*, Vol. CAS-23, pp. 73–81, February 1976.
16. T. T. Ha, *Solid State Microwave Amplifier Design*. New York: John Wiley, 1981.
17. H. Rothe and W. Dahlke, "Theory of Noisy Fourpoles," *Proc. of the IRE*, Vol. 44, pp. 811–818, June 1956.
18. H. Haus, "IRE Standards on Methods of Measuring Noise in Linear Two-Ports," *Proc. of the IRE*, Vol. 48, pp. 60–68, January 1960.
19. H. Haus and R. Adler, *Circuit Theory of Linear Noisy Networks*. New York: John Wiley, 1959.
20. C. A. Liechti and R. L. Tillman, "Design and Performance of Microwave Amplfiiers with GaAs Schottky-Gate Field-Effect Transistors," *IEEE Trans. on Microwave Theory and Techniques*, Vol. MTT-22, pp. 510–517, May 1974.
21. G. L. Matthaei, L. Young, and E. M. T. Jones, *Microwave Filters, Impedance-Matching Networks, and Coupling Structures*. New York: McGraw-Hill, 1964.
22. G. P. Young and S. O. Scanlan, "Matching Network Design Studies for Microwave Transistor Amplifiers," *IEEE Trans. on Microwave Theory and Techniques*, Vol. MTT-29, pp. 1027–1035, October 1982.
23. E. A. Guillemin, *Synthesis of Passive Networks*. New York: John Wiley, Chapter 5, 1957.

Chapter 17
Present and Future Developments

17.1 REVIEW OF MICROWAVE TECHNOLOGY

This book has provided design techniques for a large number of passive and active microwave circuits. The text included designs for passive elements such as attenuators, filters, impedance transformers, and directional couplers. These circuits may be realized in a variety of transmission line media, and formulas for their implementation were provided. The application of the digital computer to microwave design, manufacture, and testing was discussed. The future automation in these three areas will undoubtedly have a profound effect on the field of microwave engineering. Some general properties of amplifiers, oscillators, and noise were offered, along with some basic theory of semiconductors. The remaining material considered the theory of operation of some of the most widely used microwave semiconductor devices and their applications in circuits. The discussion on diodes included the varactor, step recovery, Schottky barrier, PIN, IMPATT, and Gunn devices. Finally, bipolar and field effect transistor circuits were considered.

This list of topics by no means exhausts the areas of concern for microwave engineering. The subjects of antenna design, wave propagation through the atmosphere, and scattering theory are necessary ingredients in building radar systems, overland microwave links, and satellite communication networks. These topics are distinct from circuit and device design. The subject of microwave tubes was also not included in the discussion. This does not imply that microwave tubes have outlived their usefulness. The high power magnetron is likely to remain the active source in microwave ovens. The traveling wave tube amplifier will not soon be replaced, since this tube can supply simultaneously more power, gain, and bandwidth than any presently available semiconductor.

There are some additional areas of concern to the microwave engineer that may be explored. One of these areas is the class of ferrite components, which include nonreciprocal isolators, circulators, phase shifters, and tunable filters. These circuits use the material, yttrium iron garnet (YIG), which has a permeability that can be changed with the application of a DC bias current. Thus, a current-controlled inductance can be obtained using YIG, and it can be used in a fashion somewhat analogous to the voltage-controlled capacitance of a varactor diode.

Microwave accoustic devices have been found useful in frequency control of oscillators, data storage elements, time delay networks, and pulse compression circuits. While these devices operate largely at frequencies no higher than UHF, they have been found particularly useful in signal processing applications.

Another circuit element not considered in this book is the dielectric resonator. This is usually a small dielectric cylinder that may be attached near a transmission line in a microstrip circuit. Energy at the resonant frequency of the dielectric cylinder is coupled into it. Since the dielectric material is chosen for its low loss and temperature insensitivity, it can be used to stabilize microwave semiconductor oscillators mounted in microstrip circuits. Since the dielectric constant of the resonator is high, it is particularly useful for compact microstrip circuits.

Several specialized devices were not discussed in this book. Among them are the transit time or avalanche devices related to the IMPATT diode. The BARITT diode (BARrier Injected Transit Time) is one from this class in which carriers are injected into a drift region without producing an avalanche. The negative resistance produced by the BARITT diode can provide oscillations with much less noise and efficiency than the IMPATT. The BARITT diode has been found useful in low level doppler radars. The three terminal CATT (Controlled Avalanche Transit-time Triode) is another avalanche device in which the avalanche mechanism is used to multiply the charge injected into the collector region.[1] This device has a noise and power characteristic intermediate between the IMPATT diode and the FET.

The tunnel diode was one of the early microwave solid state amplifiers. It provides low noise amplification over a broad frequency range. Its noise figure usually lies somewhere between that achieved by a parametric amplifier and a bipolar transistor. Its major shortcoming is that it saturates at low power levels and is easily burned out by moderately high signals. Consequently, the circuit design needed to make a practical tunnel diode amplifier can be quite complex. The circuit must provide optimum impedance match to the diode and reject out-of-band signals. For these reasons, the FET has replaced the tunnel diode in most applications.

A device closely related to the tunnel diode is the backward diode. This device has been found useful as a zero bias mixer and detector, and it has been found to be less temperature sensitive than Schottky-barrier diodes. However, the backward diode has been found difficult to fabricate and suffers from reliability problems.

17.2 FUTURE DEVELOPMENTS

The above enumerated list of topics has been the subject of a great deal of research in the past. Some of these investigations have been pressed into the service of many practical applications, while other technologies have been laid aside because of one or more critical shortcomings. This process of discovering what will and what will not be found useful continues today. Extrapolating the present circumstances into the future leads to two categories that will show growth: new technology and better quality control.

The two major new technological areas in microwaves seem to be monolithic circuits and millimeter waves. Monolithic circuits are usually designed on a thin semi-insulating GaAs substrate. By definition a monolithic circuit is formed completely by a deposition method such as liquid phase epitaxy, vapor phase epitaxy, ion implantation, evaporation, or sputtering. The circuit elements that can be formed on monolithic circuit boards include transmission lines, lumped overlay or interdigital capacitors, lumped spiral inductors, thin film resistors, FETs, and when absolutely necessary, transmission line stubs. Since space is at a premium, lumped elements are preferred to distributed elements. The preferred transmission lines are microstrip and coplanar waveguide. Of these two, microstrip has less loss and can be made with a characteristic impedance as high as 90 ohms on a 125 μm GaAs substrate. Since the complete circuit is manufactured using automated techniques, no final adjustments can be made on it after it has been produced. The circuit design must rely heavily on computer modeling and optimization. The lumped overlay capacitor can provide between 0.5 pF to as high as 70 pF. For a 1 pF capacitor, at 10 GHz the Q would lie between 50 and 100. A single-loop inductor can provide 2 nH of inductance up to 18 GHz. At 10 GHz a single loop inductor can be made with a Q between 40 and 80. In monolithic circuits, a FET is relatively easy to make. Thus FETs are used for a wide variety of purposes including amplifiers, mixers, frequency multipliers, and resistors.

Monolithic circuits are very likely to find their place in microwave applications requiring large numbers of compact units. Although they are at present expensive to design, they offer low manufacturing cost and good unit to unit repeatability. Ion implantation techniques are expected to find greater use as monolithic circuit designs become more complex.

Millimeter wave circuit design is usually approached by extrapolating from design techniques learned from lower frequency microwave engineering. However, this approach must be used with care, since at millimeter wave frequencies, mechanical dimensions are much more critical, surfaces must be made much smoother, and transmission lines are found to be much more lossy. As a result, design and fabrication of millimeter wave circuits must be done with more care. Because of the difficulty of modeling all the important effects, millimeter circuit design remains to a large degree an empirical science. The advantages offered by using millimeter waves are the high resolution radars that can be obtained with small antennas, wideband communication channels, and narrow antenna beam widths that are especially useful in space communications between satellites. The reawakening of interest in the gyrotron, a high power millimeter wave tube, not only has potential applications as a transmitter, but also may be useful in obtaining energy from fusion. The future of the microwave industry will depend on its ability to economically produce miniaturized and higher frequency circuits.

REFERENCE

1. S. P. Yu, W. Tantraporn, and J. R. Eshbach, "Theory of a New Three-Terminal Microwave Power Amplifier," *IEEE Trans. on Electron Devices*, Vol. ED-23, pp. 332–343, March 1976.

Appendix A
Read Diode Avalanche Current Equation

In chapter 14 reference was made to the expression for the avalanche current with unequal ionization coefficients and unequal saturated velocities in an n^+pip^+ Read type diode. The derivation of this expression begins by eliminating the hole and electron current densities by adding together the continuity equations (14.23) and (14.24).

$$\frac{\partial}{\partial t}\left(\frac{J_p}{v_p} + \frac{J_n}{v_n}\right) + \frac{\partial}{\partial x}(J_p - J_n) = 2(\alpha_p J_p + \alpha_n J_n) \quad (A.1)$$

From (14.25) and (14.26) the hole and electron currents may be found in terms of the total conduction current and the induced space charge field.

$$J_p = \frac{v_p}{v_p + v_n}\left[J - \epsilon\frac{\partial E_{sc}}{\partial t} + v_n\epsilon\frac{\partial E_{sc}}{\partial x}\right] \quad (A.2)$$

$$J_n = \frac{v_n}{v_p + v_n}\left[J - \epsilon\frac{\partial E_{sc}}{\partial t} - v_p\epsilon\frac{\partial E_{sc}}{\partial x}\right] \quad (A.3)$$

Substitution of (A.2) and (A.3) into (A.1) yields an expression for the total conduction current.

$$\left(-\frac{\partial}{\partial t} + \alpha_n v_n + \alpha_p v_p\right)J = v_n v_p\left(\frac{\partial}{\partial x} + \alpha_n - \alpha_p\right)\epsilon\frac{\partial E_{sc}}{\partial x}$$
$$+ \left[(v_n - v_p)\frac{\partial}{\partial x} - \frac{\partial}{\partial t} + \alpha_n v_n + \alpha_p v_p\right]\epsilon\frac{\partial E_{sc}}{\partial t} \quad (A.4)$$

If this equation is spatially integrated, then only a time dependent expression remains to be solved. This integration results in (14.27), and the procedure begins by multiplying (A.4) by the integration factor $e^{R(x,x_a)}$ where

$$R(x_1, x_2) \stackrel{\Delta}{=} \int_{x_2}^{x_1} (\alpha_p - \alpha_n)\, dx. \quad (A.5)$$

It was shown in chapter 14 that the following relationships hold.

$$\epsilon \frac{\partial E_{sc}}{\partial x} = \frac{J_p}{v_p} - \frac{J_n}{v_n} \qquad (A.6)$$

$$J = J_p + J_n + \epsilon \frac{\partial E_{sc}}{\partial t} \qquad (A.7)$$

$$\frac{\tau_n}{\tau_p} = \frac{M_p}{M_n} = e^{R(0,x_a)} \qquad (A.8)$$

$$\tau_n = \frac{1}{v_p + v_n} \int_0^{x_a} e^{R(x,x_a)} \, dx \qquad (A.9)$$

$$1 - \frac{1}{M_n} = \int_0^{x_a} \alpha_n e^{R(x,x_a)} \, dx \qquad (A.10)$$

It may easily be shown as well that

$$\frac{M_p}{M_n} - 1 = \int_0^{x_a} (\alpha_p - \alpha_n) e^{R(x,x_a)} \, dx \qquad (A.11)$$

$$\frac{1}{M_n}(M_p - 1) = \int_0^{x_a} \alpha_p e^{R(x,x_a)} \, dx. \qquad (A.12)$$

Multiplying (A.4) by the above mentioned integration factor and using (A.9), this equation becomes

$$\frac{\partial J}{\partial t}(v_n + v_p)\tau_n - Jv_n\left(1 - \frac{1}{M_n}\right) - \frac{Jv_p}{M_n}(M_p - 1)$$
$$= F_1 + F_2 + F_3 \qquad (A.13)$$

where

$$F_1 \triangleq -v_n v_p \int_0^{x_a} \frac{\partial}{\partial x}\left(e^{R(x,x_a)} \frac{\partial E_{sc}}{\partial x}\right) dx, \qquad (A.14)$$

$$F_2 \triangleq -\int_0^{x_a} (v_n - v_p) \epsilon \frac{\partial^2 E_{sc}}{\partial x \partial t} e^{R(x,x_a)} \, dx, \qquad (A.15)$$

$$F_3 \triangleq - \int_0^{x_a} \left[\alpha_n v_n + \alpha_p v_p - \frac{\partial}{\partial t} \right] \epsilon \frac{\partial E_{sc}}{\partial t} e^{R(x,x_a)} \, dx. \qquad (A.16)$$

The indicated integration of F_1 is straightforward.

$$F_1 = -v_n v_p \left[\epsilon \frac{\partial E_{sc}(x_a,t)}{\partial x} - \epsilon \frac{\partial E_{sc}(0,t)}{\partial x} e^{R(0,x_a)} \right] \qquad (A.17)$$

The spatial derivatives of the space charge field may be eliminated with the help of (A.2) where $J_p(0) = J_{ps}$ and with the help of (A.3) where $J_n(x_a) = J_{ns}$. This substitution puts F_1 in the desired form.

$$F_1 = v_n v_p \left\{ e^{R(0,x_a)} \left[\frac{v_p + v_n}{v_p v_n} J_{ps} - \frac{J}{v_n} + \frac{v_p + v_n}{v_p v_n} J_{ns} \right. \right.$$
$$\left. \left. - \frac{J}{v_p} + \frac{\epsilon}{v_p} \frac{\partial E_{sc}}{\partial t}(x_a,t) \right] \right\} \qquad (A.18)$$

or

$$F_1 = (v_p + v_n) \left(\frac{J_{ps} \tau_n}{\tau_p} + J_{ns} \right) - J \left(\frac{v_p \tau_n}{\tau_p} + v_n \right)$$
$$+ \epsilon \tau_n \left[\frac{v_p}{\tau_p} \frac{\partial E_{sc}(0,t)}{\partial t} + \frac{v_n}{\tau_n} \frac{\partial E_{sc}(x_a,t)}{\partial t} \right] \qquad (A.19)$$

The function F_2 is reduced by the standard formula for integration by parts.

$$\int u \, dv = uv - \int v \, du \qquad (A.20)$$

If $u = e^{R(x,x_a)}$ and $dv = \frac{\partial}{\partial x}\left(\frac{\partial E_{sc}}{\partial t}\right) dx$, then

$$F_2 = -(v_n - v_p)\epsilon \left[\frac{\partial E_{sc}(x_a,t)}{\partial t} - \frac{\partial E_{sc}(0,t)}{\partial t} e^{R(0,x_a)} \right]$$
$$- (v_n - v_p) \int_0^{x_a} (\alpha_p - \alpha_n) J e^{R(x,x_a)} \, dx$$
$$+ (v_n - v_p) \int_0^{x_a} (\alpha_p - \alpha_n)(J_p + J_n) e^{R(x,x_a)} \, dx \qquad (A.21)$$

The third term in (A.21), which will be called G_2, is again integrated by parts using $u = J_p + J_n$ and $dv = (\alpha_p - \alpha_n)e^R \, dx$.

$$G_2 = (v_n - v_p) \left[-(J_p(x_a) + J_n(x_a)) + (J_p(0) + J_n(0)) \frac{M_p}{M_n} \right.$$
$$\left. - \frac{\partial}{\partial t} \int_0^{x_a} \epsilon \frac{\partial E_{sc}}{\partial x} e^{R(x,x_a)} \, dx \right] \quad (A.22)$$

The integral in G_2 is again evaluated by parts using $u = e^{R(x,x_a)}$ and $dv = \epsilon \frac{\partial E_{sc}}{\partial x} dx$. Also, (A.7) is used to eliminate $(J_p + J_n)$ in the first part of G_2.

$$G_2 = (v_n - v_p) \left[J \left(\frac{M_p}{M_n} - 1 \right) \right.$$
$$\left. - \frac{\partial}{\partial t} \int_0^{x_a} \epsilon E_{sc}(x,t) \cdot (\alpha_p - \alpha_n) e^{R(x,x_a)} \, dx \right] \quad (A.23)$$

The function F_2 is therefore

$$F_2 = \epsilon \tau_n (v_n - v_p) \left[\frac{1}{\tau_p} \frac{\partial E_{sc}(0,t)}{\partial t} - \frac{1}{\tau_n} \frac{\partial E_{sc}(x_a,t)}{\partial t} \right]$$
$$- (v_n - v_p) \int_0^{x_a} (\alpha_p - \alpha_n) e^{R(x,x_a)} \epsilon \frac{\partial E_{sc}}{\partial t} \, dx \quad (A.24)$$

where (A.8) has been used to eliminate $e^{R(0,x_a)}$ in (A.21) and (A.11) has been used to evaluate the second integral of (A.21). Now the terms F_1, F_2, and F_3 on the right side of (A.13) may be summed. Appropriate terms on both sides of the equation can be cancelled and the integrands of F_2 and F_3 combined to yield the equation for the avalanche conduction current.

$$\left(\frac{\partial}{\partial t} + \frac{1}{M\tau} \right) J = \frac{J_s}{\tau} + \frac{\epsilon}{v_p + v_n} \left[\frac{v_n}{\tau_p} \frac{\partial E_{sc}(0,t)}{\partial t} + \frac{v_p}{\tau_n} \frac{\partial E_{sc}(x_a,t)}{\partial t} \right.$$
$$\left. - \frac{1}{\tau_n} \int_0^{x_a} \left(\alpha_n v_p + \alpha_p v_n - \frac{\partial}{\partial t} \right) \epsilon \frac{\partial E_{sc}}{\partial t} e^{R(x,x_a)} \, dx \right] \quad (A.25)$$

Appendix B
Quasistatic Approximation for the Avalanche Current

The large signal analysis given in chapter 14 and Appendix A can be greatly simplified by using the quasistatic approximation.[1] By this is meant that the ratios of the AC to the total current and the DC to the total current are equal.

$$r_p(x) \triangleq \frac{J_p}{J} = \frac{J_{po}}{J_o} \tag{B.1}$$

The subscript o designates the DC component. The time independence of the ratio $r_p(x)$ is exploited in simplifying (14.27) to the quasistatic expression (14.35). In addition since the second-order time derivative in (14.27) is proportional to the square of the frequency, this term becomes negligible in the quasistatic approximation. Hence, to prove (14.35) it must be shown that

$$(1 - \kappa)\frac{\partial J}{\partial t} = \frac{\epsilon}{v_p + v_n}\left[\frac{v_n}{\tau_p}\frac{\partial E_{sc}(0,t)}{\partial t} + \frac{v_p}{\tau_n}\frac{\partial E_{sc}(x_a,t)}{\partial t}\right]$$
$$- \frac{1}{\tau_n}\int_0^{x_a}(\alpha_n v_p + \alpha_p v_n)\frac{\partial E_{sc}}{\partial t}e^{R(x,x_a)}\,dx \tag{B.2}$$

where the function $R(x_1,x_2)$ is defined as

$$R(x_1,x_2) = \int_{x_2}^{x_1}(\alpha_p - \alpha_n)\,dx. \tag{B.3}$$

Poisson's equation (14.25) can be rewritten as

$$\epsilon\frac{\partial E_{sc}}{\partial x} = \frac{J_p}{v_p} - \frac{J_n}{v_n}$$
$$= Jg(x) \tag{B.4}$$

where

$$g(x) = \left(\frac{1}{v_p} + \frac{1}{v_n}\right) r_p(x) - \frac{1}{v_n}. \tag{B.5}$$

If (B.4) is integrated with respect to x and differentiated with respect to t, then

$$\epsilon \frac{\partial E_{sc}}{\partial t} = \frac{\partial J}{\partial t} \int_0^x g(x') \, dx'. \tag{B.6}$$

This can then be substituted into (B.2) and the $\partial J/\partial t$ terms cancelled on both sides of the equation to give the expression for κ that is to be evaluated.

$$\kappa = 1 + \frac{1}{v_n \tau_n}\left[-\int_0^{x_a} r_p(x) \, dx + \frac{x_a v_p}{v_p + v_n} \right.$$

$$+ \frac{1}{v_p} \int_0^{x_a} (\alpha_n v_p + \alpha_p v_n) e^{R(x,x_a)} \int_0^x r_p(x') \, dx' \, dx$$

$$\left. - \frac{1}{v_p + v_n} \int_0^{x_a} x(\alpha_n v_p + \alpha_p v_n) e^{R(x,x_a)} \, dx \right] \tag{B.7}$$

Let the last integral in (B.7) be designated as F_1. This integral can be simplified by adding and subtracting $\alpha_n v_n/(v_p + v_n)$ to the integrand.

$$F_1 = -\frac{1}{\tau_n v_n}\left[\int_0^{x_a} \alpha_n x e^{R(x,x_a)} \, dx + \frac{v_n}{v_p + v_n} \int_0^{x_a} x(\alpha_p - \alpha_n) e^{R(x,x_a)} \, dx \right] \tag{B.8}$$

Reduction of κ to that given by (14.36) requires the repeated use of the integration by parts formula.

$$\int u \, dv = uv - \int v \, du \tag{B.9}$$

In particular for the two integrals in (B.8) let $u = x$ and dv be the remainder of the integrand in each of the two cases. Then (B.8) becomes

QUASISTATIC APPROXIMATION FOR THE AVALANCHE CURRENT 407

$$F_1 = -\frac{1}{\tau_n v_n}\left[x \int_0^x \alpha_n e^{R(x',x_a)}\, dx' \bigg|_0^{x_a} - \int_0^{x_a}\int_0^x \alpha_n e^{R(x',x_a)}\, dx'\, dx \right.$$

$$\left. + \frac{v_n}{v_p + v_n}\left\{ -xe^{R(x,x_a)}\bigg|_0^{x_a} + \int_0^{x_a} e^{R(x,x_a)}\, dx \right\} \right] \quad \text{(B.10)}$$

The first integral in (B.10) is obtained from (A.10) and the last from (A.9).

$$F_1 = -\frac{1}{\tau_n v_n}\left[\frac{x_a v_p}{v_p + v_n} - \frac{x_a}{M_n} - \int_0^{x_a}\int_0^x \alpha_n e^{R(x',x_a)}\, dx'\, dx \right] - 1. \quad \text{(B.11)}$$

The remaining integral in (B.11) can be rewritten as

$$\int_0^{x_a} e^{R(x,x_a)} \int_0^x \alpha_n e^{R(x',x)}\, dx'\, dx \quad \text{(B.12)}$$

The interior term can be simplified by (14.37) which is rewritten below.

$$\int_0^x \alpha_n e^{R(x',x)}\, dx' = r_p(x) - \frac{e^{R(0,x)}}{M_p} \quad \text{(B.13)}$$

Hence, F_1 in (B.11) becomes

$$F_1 = -\frac{1}{\tau_n v_n}\left[\frac{x_a v_p}{v_p + v_n} - \frac{x_a}{M_n} - \int_0^{x_a} r_p(x) e^{R(x,x_a)}\, dx \right.$$

$$\left. + \frac{1}{M_p} \int_0^{x_a} e^{R(0,x)} e^{R(x,x_a)}\, dx \right] - 1$$

$$= -\frac{1}{\tau_n v_n}\left[\frac{x_a v_p}{v_p + v_n} - \frac{x_a}{M_n} - \int_0^{x_a} r_p(x) e^{R(x,x_a)}\, dx + \frac{1}{M_p}\frac{M_p}{M_n} x_a \right] - 1 \quad \text{(B.15)}$$

where use has been made of the identity $R(0,x) + R(x,x_a) = R(0,x_a)$ and (A.8). The expression for κ in (B.7) can now be expressed as follows.

$$\kappa = \frac{1}{\tau_n v_n} \left[- \int_0^{x_a} r_p(x)(1 - e^{R(x,x_a)}) \, dx \right.$$

$$\left. + \frac{1}{v_p} \int_0^{x_a} (\alpha_n v_p + \alpha_p v_n) e^{R(x,x_a)} \int_0^x r_p(x') \, dx' \, dx \right] \quad \text{(B.16)}$$

The first term is integrated by parts using $u = 1 - e^{R(x,x_a)}$ and $dv = r_p \, dx$. Since the uv term is zero at both end points the first integration in (B.16) reduces to

$$\frac{1}{\tau_n v_n} \int_0^{x_a} (\alpha_p - \alpha_n) e^{R(x,x_a)} \int_0^x r_p(x') \, dx'. \quad \text{(B.17)}$$

This is readily combined with the second integral in (B.16) to give

$$\kappa = \frac{v_p + v_n}{\tau_n v_n v_p} \left[\int_0^{x_a} (\alpha_p - \alpha_n) e^{R(x,x_a)} \int_0^x r_p(x') \, dx' \, dx \right.$$

$$\left. + \int_0^{x_a} \alpha_n e^{R(x,x_a)} \int_0^x r_p(x') \, dx' \, dx \right] \quad \text{(B.18)}$$

These two integrals in (B.18) are evaluated again by parts by letting $u = \int r_p(x') \, dx'$ and dv be the remaining part of the integrands. Making use of the formula for the electron multiplication factor (A.10), the expression for κ becomes

$$\kappa = \frac{v_p + v_n}{\tau_n v_p v_n} \left[-\frac{1}{M_n} \int_0^{x_a} r_p(x) \, dx + \int_0^{x_a} r_p(x) e^{R(x,x_a)} \cdot \right.$$

$$\left. \left(1 - \int_0^x \alpha_n e^{R(x',x)} \, dx' \right) dx \right] \quad \text{(B.19)}$$

Using (B.13) to remove the α_n term in (B.19) the equation for κ becomes

$$\kappa = \frac{v_p + v_n}{\tau_n v_p v_n} \left[-\frac{1}{M_n} \int_0^{x_a} r_p(x) \, dx \right] \quad \text{(B.20)}$$

$$+ \int_0^{x_a} r_p(x)(1 - r_p(x))e^{R(x,x_a)}\,dx + \frac{1}{M_p}\int_0^{x_a} r_p(x)e^{R(0,x_a)}\,dx \bigg] \quad \text{(B.20)}$$

Since from (A.8) the first and third terms cancel one another, the remaining portion of κ in (B.20) is given in its final form.

$$\kappa = \frac{v_p + v_n}{\tau_p v_p v_n} \int_0^{x_a} r_p(x)(1 - r_p(x))e^{-R(0,x)}\,dx \quad \text{(B.21)}$$

REFERENCE

1. D. R. Decker, "IMPATT Diode Quasi-Static Large-Signal Model," *IEEE Trans. on Electron Devices*, Vol. ED-21, pp. 469–479, Aug. 1974.

Index

ABCD matrix, 4–6, 9–10, 20–23, 51, 53, 76–77, 85–86, 95–96, 106, 282–283
Acoustic device, 396
Accumulation layer, 339–341, 364–365
Accumulation layer mode, 333, 337, 339
Admittance inverter. *See* Impedance inverter
Allpass circuit. *See* C-section
Alpha cutoff frequency, 360, 362
Alumina, 129, 152
Antinomy, 198
Attenuation constant, 117–118, 122
Attenuator, 29–31
Auto correlation, 170
Automatic network analyzer, 148
Avalanche
 current, 401–409
 frequency, 307, 310
 region, 295, 297–304, 306–309, 314

Back diode, 397
Balun, 28, 30
Bandwidth, 156–157, 163, 165, 176, 251–252
BARITT diode, 397
Barrier height, 245
Bipolar transistor, 154, 162, 355–362
Black body, 171, 186, 196, 201–202
Bloch function, 327
Boltzmann
 distribution, 194, 196, 344
 equation, 342
 factor, 160, 171, 186–187, 189
Bose-Einstein distribution, 172, 186, 192, 195–196
Breakdown electric field, 296, 309

CAD, 136–137
CAM, 136, 145–146
Capacitance matrix, 62–63
Capacitance
 static, 117
 stray, 122
CAT, 137, 146
Cauchy product, 142

Characteristic impedance, 5, 7, 20, 65, 114–117, 126–128, 131, 350
Chebyshev polynomial, 39–41, 45, 107
Circulator, 32–34, 154
Class A, 162–163
Class AB, 163
Class B, 162–163
Class C, 163
Coaxial line, 65, 74–76, 117–120, 135, 152
Combiner/divider
 cavity, 317–323
 Gysel, 28, 317
 Wilkinson, 315–316
Compact, 138
Computer aided design. *See* CAD
Computer aided manufacture. *See* CAM
Computer aided test. *See* CAT
Conduction band, 201–202, 204, 243–244, 328, 364
Conduction valley, 325, 328–330, 342, 344
Connector, 151
Continuity equation, 206, 209, 235, 300, 304, 401
Conversion loss, 254–255, 259–260
Coplanar line, 88
Coupler. *See* Directional coupler
Coupling coefficient, 92, 97, 320, 322
Critical electric field, 325
Crystal momentum, 327
C-section, 56, 280–282
Current gain, 360, 362
Cutoff frequency, 360, 362, 367–368

Delay mode. *See* Inhibited mode
Depletion
 capacitance, 205–207, 209, 211, 215–218, 225, 233, 245, 264–265, 270
 layer, 339–341, 348, 363, 368–369
Detector, 243–244, 248–253
Dielectric
 relaxation time, 336
 resonator, 397

Diffusion
  capacitance, 208–209, 212–217, 234
  length, 215, 248, 267, 357, 359
  theory, 346–347
Dipole domain, 330–334, 342, 347
Dipole domain mode, 332–333, 337–338
Directional coupler:
  assymmetric. *See* Directional coupler, 180°
  branched line, 19–25
  rat race, 20, 23–29
  symmetric. *See* Directional coupler, 90°
  180°, 100–106
  8.34 dB, 18–19, 109–110
  90°, 16, 18, 106–110
Directivity, 15, 97
Discontinuity, 116, 121–122, 128–130, 132–134
  reactance (susceptance), 118–120
Discriminator, 183
Dispersion, 112
Dissipation, 121
Distribution function, 190, 192
Domain
  charge, 348
  current, 347
  width, 341, 347
Drift region, 295, 297, 304, 307, 309, 314
Dynamic range, 156, 163–164, 249

Effective dielectric constant, 126–127, 131–132
Effective mass, 327–328
Efficiency, 161, 163, 166–167, 233–235, 240–241, 313–315, 326, 336
Einstein relationship, 209
Electron affinity, 243–244
Emitter efficiency, 359–360
End effect, 122, 126, 128
Entropy, 189–190, 196–198
Equal apriori probability, 187–188
Equal areas rule, 338
Equipartition theorem, 174, 191
Ergodic hypothesis, 197
Error
  criteria, 137
  function, 107, 138–139, 159
Even/odd mode
  capacitance, 69, 132
  impedance, 69–70, 86–88, 97, 101–102, 106, 121, 124, 282

Fermi-Dirac distribution, 186, 192, 194–195, 203
Fermi level, 195, 243–244, 362
Ferrite, 31, 34, 219–220
FET. *See* Field effect transistor
Field effect transistor, 162–163, 355, 362–373
Filter
  bandpass, 48, 59–60, 71, 76, 90
  bandstop, 48, 56, 59
  Butterworth, 44–47, 51, 55–56, 58, 68, 385
  Chebyshev. 44–47, 55–56, 59–60, 62, 89–90, 385–386
  coupled line, 78–92
  highpass, 48, 56–57, 59–60, 79, 84
Flat profile IMPATT, 297
Flow graph, 2, 11–12, 149, 376–378
Forbidden gap, 202, 328
Forward bias, 208, 214, 246–247, 266, 269–270, 273
Foster reactance theorem, 2
Fourier
  series, 140–145, 210–211, 225, 237, 239
  transform, 3, 305
Free energy level, 244
Frequency domain, 140, 159, 305
Frequency multiplier. *See* Multiplier

GaAs. *See* Gallium Arsenide
Gain
  available, 156
  compression, 163
  exchangeable, 156
  insertion, 156
  power, 155
  transducer, 44, 155, 164–167, 175, 178, 227, 229–232, 374–379
Gallium arsenide, 160–161, 325, 329, 341, 363–364
Gauss's law, 206–207
Gunn diode, 325, 329–330, 332, 338, 345–346, 350, 353

Harmonic balance, 139–140, 210, 223–224
Harmonic frequencies, 141, 159, 212–213, 222–223, 225, 233–234, 237, 239, 249, 259, 305, 315
Heat
  sink, 317, 373
  transfer, 160
Heisenberg uncertainty principle, 187

Hybrid mode, 333, 337
Hyperabrupt junction, 207–208

Idler circuit, 234
IEEE 488-1978, 148
Image
 frequency, 254, 258
 impedance, 4–5, 14, 29, 31, 78, 259
IMPATT diode, 160–161, 294–315, 325–326
Impedance
 inverter, 71–87
 match, 37, 47, 92, 104, 386–394
Inhibited mode, 333–335
Intercept point, 164
Intermodulation, 164
Ionization coefficient, 297, 313, 401
Isolation, 15–16
Isolators, 16, 32

Jacobian, 107, 139
J inverter. *See* Impedance inverter

K inverter. *See* Impedance inverter
Kirchoff current law, 31, 145
Klopfenstein tapper, 102–104
Kuroda identities, 49–50, 52–54, 83–84, 90
Kurokawa stability criterion, 220–221

Launcher, microstrip, 152–153
Limiter, 272, 274
Linear circuit, 137, 140
Locked oscillator. *See* Oscillator
Loss, 14, 18, 27, 29–30, 44–45, 117, 121–122, 126–127, 273
LSA mode, 333, 335–337, 349
LSUC. *See* Upconverter, lower sideband
Lumped prototype, 71–72, 79–82, 89

Magic tee, 183–184
Manley-Rowe equations, 223, 232, 254
Maxwell's equations, 2, 113
MESFET. *See* Field effect transistor
Microstrip, 120, 126–132, 145, 152, 253
Millimeter waves, 314, 398
Minority carrier lifetime, 208–209, 213, 216–218, 234, 237, 243, 247, 269, 356
Mixer, 253–262
 balanced, 183, 255–256
 double balanced, 255, 257
 image enhanced, 261

single balanced. *See* Mixer, balanced
 single ended, 255, 258
Mobility, 297, 330, 332, 336, 342
Model, device circuit, 345–348
Modulation, 163, 181
Monolithic circuits, 398
Multiplication factor, 298–303
Multiplier, 207, 209, 219, 222, 232–235, 237–240

Negative conduction mode. *See* Accumulation layer mode
Negative resistance, 16, 154, 156, 175, 179, 222, 226, 230–232, 310, 325–326, 337–338, 343, 345, 349
Newton's method, 61, 138–139, 145, 277
Nonlinear
 analysis, 137, 139–140, 144–147, 164, 223
 device, 141, 144, 163, 212, 223, 225, 233, 253, 311
Non-redundant synthesis, 35, 54, 60
Noise
 AM, 181–184, 345
 equivalent power, 251
 figure, 164–166, 175–176, 178–180, 227–232, 254, 312–313, 326, 371–373, 384
 flicker, 248, 253–254
 FM, 181–184, 345
 measure, 176, 384
 oscillator, 179–180
 power, 164, 172–173, 177–178, 183
 shot, 164, 170, 173–175
 spot, 176
 temperature, 176
 thermal, 164, 170, 172, 174, 177, 181
Nyquist formula, 165, 174–175

Objective function, 138
Offset coupled lines, 124–126
Ohmic contact, 330, 363
Ohm's law, 141, 160
One-port, 154–155, 167
Optimization, 107, 137–138, 159, 275
Oscillator, 35
 locked, 15

Package, 151, 157–159, 234, 353
Parametric amplifier. *See* Reflection parametric amplifier
Parasitic reactance (susceptance), 273, 384
Pauli exclusion principle, 195, 202

Penrose pseudo-inverse, 108
Permeability, 32
Phase shifter, 264, 277, 278
   loaded line, 278, 280–281, 286–290
   lowpass/highpass, 278, 280–285
   reflection, 278, 280, 290–292
   switched line, 278, 280–281
Point contact diode, 248
Poisson's equation, 206, 338
Power, 160
Power handling, 159, 233–234, 241, 264, 269–270, 272, 281, 315, 373
Premature collection mode, 314
Propagation, 31
Propagation constant, 78, 113
Prototype. *See* Lumped prototype
Pulse, 160–162, 174
Punch through, 265

Quasistatic, 302–303, 405–409
Quenched domain mode, 333, 335, 337

Radiation loss, 121
Read diode, 294–295, 297–310
Reciprocity, 4, 16
Redundant
   filter synthesis, 35, 49, 53
   unit elements, 68, 90
Reflection
   circuit, 15–16
   coefficient, 9, 11, 20–21, 27, 37–40, 42, 95–96, 98–101, 108, 132–133, 149–150, 154, 290–291
   parametric amplifier, 156, 168, 170, 207, 219, 222, 226–230
Reliability, 160
Resistive loss, 113
Response time, 264, 301–302
Reverse bias, 205, 222, 245, 264–265, 269–270, 273, 275
Richard's theorem, 55, 60–61, 282
Richard's transformation, 35, 49
Rise time, 249
RPA. *See* Reflection parametric amplifier
RWH mechanism, 325–326, 329–330

Satellite valley. *See* Conduction valley
Saturated drift velocity, 297, 370
Scattering
   matrix, 2, 4, 9–10, 16–18, 32–34, 56, 100, 102, 374–384
   parameters, 95, 101, 137, 149–150

Schottky barrier, 200–201, 243–248, 254, 362–363
Schrödinger equation, 171
Secondary avalanche mode, 314
Self-aligned flip chip, 373
Slope parameter, 73
Small signal analysis, 222–223, 306, 365–366
S-matrix. *See* Scattering matrix
Source via hole, 373
Space charge, 308
S-parameters. *See* Scattering parameters
SPDT switch, 275, 279, 281
S-plane, 50, 53–56, 61, 71, 83, 394
Spline, 106
Spurious oscillations, 311
Square law, 249
Stability
   circles, 383
   factor, 379–383
   Kurokawa criterion. *See* Kurokawa stability criterion
Step recovery diode, 200, 219, 233–235, 237
Storage time, 236, 248
Stored charge, 234–237, 240, 253
Stripline, 120–126, 145
Sum frequency, 254
Super Compact, 137–138
Superheterodyne, 181, 183
Surface
   roughness, 114
   states, 245
Switch, 272–277, 290
Switching speed, 270, 356
Symmetrical circuit 14, 19, 20

Tangential sensitivity, 251
Temperature, 160, 165, 172, 174, 177, 189–191, 197, 203, 228, 269, 311, 344–345
TE waves, 114
TEM waves, 112, 116, 126
Thermal
   equilibrium, 190, 245
   relaxation time, 345
   resistance, 160
Thermionic theory, 246
Three-port, 32
Three terminal, 2, 200
Threshold field, 330–332, 334–335
Time delay, 277–278, 281, 290
Time domain
   integration, 139, 305
   measurement, 151–153, 159

INDEX

TM waves, 114
Transconductance, 370
Transformer, 35–43, 62–63, 68, 71, 83, 90, 103, 106, 108–109, 122, 134
Transient solution, 140
Transition time, 237
Transit time, 308, 314, 325, 361–362, 366
Transit time mode. *See* Dipole domain mode
Transmission
 circuit, 15–16
 coefficient, 9–11, 20–21, 37–38, 58, 95–96, 102, 132–133, 283
 equation, 5–8, 36–37, 48
 phase, 44
Transport factor, 359–360, 362
TRAPATT diode, 375
Tube, 396
Tunnel diode, 397
Two-port, 12, 21, 43, 94, 154, 167
Two-terminal, 200
Two-wire line, 112

Ultraviolet catastrophe, 171, 201
Unit element, 35, 49, 50–51, 53–55, 60, 62, 68–71, 90, 394
Upconverter
 lower sideband, 228–232
 parametric, 156, 168, 219, 222
 upper sideband, 229, 231–232
USUC. *See* Upconverter, upper sideband

Valance band, 201–202, 204, 243
Varactor diode, 145–147, 158, 160, 200, 207, 217, 219–223, 226, 228, 233, 254
VCO. *See* Voltage controlled oscillator
Voltage controlled oscillator, 219–222
Volterra series, 140

Waveguide, 112–115, 157
Waveguide mount, 350–353
Wave impedance, 115
Work function, 243

YIG. *See* Ferrite
Y-factor, 176–177